INVESTIGATING HUMAN DISEASES WITH THE MICROBIOME: METAGENOMICS BENCH TO BEDSIDE

INVESTIGATING HUMAN DISEASES WITH THE MICROBIOME: METAGENOMICS BENCH TO BEDSIDE

Huijue Jia, PhD
Director, Shenzhen Key Laboratory of Human Commensal Microorganisms and Health Research, BGI, Shenzhen, China
Mentor for graduate students, BGI Education Center, University of Chinese Academy of Sciences, Shenzhen, China

Academic Press is an imprint of Elsevier
125 London Wall, London EC2Y 5AS, United Kingdom
525 B Street, Suite 1650, San Diego, CA 92101, United States
50 Hampshire Street, 5th Floor, Cambridge, MA 02139, United States
The Boulevard, Langford Lane, Kidlington, Oxford OX5 1GB, United Kingdom

Library of Congress Cataloging-in-Publication Data
A catalog record for this book is available from the Library of Congress

British Library Cataloguing-in-Publication Data
A catalogue record for this book is available from the British Library

ISBN 978-0-323-91369-0

For information on all Academic Press publications
visit our website at https://www.elsevier.com/books-and-journals

Publisher: Stacy Masucci
Editorial Project Manager: Mica Ella Ortega
Production Project Manager: Selvaraj Raviraj
Cover Designer: Christian Bilbow
Cover art by Yuan Fang

Typeset by STRAIVE, India

Contents

Acknowledgment ix

Chapter 1 The supraorganism 1

1.1 New discoveries with new technology—A historical account 1

1.2 How many microbial cells can a human body have? 5

1.3 Viral particles in the human body 12

1.4 Microbiome in other species 13

1.5 Microbiome from ancient times 13

1.6 Summary 17

References 17

Chapter 2 Microbiota 21

2.1 Trophic levels in macroecology 21

2.2 Microbiome stability, diversity, and richness 21

2.3 De novo assembly of microbiota and robustness against invasions 26

2.4 Types of habitats for the skin microbiome 29

2.5 Forces shaping the oral microbiome 31

2.6 A stable gut microbiome 33

2.7 "Enterotypes" and the Serengeti rules? 38

2.8 Summary 46

References 46

Chapter 3 Collecting samples for metagenomics 57

3.1 Nonmicrobial components in the sample that could influence DNA extraction and sequencing amount 57

3.2 Beware of contamination in each step, from stools to low-biomass samples 59

3.3 Reagents that prevent microbial growth after sampling 65
3.4 DNA extraction for metagenomic samples 67
3.5 Sequencing amount 68
3.6 Taxonomic and functional profiles, absolute abundance 69
3.7 Sample size for metagenome-wide association studies 71
3.8 Summary 73
References 76

Chapter 4 Epidemiology in the human body 83

4.1 Analogy to COVID-19 83
4.2 Sources of potential pathogens in the infant gut 84
4.3 Ectopic presence of commensal microbes 84
4.4 Get to where it matters for the disease 88
4.5 Interkingdom interactions in the microbiome in diseases 94
4.6 Other omics data that hint at a difference in microbiome 97
4.7 Summary 101
References 101

Chapter 5 The evolving microbial taxonomy 109

5.1 Approaching a closed reference set for routine applications 109
5.2 Sparser data with increasing taxonomic resolution. 118
5.3 Evolutionary history below the species level 121
5.4 Whole-cell modeling to predict functional differences from genomic differences? 124
5.5 Summary 129
References 129

Chapter 6 Blurring the line between opportunistic pathogens and commensals 133
6.1 Causal reasoning 101. 133
6.2 Levels of existing evidence for the human microbiome and diseases. 138
6.3 From microbes to molecules 143
6.4 Summary 149
References 149

Chapter 7 Metagenomics from bench to bedside and from bedside to bench. 157
7.1 Metagenomics for decision-making in diagnosis and treatment 157
7.2 Further research to be inspired by clinical practice 172
7.3 Potential to modify existing categorization of diseases with knowledge of the microbiome. 175
7.4 Summary 177
References 177

Chapter 8 A microbiome record for life. 189
8.1 Proactive sampling of the microbiome at important time periods. 189
8.2 From genetic risk to the prevention of diseases. 200
8.3 Summary 204
References 204

Index. 215

Acknowledgment

As scientists get used to online conferences in the past year, it is good to be able to spend some time thinking.

My teacher on microbiology from Fudan University, Prof. Baolong Hu, described a thorough bathing and scrubbing of oneself as "spread plate, spread plate," years before the launch of the Human Microbiome Project (HMP) and the Metagenome of the Human Intestinal Tract (MetaHIT) Consortium. The textbook written by his teacher, Prof. Deqing Zhou, is widely in use by college and graduate students in China. It is my great honor to take on this opportunity, made possible by Elsevier, to put together a book on the human microbiome, which is a field both young and old.

The virology class taught by Prof. Jiang Zhong opened my eyes to the fascinating things viruses can do. The zoology class led by Prof. Yanyun Yang was lively and comprehensive, and we had elegant charts from some missionaries a century ago. My biochemistry teacher, Prof. Weida Huang was teaching according to *Lehninger Principles of biochemistry*, my first biology textbook in English.

Much of the metagenomic studies on the microbiome depend on bioinformatics. My teacher on bioinformatics at Fudan University, Prof. Yang Zhong (who also contributed to developments in Tibet University, ShanghaiTech University, etc.), has unfortunately passed away in a car accident in the Inner Mongolia Autonomous Region.

I was very fortunate to be able to study for my Ph.D. at Case Western Reserve University. Cleveland is a very affordable city with plenty of activities. I enjoyed a very nurturing environment in the laboratory of Prof. Eckhard Jankowsky, with fun and collaborative people in the lab and in the department at large. My brief time in Prof. Yi Zhang's lab then at Chapel Hill, and as an editor in London, made me further unfit for a conventional narrow path in academia.

I stumbled into the metagenomic field at the end of the year 2012, after meeting many young people at BGI, and as Mr. Qiang Feng and Prof. Jun Wang insisted. Despite being strangled by Illumina after BGI purchased Complete Genomics and rushing to upgrade the sequencing technology (led by Prof. Xun Xu), we did OK in the metagenomic field, never forgot our initial mission to benefit human health.

Academician Huanming (Henry) Yang has high hopes for metagenomics. He, a few years ago, likened the metagenomics technology to the invention of microscopes, as a major advance in microbiology. Prof. Jian Wang is pushing for large-scale screening of colorectal cancer. After successful protection of infants, longevity for a population breaks down into how well we control the major diseases that come with age, without much decay in physical and cognitive functions.

I sincerely thank the Elsevier staff, Stacy, Mica, Kavitha, Mohanraj, Selva, and probably many more people I didn't get to know. Thank my colleagues for their excellent studies over the years. In particular Chen Chen, Lilan Hao, Xin Tong, Youwen Qin, Yanmei Ju, Jie Zhu, Yanzheng Meng, Weiting Liang, and Zhuye Jie, for checking the text, and for improving the worked samples and the high-resolution display items for publication in this book. Thank my dear husband, Dong, for support and inspiration.

And thank you for being interested in this book. This has been a great opportunity for me to learn, and to put together discrete information for a more coherent picture. I apologize that it is increasingly difficult to keep track of every publication. But it is high time that we go beyond fecal samples and crude questionnaires, and carefully research in the microbes that show up in "sterile" body sites. I do hope that there would be better examples to be included into subsequent editions, for clinical practice, new research, and education.

1

The supraorganism

1.1 New discoveries with new technology—A historical account

Antoni van Leeuwenhoek looked at all kinds of things he could think of using his mysterious single-lens microscope, which had an impressive resolution of around 1 μm (10^{-6} m, Fig. 1.1) [1,2]. From the dental plaques of a few people, he described with astonishment several different bacteria, despite paying more attention to oral hygiene than his contemporaries (Box 1.1) [3]. For size comparison, Leeuwenhoek mentioned sandgrains (Box 1.1), the diameter of which are in the submillimeter range, and the finer ones are indeed at the resolution of the human eye (100–200 μm).

Acceptance of Leeuwenhoek's various works grew after Robert Hooke recapitulated one of Leeuwenhoek's pepper-water experiments with his not as high-resolution but more accessible two-lens microscope (compound microscope) [1]. Nonetheless, Carolus Linnaeus did not give microbes a slot in his 10th edition of *Systema Naturae* (1758). His tree of life was only about plants and animals. Ernest Haeckel added "Protista" (unicellular organisms) in 1866. The taxon "Monera" for unicellular organisms that lack a nucleus, such as bacteria, was proposed as a phylum, and later elevated among the kingdoms. Robert Whittaker added the fungi kingdom to the tree of life in 1969, until Carl Woese revamp of the tree with archaea, according to sequence comparisons in the 16S ribosomal RNA [4].

Developments in metagenomics in the past two decades represent another major leap in technology that has allowed us to better appreciate the microbial world (Fig. 1.2). Also with curiosity and excitement, we applied the technology to all kinds of samples. However, "messy" or "shitty" the host-associated (animals or plants are the hosts for the microbes) or environmental sample is, we can sequence all the DNA in the

Investigating Human Diseases with the Microbiome: Metagenomics Bench to Bedside.
https://doi.org/10.1016/B978-0-323-91369-0.00009-1

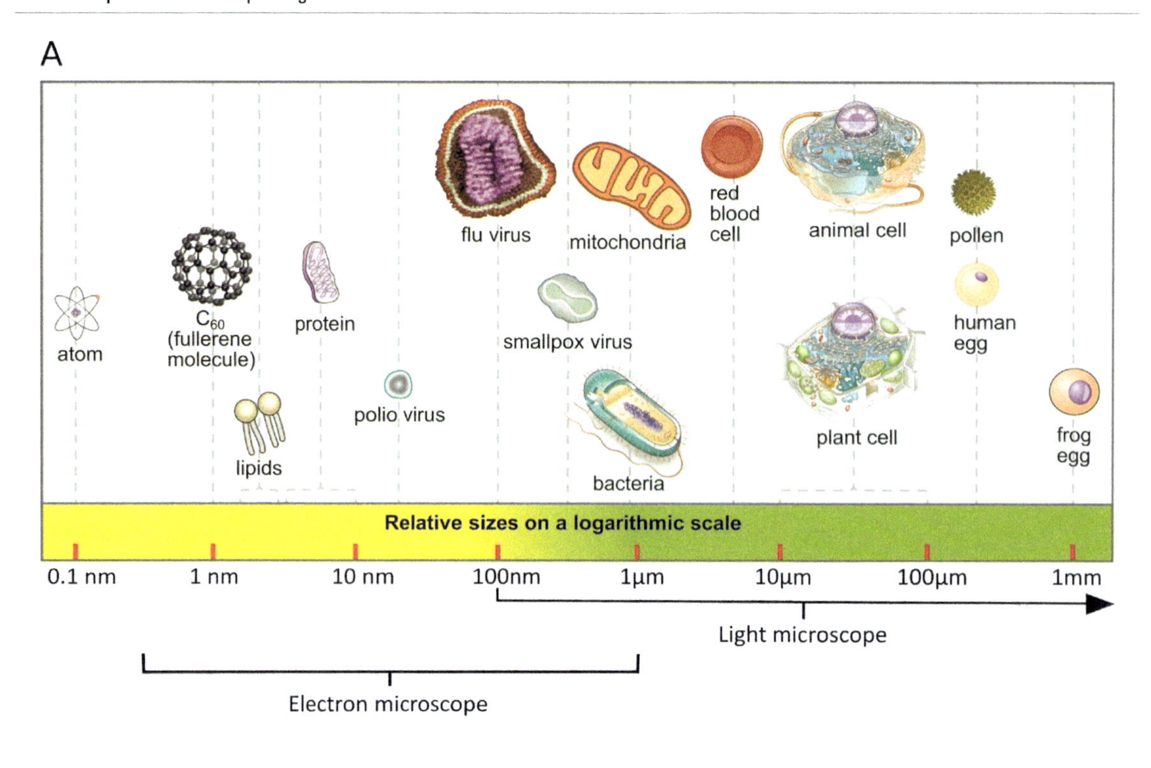

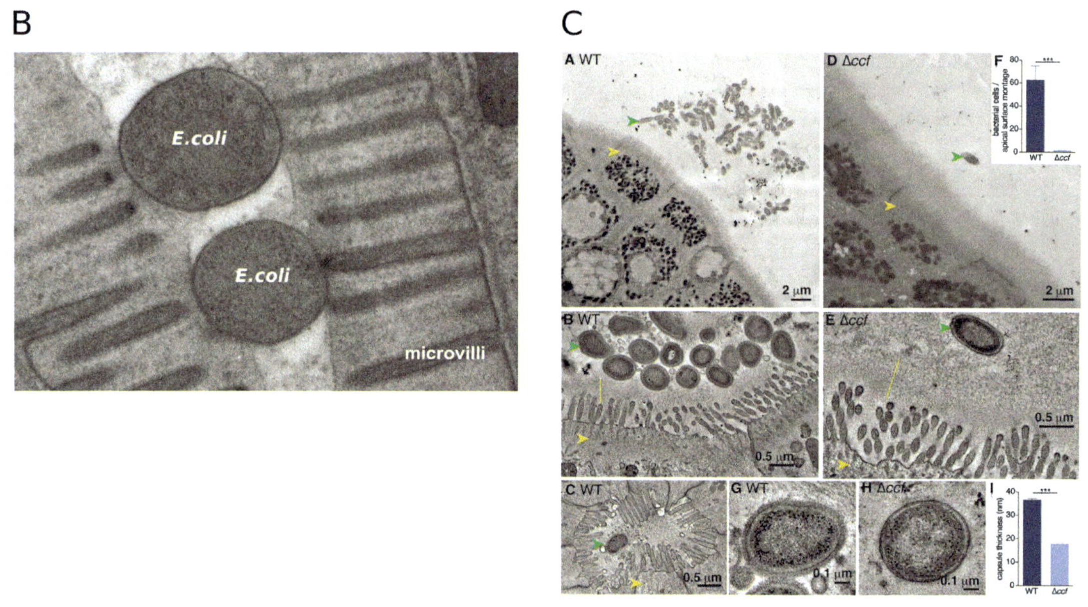

Fig. 1.1 Most bacteria are smaller than their host cells by 1 order of magnitude. (A) Size chart for biomolecules and cells. (B) *Escherichia coli* cells are like tapioca pearls in a bubble tea straw when in the intestine of *Caenorhabditis elegans*. The cross-section of *E. coli* is about 400 nm. (C) Wildtype and Δ*ccf* mutant *Bacteroides fragilis* in the colon of mice monocolonized with the bacterium. Note the thinner cell wall of the Δ*ccf* mutant. Credit: (A) https://courses.lumenlearning.com/microbiology/chapter/types-of-microorganisms/, (B) photo Shigeki Watanabe and Erik Jorgensen, and (C) Fig. 1 of http://science.sciencemag.org/content/360/6390/795.long.

Box 1.1 Dental bacteria observed by Leeuwenhoek in 1683

Although saliva is somehow free of 'animalcules', clearly mobile microorganisms referred to as little animals, Mr. Antoni van Leeuwenhoek's study on the bacteria in dental plaques, as described in his September 17th, 1683 letter to the Royal Society of London [3], showcases his scientific rigor.

Oral hygiene, and Leeuwenhoek's dental sample were examined multiple times:

"I am in the habit of rubbing my teeth with salt in the morning, and then rinsing my mouth with water. After eating I usually pick my molars with a tooth-pick and also rub them with a cloth quite vigorously. This keeps my teeth and grinders so clean and white that only few people of my age can compare with me. Also when I rub my gums with hard salt, they will not bleed. Yet all this does not make my teeth so clean but that I can see, looking at them in a magnifying glass that something will stick or grow between some of the molars and teeth, a little white matter, about as thick as batter. Observing it I judged that, although I could not see anything moving in it, there were yet living animalcules in it. I then mixed it several times with pure rain-water, in which there were no animalcules, and also with saliva that I took from my mouth after eliminating the air-bubbles lest these should stir the spittle. I then again and again saw to my great astonishment, that there were many very small living animalcules in the said matter, which moved very prettily. The big sort had the shape of fig. A; these had a very strong and swift motion, and shot through the water or spittle like a pike through the water. These were mostly few in number. The second sort had the shape of fig. B. These often spun round like a top and every now and then took a course like that shown between C and D. These were far more in number. I could not make out the shape of the third sort, for at one time they seemed to be long and round while at another time they appeared to be round. These were so small that I could see them no bigger than fig. E and therewithal they went forward so rapidly and whirled about among one another so densely that one might imagine to see a big swarm of gnats or flies flying about together. These last at times appeared to me so numerous that I judged that I saw several thousands of them in a quantity of water or spittle (mixed with the aforesaid matter) no bigger than a sand-grain, although there were quite nine parts of water or spittle to one part of the matter taken from between my front-teeth and grinders. Furthermore the matter consisted for the greater part of a great number of fibres, some greatly differing from others in their length, yet of one and the same thickness, some bent crooked, some straight as in fig. F and which lay about in disorderly confusion. And because I had formerly seen in water live animalcules that had the same figure, I made every endeavour to see if there was any life in them, but I could not make out the least motion in any of them that at all looked like life."

Dental samples from other people, of different gender, age, oral hygiene, and drinking/smoking habits:

"I also took spittle from the mouths of two different women, who, I am convinced, daily cleaned their mouths, and I examined it as closely as I could. But in this, I could not discern any living animalcules. I then mixed the same saliva with a little of the matter that I picked with a needle from between their teeth and then discovered as many living animalcules and also the long particles, as before related.

I have also examined the spittle of a child about 8 years old, but there also could not discover any living animalcules; and after that I mixed the spittle with some of the matter taken from between the child's teeth and discovered as great a number of animalcules and other particles as mentioned before."

"While an old man who leads a sober life and never drinks aqua vitae (ethanol solution) or tobacco and very seldom any wine was talking to me, my eye fell on his teeth, which were all coated over; this made me ask him when he had

Continued

Box 1.1 Dental bacteria observed by Leeuwenhoek in 1683—cont'd

last cleaned his mouth and the reply was, that he had never washed his mouth all his life. So I took spittle from his mouth and examined it, but could not find in it anything but what I had seen in my own spittle or that of the others.

I took also the matter that stuck between and against his teeth; on mixing this with clean water in which there were no animalcules, and also with his spittle, I observed an incredible number of living animalcules, swimming more nimbly than I had ever seen up to this time. The big sort which were very plentiful, bent their body into curves while going forward, as in fig. G. Furthermore the other animalcules were so excessively numerous that all the water seemed to live, although only very little matter - taken from the teeth - had been mixed with it. The long particles, mentioned before, were also numerous.

I also took the spittle and the white matter, lodged upon and between his teeth from an old man who is in the habit of taking aqua vitae in the morning and of drinking wine and tobacco in the afternoon, wondering whether the little animals could live in spite of this continual drinking. I judged that this man, because his teeth were so uncommonly dirty, would not clean his mouth; when I asked him, he answered: never in all my life with water, but every day by flushing it with aqua vitae and wine. Yet I could not find anything in his spittle in addition to what I found in other saliva. I also mixed his spittle with the matter sticking to the front side of his teeth, but did not find anything in it save a few only of the smallest sort of living animalcules repeatedly mentioned heretofore. However, in the matter which I had taken from between his front-teeth (for he had not a back-tooth in his mouth) I saw many more animalcules, consisting of two of the smallest sort."

Intervention on his own dental sample:

"I did not clean my mouth on purpose for three days and then took the matter that, in a small quantity, had stuck to the gum above my front-teeth; this I mixed both with spittle and with clean water and discovered a few living animalcules in it."

"Furthermore I took some strong wine-vinegar into my mouth, set my teeth, and let the vinegar run between them several times; after this I rinsed three times with clean water. I then once more took some of the foresaid matter both from between my front-teeth and my grinders, mixing it as before several times with spittle as well as with clean rain-water; nearly always I discovered an incredible number of living animalcules, but mostly in the matter which I took from between my back-teeth. Few, however, had the shape of fig. A. I also mixed a little wine-vinegar with the mingled spittle and with the water; the little animals therein died at once. From this I drew the conclusion that the vinegar which I had in my mouth did not penetrate through all the matter which was firmly lodged between and against my front-teeth and my grinders, and only killed those animalcules that were in the outermost parts of the white matter."

On the number of oral bacteria:

"I have had several gentlewomen in my house, who were eager to see the little eels in vinegar. Some of them were so disgusted at what they saw that they resolved never to take vinegar again. But what if in future one should tell such people that there are living more animals in the unclean matter on the teeth in one's mouth than there are men in a whole Kingdom? Especially in those who never clean their mouths, owing to which such a stench comes from the mouth of many that one can hardly bear talking to them. Many call this a stenching breath, but actually it is in most cases a stinking mouth. For my part, I judge from my own case, although I clean my mouth in the manner heretofore described, that there are not living in our United Netherlands so many people as I carry living animals in my mouth this very day. For when I saw that one of my back-teeth was coated against the gum with the said matter about the thickness of a horse-hair, where to all appearance the salt had not scoured this matter for a few days, there were so enormous a number of living animalcules, that I imagined that I could discern as many as 1000 living little animals in a quantity of this matter no bigger than 1/100 part of a sandgrain." [3] (Remember the cube in volume calculations.)

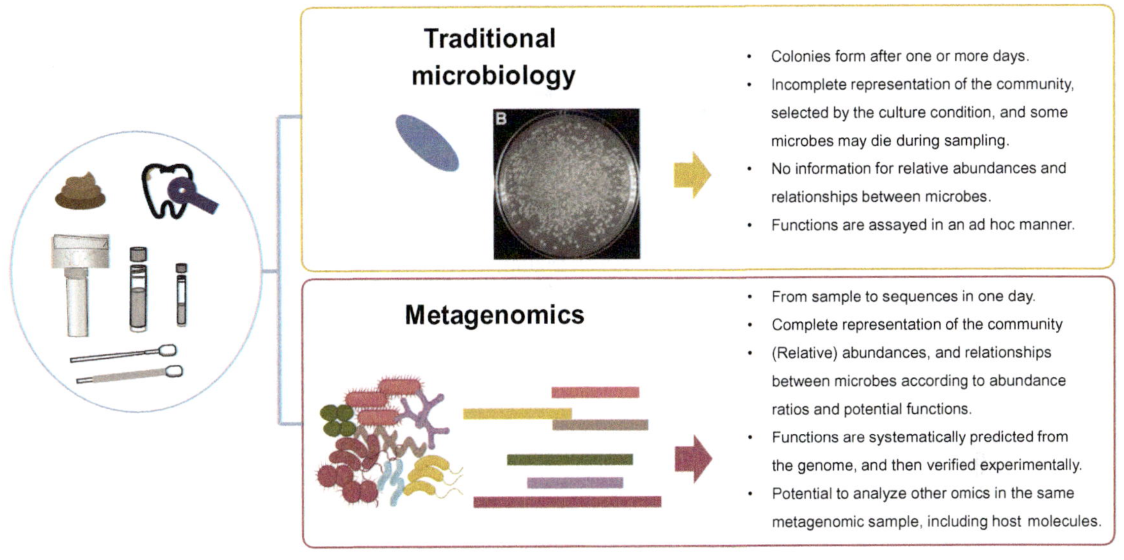

Fig. 1.2 Metagenomics vs traditional microbiology. Absolute abundances can be obtained if quantitatively performed at each step (more in Chapter 3). Relative abundances are important for ecological studies (Chapter 2). Credit: Huijue Jia.

sample, piece together genomic information for individual microbes, and quantify how abundant each microbe is in the sample. Before the advent of high-throughput sequencing technologies, impressive work has been done using traditional microbiology, and later using molecular biology techniques. Without having to guess on the culture conditions and grow out each microbe in a plate, metagenomics make it possible for researchers and clinicians to know all the microbes that are there and what they could do. Sequencing platforms both large and small, along with developments in bioinformatics, are making metagenomic studies more accessible for researchers and clinicians who have different needs for the cost and for the time to results.

Our quest to understand the human microbiome (Box 1.2) will bring about more technology. Taking better care of the microbes that live inside or on the surface of the human body will help us better cope with diseases in populations around the world.

1.2 How many microbial cells can a human body have?

Bacteria are typically in the low micrometer range (Fig. 1.1), if not in longer filaments. Fungi and viruses span a wider range on the larger and the smaller sides. At the smallest volume, free-living bacteria are limited by the volume of DNA and proteins inside; at the largest

Box 1.2 "Metagenome", "microbiota" and "microbiome"

Metagenome:

Dr. Jo Hendelsman introduced the word "metagenome" in a 1998 publication on the soil microbiome [5], for which only a fraction of the microbes had been cultured in isolation. The article advocates for direct cloning of metagenomic DNA from soil samples into ***E. coli*** BAC (Bacterial Artificial Chromosome) libraries for further analyses on the gene functions, without culturing each microbe [5]. Now with high-throughput sequencing and bioinformatic analyses, we use the term "metagenomics" to refer to unbiased direct sequencing of the microbial community in any sample.

Microbiota or microbiome:

A popular assumption is that the Nobel Laureate Joshua Linderburg coined the term "microbiome" in 2001. At face value, the terms "microbiota" or "microflora" are more about the microbial ecosystem. The term "Microbiome," however, is not a product of the genomics era and already includes everything in the ecosystem, a "biome" after all [6]. The words for microbial communities in fact date back earlier [6,7].

Back to work in 1923, Sergei Nikolaievich Winogradsky advocated for the study of interacting microbes in their natural contexts, following his discovery by the end of the 19th century, of how aerobic nitrification (the oxidation of ammonium salts to nitrites and nitrites to nitrates) bacteria locally depleted oxygen, so that the anaerobic nitrogen-fixer ***Clostridium pasteurianum*** could live in the niche created by its neighbors and finish the two-step nitrification process.

Some interesting comments on pure culture from a 1949 paper:

"... condition of pure culture in an artificial environment is never comparable to that in a natural environment ... one cannot challenge the notion that a microbe cultivated sheltered from any living competitors and luxuriously fed becomes a hot-house culture, and is induced to become in a short period of time a new race that could not be identified with its prototype without special study" [6,8].

Specific pathogen-free and germ-free animal models become a common laboratory practice in 1960s. To determine the selected microbiota that is compatible for the sustained health of the animal models has been a goal:

"to endow them, in short, with a selected microbiota compatible with sustained health and conferring some ability to withstand the assaults of other micro-organisms that would almost inevitably be encountered outside the germ-free environment" wrote Lane-Petter in 1962 [9]. In 1986, Linda R. Hegstrand and Roberta Jean Hine discovered a difference in hypothalamic histamine levels between germ-free and conventionally raised animals, an early example of the gut-brain axis [7].

When studying plant diseases, John M. Whipps wrote in 1988:

"A convenient ecological framework in which to examine biocontrol systems is that of the microbiome. This may be defined as a characteristic microbial community occupying a reasonably well defined habitat which has distinct physio-chemical properties. This term thus not only refers to the microorganisms involved but also encompasses their theatre of activity" [6,10].

volume, bacteria are limited by the number of ribosomes required for maintaining such a size, which would be too much to fit into the cell volume (Fig. 1.3) [11]. The cytoplasmic volume of *E. coli* shrank by 17% upon nutrient starvation [12]. The same principles should also apply to archaea. The smallest observed archaea are comparable to the smallest bacteria, with a volume of about 3.41×10^{-20} m^3, 0.5 megabases (Mb) genome, and about 92 ribosomes per cell [11].

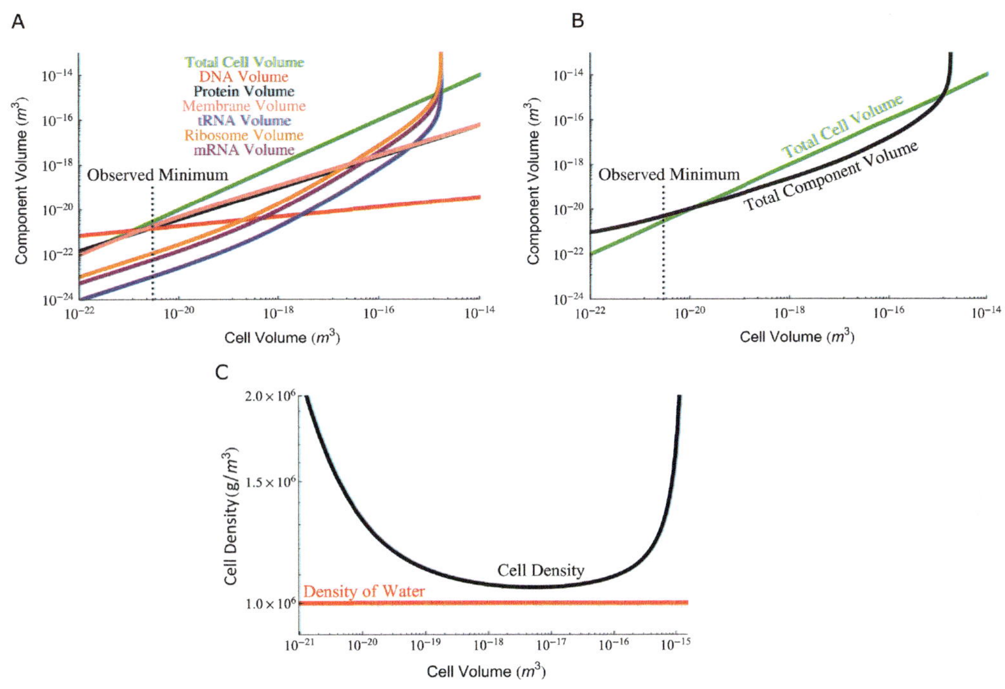

Fig. 1.3 Theoretical limitations for free-living bacteria and archaea to not go much smaller or larger. For example, a typical *E. coli* cell is 2 μm long, 0.5 μm in diameter. (A) The volume-dependent scaling of each of the major cellular components for bacteria. (B) The total cell volume was compared with the volume of all cellular components as a function of cell size. (C) The volume-dependent scaling for the calculated total cellular density. The *black curve* is the calculated density and the *red curve* is the reference value for the density of water. Credit: (A and B) Fig. 3a,b, (C) Fig. S5 of Kempes CP, Wang L, Amend JP, Doyle J, Hoehler T. Evolutionary tradeoffs in cellular composition across diverse bacteria. ISME J 2016;10:2145–57. https://doi.org/10.1038/ismej.2016.21.

We like to talk about tigers, elephants, and whales, but humans are actually among the large animals on earth. There is an interesting linear relationship between the logarithm of the bodyweight of an animal and the logarithm of the number of microbes (Fig. 1.4A). Each gram of an animal averages approximately 3.4×10^9 of associated prokaryotes (bacteria or archaea), and animals are approximately 0.34% prokaryotes by weight [13]. Animal-associated microbes total about 2.1–2.3×10^{25} of the 9.2–31.7×10^{29} prokaryotic cells on earth. Total gut volume has been estimated to scale with animal body mass with exponents of 1.0–1.08, and the surface area of the gut scales with body mass with an exponent of 0.75. However, counts of microbes per unit volume or mass of gut contents vary over several orders of magnitude, and the proportion of the gut (gut in zoology sense, the entire alimentary tract) devoted to intensive microbial activities also varies extensively among animals.

Most of the 3.8×10^{13} microbial cells for a healthy adult human resides in the colon (Figs. 1.4B and 1.5) [14,15]. The number of microbial

A

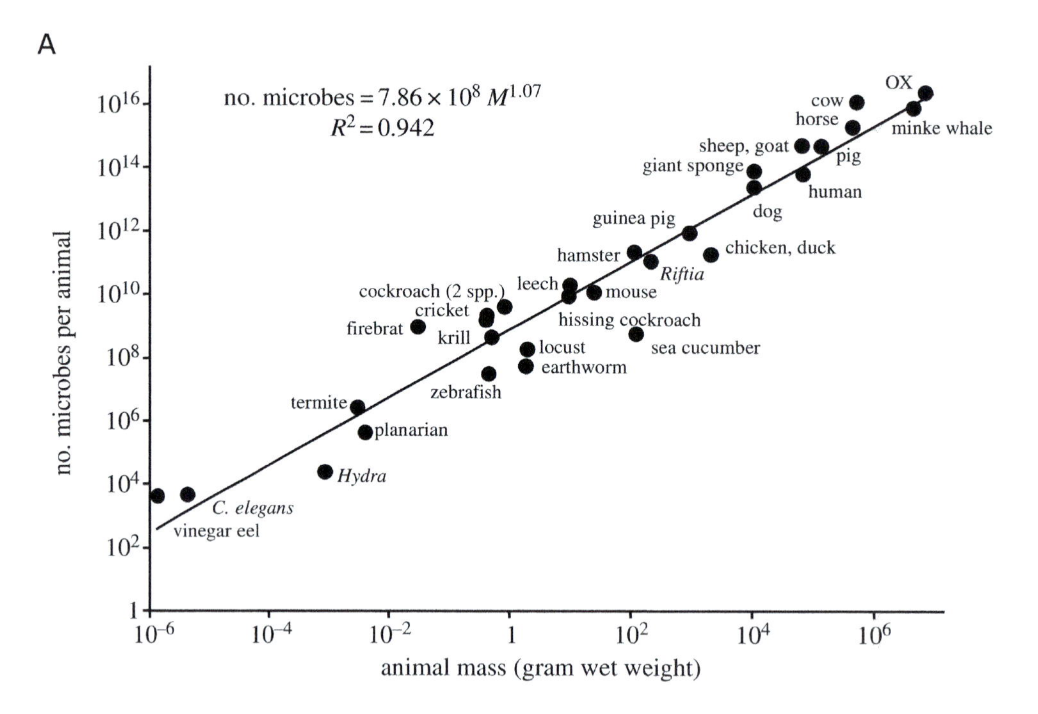

B

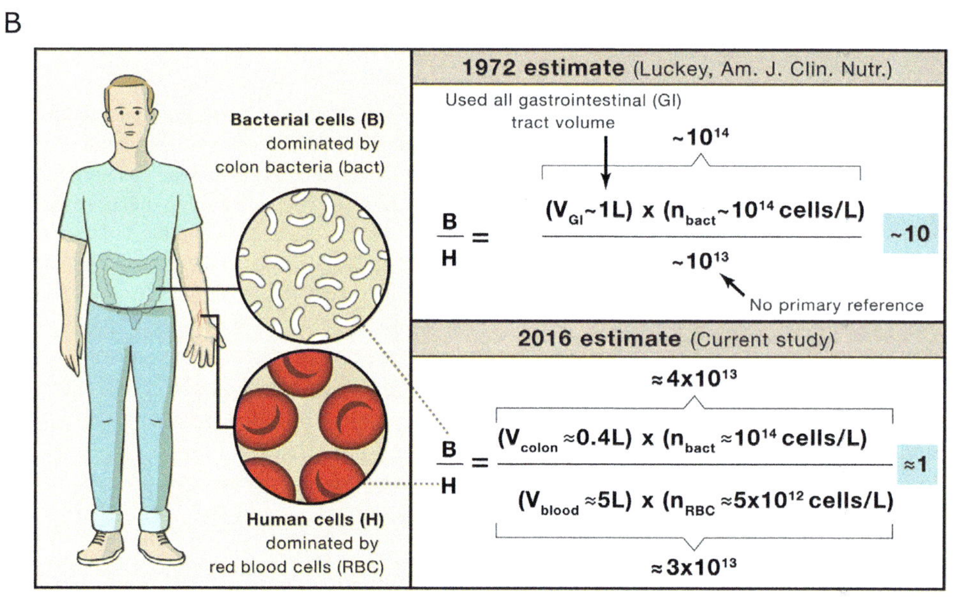

Fig. 1.4 Current estimate for the number of microbes in an adult human, and the general trend in animals of different sizes. (A) Counts of microorganisms per individual animal versus individual animal body mass (M, wet weight in gram), log-log plot. (B) The total number of bacterial cells was estimated according to the typical volume of the adult human colon, instead of volume of the entire gastrointestinal tract, as the bacterial community is much denser in the colon. Credit: (A) Fig. 1 of Kieft TL, Simmons KA. Allometry of animal–microbe interactions and global census of animal-associated microbes. Proc R Soc B Biol Sci 2015;282:20150702. https://doi.org/10.1098/rspb.2015.0702. (B) Fig. 1b of Sender R, Fuchs S, Milo R. Are we really vastly outnumbered? revisiting the ratio of bacterial to host cells in humans. Cell 2016;164:337–40. https://doi.org/10.1016/j.cell.2016.01.013.

species in the human colon is over 600, and their number of cells could range from 1 to more than 10^{13}, corresponding to relative abundances of less than 10^{-13} to over 0.3 (The relative abundances of all species sum up to one). The skin has a large surface area, but has relatively simple microbial communities in humans (Fig. 1.6A and B) [16,17]. The lung is topologically an outer surface, with an even larger surface area and

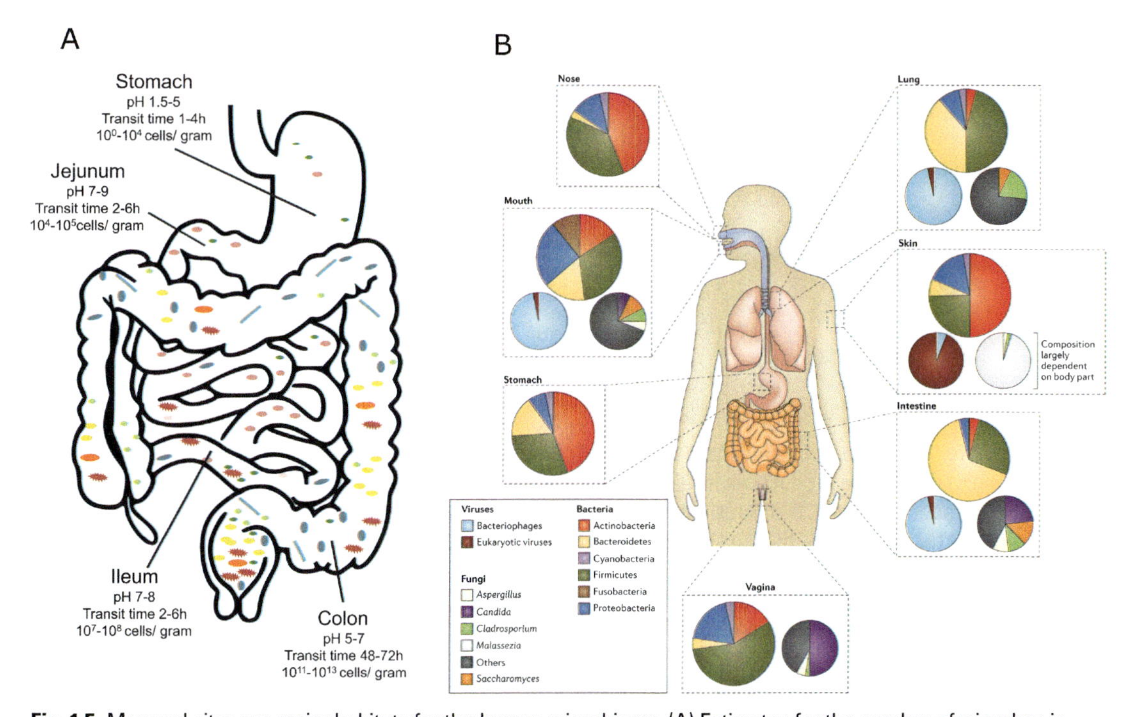

Fig. 1.5 Mucosal sites are major habitats for the human microbiome. (A) Estimates for the number of microbes in different segments of the gastrointestinal tract, together with major factors that influence the community such as pH and transit time. (B) Bacteria phyla, fungi genera, and viruses at different body sites. The figure shows the relative abundance of bacterial, fungal, and viral communities at different body sites exposed to the external environment—the nose, mouth, skin, stomach, intestinal tract, vagina, and lungs. Bacterial composition is represented by the six most commonly detected phyla—Actinobacteria, Bacteroidetes, Cyanobacteria, Firmicutes, Fusobacteria, and Proteobacteria. Fungal composition includes Aspergillus, Candida, Cladosporium, Malassezia, and Saccharomyces as the most prominent genera. Additional types of fungi are summarized as "Others." Viral composition is classified simply as bacteriophages or eukaryotic viruses. Data based on Erb-Downward JR, Thompson DL, Han MK, Freeman CM, McCloskey L, Schmidt LA, et al. Analysis of the lung microbiome in the "healthy" smoker and in COPD. PLoS One 2011;6:e16384. https://doi.org/10.1371/journal.pone.0016384; Underhill DM, Iliev ID. The mycobiota: interactions between commensal fungi and the host immune system. Nat Rev Immunol 2014;14:405–16. https://doi.org/10.1038/nri3684; Spor A, Koren O, Ley R. Unravelling the effects of the environment and host genotype on the gut microbiome. Nat Rev Microbiol 2011;9:279–90. https://doi.org/10.1038/nrmicro2540; Abeles SR, Pride DT. Molecular bases and role of viruses in the human microbiome. J Mol Biol 2014;426:3892–906. https://doi.org/10.1016/j.jmb.2014.07.002. Credit: (A) Fig. 1a of Belzer C, de Vos WM. Microbes inside—from diversity to function: the case of Akkermansia. ISME J 2012;6:1449–58. https://doi.org/10.1038/ismej.2012.6. (B) Fig. 1 of Marsland BJ, Gollwitzer ES. Host-microorganism interactions in lung diseases. Nat Rev Immunol 2014;14:827–35. https://doi.org/10.1038/nri3769.

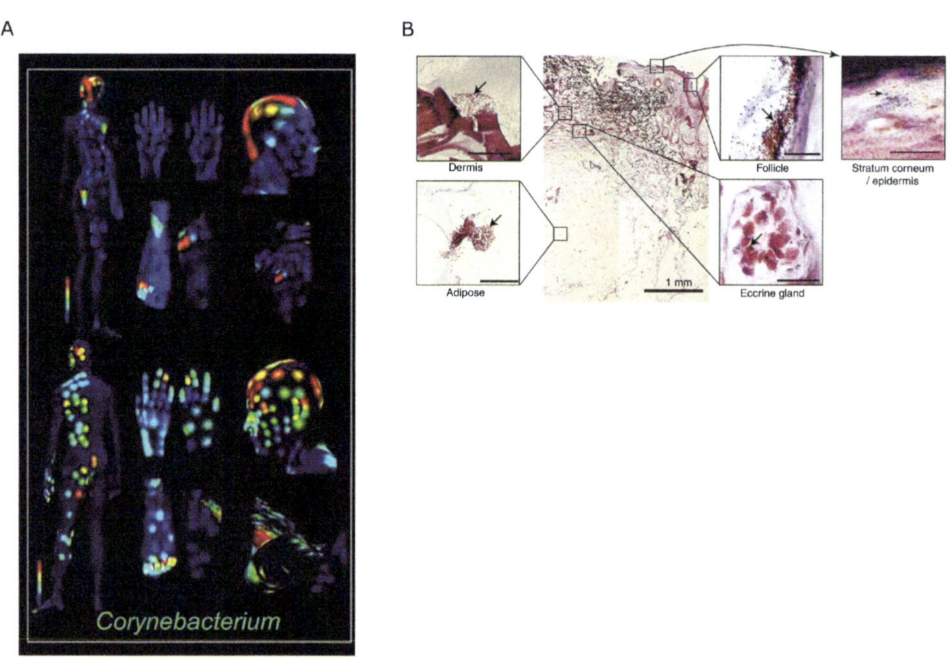

Fig. 1.6 Bacteria on and under the skin. (A) Distribution of *Corynebacterium* spp. on the human skin. (B) Gram stain for bacteria inside the skin. Cut from normal areas of melanoma samples. Credit: (A) Cropped from Fig. 3 of Bouslimani A, Porto C, Rath CM, Wang M, Guo Y, Gonzalez A, et al. Molecular cartography of the human skin surface in 3D. Proc Natl Acad Sci U S A 2015;112:E2120–9. https://doi.org/10.1073/pnas.1424409112. (B) Fig. 1c of Nakatsuji T, Chiang H-I, Jiang SB, Nagarajan H, Zengler K, Gallo RL. The microbiome extends to subepidermal compartments of normal skin. Nat Commun 2013;4:1431. https://doi.org/10.1038/ncomms2441.

a sparse population of microbes [18,19]. The oral cavity and the vagina are densely populated with bacteria, along with some viruses and sometimes fungi, but the total number of microbial cells would be 1 or 2 orders of magnitude lower than the estimate for the colon, even if the bladder and the uterus are also considered (Fig. 1.7). Semen samples can contain more than 10^6–10^7 bacteria per mL (milliliter) [20]. Microbes are found in more traditionally "sterile" places, including in tumors (Fig. 1.8), but these are not large numbers. If one day we see each of the 3×10^{13} cells in the human body to contain more than 2 intracellular bacteria, we may start to worry about the total estimate (Fig. 1.4).

Worked sample 1.1

What physiological conditions do you think might change the number of microbes in a given body site?

How many microbial cells are we losing from a given body site every day? What does that predict regarding their growth rate? Or they are repleted by microbes from other places (More in Chapters 2 and 4).

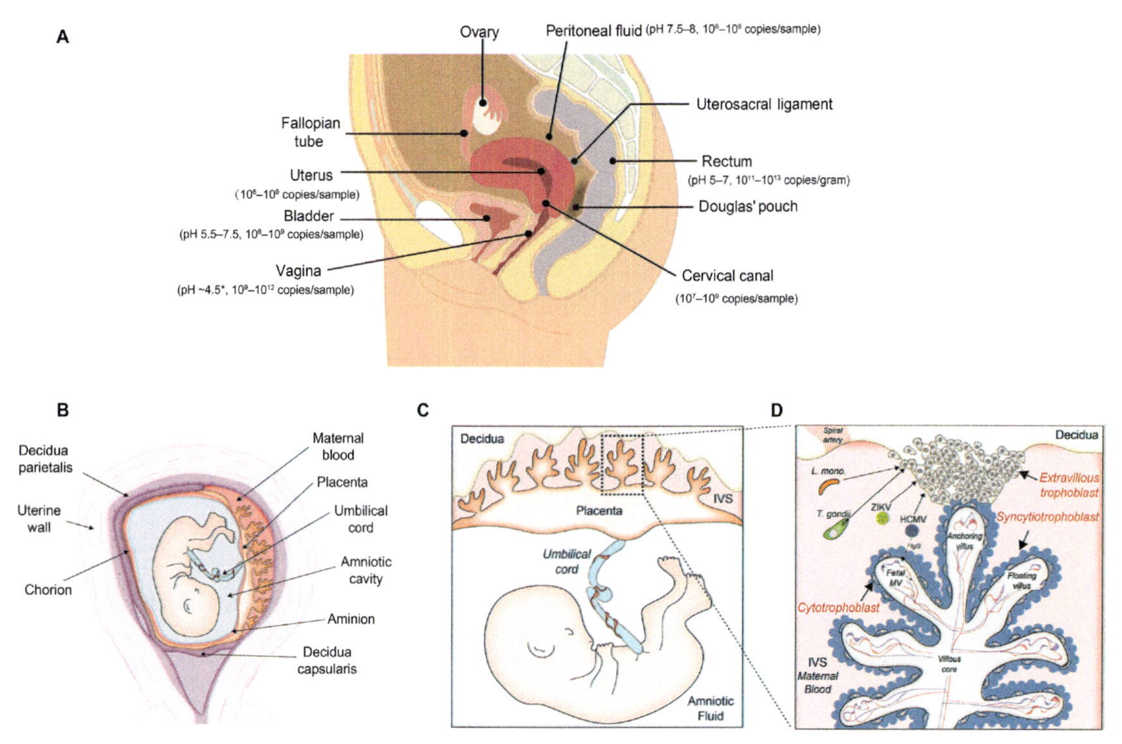

Fig. 1.7 Microbiome of the female reproductive tract. (A) Tentative estimates for the number of microbes in different regions of the female reproductive tract in nonpregnant volunteers operated for reasons not known to involve infection (e.g., uterine fibroids) [21]. Copies of total bacterial genomes were calculated according to qPCR against *Lactobacillus* species in the samples, which constitute a much smaller portion of the community in the upper reproductive tract where the pH is no longer acidic. (B–D) More research would be necessary for different regions of the placenta (more in Chapter 3). Amniotic fluid is typically only available for preterm birth. (B) Schematic of the uterine cavity during pregnancy. The developing fetus is encased within the amniotic cavity, surrounded by the chorion and amnion, and anchored to the maternal decidua by the placenta (at the site of attachment, the decidua basalis). (C) Maternal blood fills the intervillous space (IVS) via spiral arteries that bathe the surfaces of the placenta in maternal blood (once the maternal microvasculature has been established). (D) The human hemochorial placenta is formed by villous trees composed of both floating villi and anchoring villi, which attach directly to the decidua basalis by the invasion of extravillous trophoblasts (EVTs). The human placenta villous trees are covered by syncytiotrophoblasts, with a layer of cytotrophoblasts (which become discontinuous throughout pregnancy) below this layer. Several pathogens, including *Listeria monocytogenes* (*L. mono*), *Toxoplasma gondii* (*T. gondii*), human CMV (HCMV), and Zika virus (ZIKV), are thought to access the villous core following replication in EVTs. Credit: (A) Chen Chen from BGI-Shenzhen. (B–D) Fig. 1A–C of Arora N, Sadovsky Y, Dermody TS, Coyne CB. Microbial vertical transmission during human pregnancy. Cell Host Microbe 2017;21:561–7. https://doi.org/10.1016/j.chom.2017.04.007.

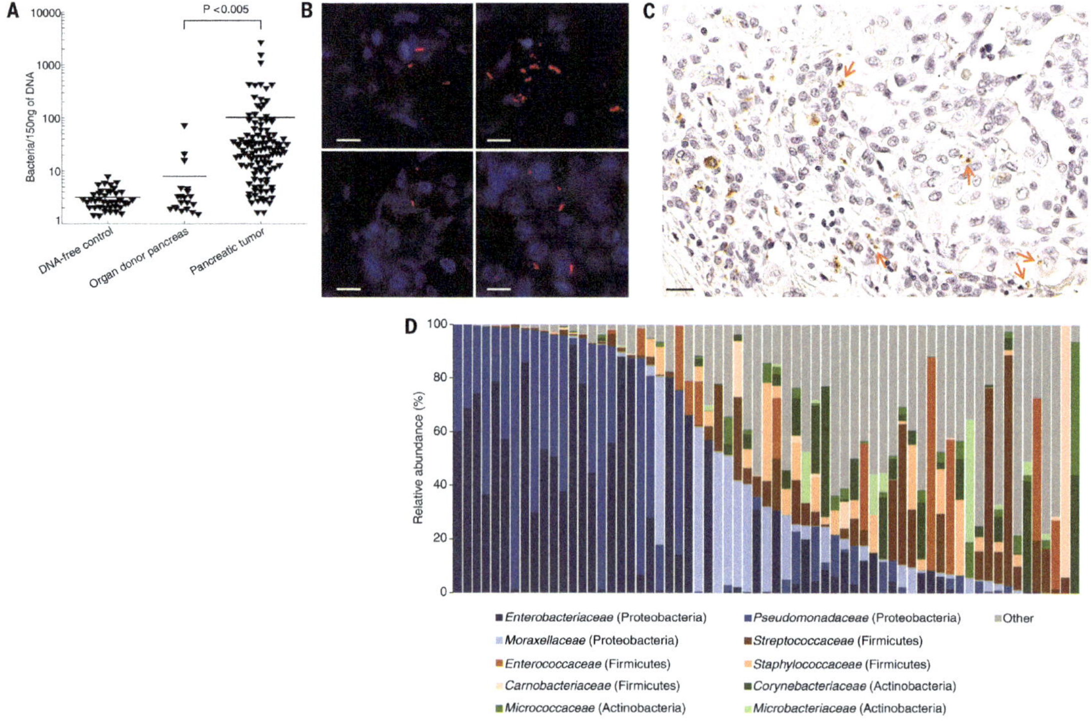

Fig. 1.8 Bacteria in pancreatic ductal adenocarcinoma (PDAC) studied by multiple methods. (A) The presence of bacteria in human pancreatic tumors or in healthy pancreatic tissue from organ donors was assessed by bacterial 16S rDNA qPCR. A calibration curve, generated by spiking bacterial DNA into human DNA, was used to estimate bacterial numbers. Bars represent the mean. (B) Fluorescence in situ hybridization was used to detect bacterial 16S rRNA sequences in a human PDAC tumor (red). Cell nuclei were stained with 4′,6-diamidino-2-phenylindole (DAPI) *(blue)*. Four sections from one tumor are presented. Scale bars, 10 mm. (C) Immunohistochemistry of a human PDAC tumor using an antibacterial LPS antibody. Arrows point to LPS staining in the tumor tissue. Scale bar, 20 mm. (D) Distribution of family-level phylotypes in 65 human PDAC tumors. Relative abundance (%) is plotted for each tumor. Credit: Fig. 4 of Geller LT, Barzily-Rokni M, Danino T, Jonas OH, Shental N, Nejman D, et al. Potential role of intratumor bacteria in mediating tumor resistance to the chemotherapeutic drug gemcitabine. Science 2017;357:1156–60. https://doi.org/10.1126/science.aah5043.

1.3 Viral particles in the human body

Viral particles have also been counted from many kinds of samples. There are 10^9 virus-like particles (VLPs) per gram of feces, 10^7 VLPs per milliliter of urine, and 10^8 VLPs per milliliter of saliva [22].

Besides bacteriophages, there are viruses for eukaryotes. Anelloviridae is a family of nonenveloped, single-stranded DNA viruses with a circular genome of 2–4 kilobases (kb). Anelloviridae is commonly detected in all the major mucosal sites, as well as in blood and semen [22]. Members of the Anelloviridae family include torque teno virus, torque teno mini virus and torque teno midi virus.

Worked sample 1.2

Do you think the total number of microbial species in the human body will be more than a few thousands?
Where do you think the current number may be underestimated?

1.4 Microbiome in other species

This book focuses on the human microbiome and diseases. With the basic principles covered, it is up to the readers to take inspirations from microbiome studies on animals or even plants (Fig. 1.9) [23], in the process of investigating a particular question.

For example, a switch of microbiome components may appear drastic in a specific lineage of the host. The insect *Cicadas* depend on the essential bacterial symbionts *Sulcia* and *Hodgkinia*, but in many Japanese *Cicadas*, *Hodgkinia* has been replaced by a fungus that is likely recruited from cicada-parasitizing *Ophiocordyceps* fungi. The fungal symbiont encodes all the pathways for B vitamins and nitrogen recycling, and could synthesize all essential and nonessential amino acids, which is more versatile than histidine and methionine synthesis provided by *Hodgkinia* [25]. Flies (*Drosophila melanogaster*) actively seek food to replenish their gut bacteria, and the taxonomic preference can be modified by early exposure to the bacteria [26].

Factors such as pH, oxygen, temperature, minerals, carbon, and nitrogen sources are well established in traditional microbiology to impact growth (Fig. 1.10) [27], and some of the bacteria can fix carbon dioxide or fix nitrogen themselves. In a community setting, and being on the surface of or inside a host presents both opportunities and challenges (more in Chapter 2). Functions such as competitive exclusion against pathogens, digestion of complex organic substrates, providing nutrients and growth factors, promoting development, and affecting behavior, are evolutionarily conserved service of the microbiome to the host, and there may be manipulation of the host for better survival of the microbes [28–31].

Compared to nutrient-poor environments such as deep ocean sediments, where a bacterium takes more than 1000 years to replicate once [33], life is fast for the human-associated microbes, despite possible differences between taxa.

1.5 Microbiome from ancient times

Paleogenomics is telling us a lot about the evolution and migration of modern humans as well as other species. Fragments of microbial DNA can also be extracted from dental calculus and coprolites (fossil feces) from ancient times (Figs. 1.11 and 1.12), which provide exciting data points to offer a glimpse into the evolution of the microbiome.

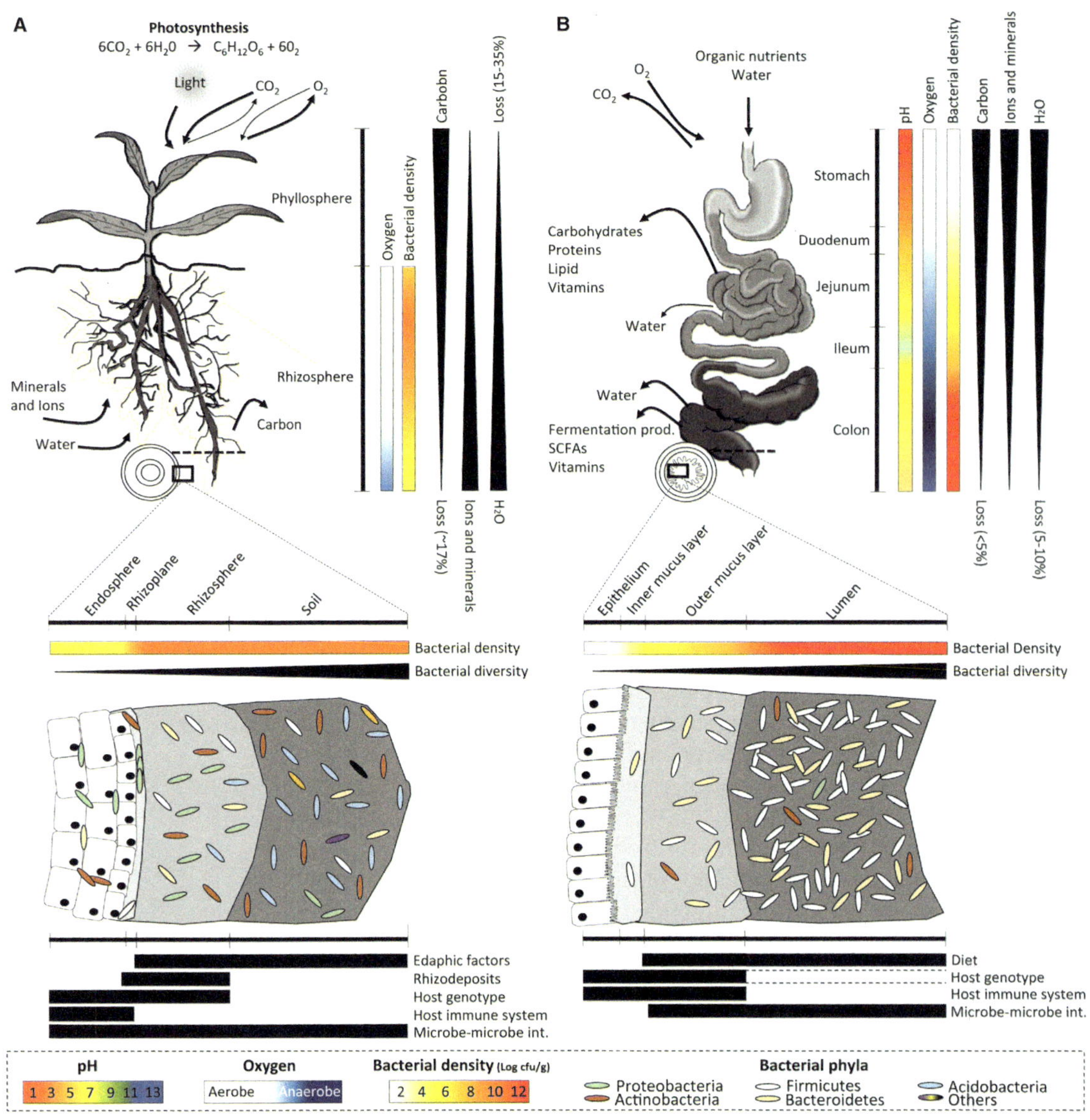

Fig. 1.9 Physiological functions of the plant roots and human gut in nutrient uptake, spatial aspects of microbiota composition, and factors driving community establishment. Spatial compartmentalization of the plant root microbiota (A) and the human gut microbiota (B). Upper panels: the major nutrient fluxes are indicated, as well as pH and oxygen gradients in relation to the bacterial density. Lower panels: compartmentalization of the microbiota along the lumen-epithelium continuum in the gut or along the soil-endosphere continuum in the root. For each compartment, the bacterial density, the bacterial diversity, and the major represented phyla are represented for both the gut and the root organs. The main factors driving community establishment in these distinct compartments are depicted with *black bars*. The gut drawing was adapted by [23] from Tsabouri et al. [24] with permission from the publisher. Credit: Fig. 1 of Hacquard S, Garrido-Oter R, González A, Spaepen S, Ackermann G, Lebeis S, et al. Microbiota and host nutrition across plant and animal kingdoms. Cell Host Microbe 2015;17:603–16. https://doi.org/10.1016/j.chom.2015.04.009.

Taxonomic rank

Phylum Order Family Genus Species

Trait	Organisms
Methanogenesis	All prokaryotes
pH	Acidobacteria, AOAs
Salinity	Verrucomicrobia, Caulobacter Planctomycetes, AOBs, SAR11
Sulfate reduction	All prokaryotes
Drought	Leaf litter comm.
Cellulase	All prokaryotes
N_2 fixation	All prokaryotes
Denitrification	All prokaryotes
Simple C substrates	Cultured prokaryotes
Temperature	Cyanobacteria, Actinobacteria, *E. coli*
Organic P acquisition	All prokaryotes
Virus resistance	Bacteria

0.15 0.1 0.05 0

Trait depth (genetic distance)

Fig. 1.10 Prokaryotic traits are conserved at different phylogenetic depths. A box plot of the depth of clades within which taxa consistently share a trait measured as the genetic distance to the root node of a clade (bottom axis; usually of the 16S rRNA gene). For some traits, the distribution was based on several studies, each with one estimate. For other traits, the authors of the original figure [32] reported the distribution calculated by a single study. For comparison, the authors of the original figure showed rough taxonomic levels on the top axis. Credit: Fig. 3 of Martiny JBH, Jones SE, Lennon JT, Martiny AC. Microbiomes in light of traits: a phylogenetic perspective. Science 2015;350:aac9323. https://doi.org/10.1126/science.aac9323.

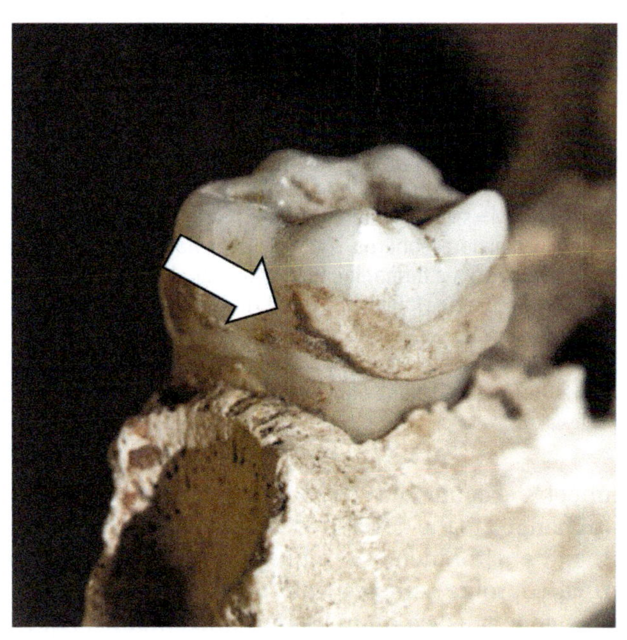

Fig. 1.11 Supra-gingival dental calculus is identifiable in a concave ring on a lower molar from a Medieval specimen, York, United Kingdom. Credit: Fig. 1 of Weyrich LS, Dobney K, Cooper A. Ancient DNA analysis of dental calculus. J Hum Evol 2015;79:119–24. https://doi.org/10.1016/j.jhevol.2014.06.018.

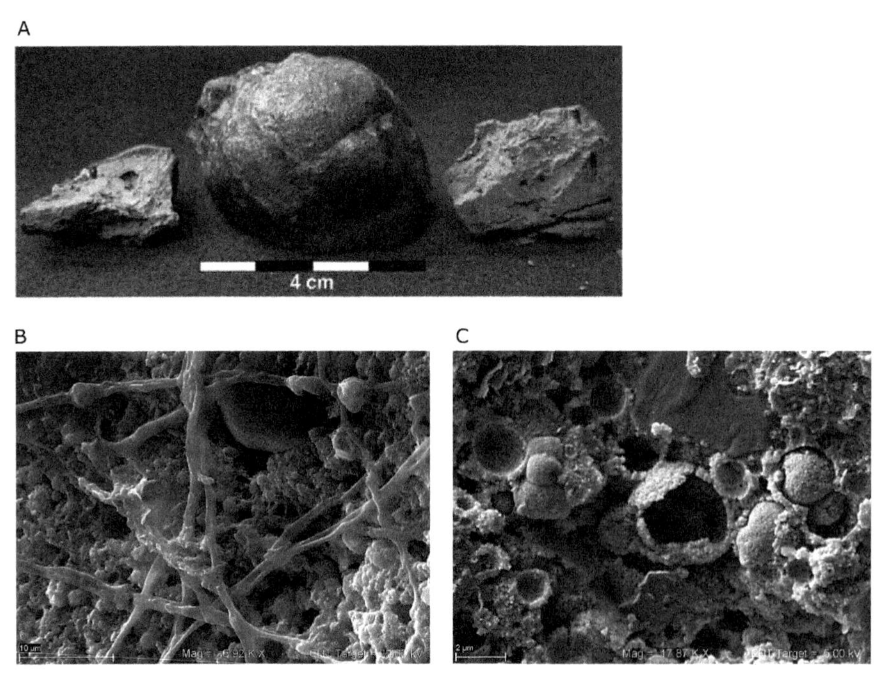

Fig. 1.12 Microbe-like structures in a coprolite (fossil feces) from the Cretaceous period, even though DNA would not have survived that long. (A) The Coprolite from a *Triceratops* dinosaur (Sterrholophus Marsh) site preserved in fluvial deposits in the Upper Cretaceous Hell Creek Formation of eastern Montana, United States. The host animal is unknown. The coprolite contains small quantities of minute bone or tooth fragments, kerogenized plant residues (pollen, spores, sporangia, cuticle, etc.), hyphae of probable fungal origin, and small detrital mineral grains in a fine-grained, highly porous matrix. (B) Branching, fungal or bacterial hyphae, and numerous small spherical objects resembling bacteria in size and shape; scale bar 10 μm; M = silicate mineral grain. (C) Structure of the coprolite matrix showing hollow, thin-walled mineral spheres, some of which have double shells with a thin void space between that may be the former location of bacterial cell walls; scale bar 2 μm; M = silicate mineral grain. Credit: Fig. 1 and Fig. 3D,F of Hollocher KT, Hollocher TC, Rigby JK. A phosphatic coprolite lacking diagenetic permineralization from the upper cretaceous hell creek formation, northeastern montana: importance of dietary calcium phosphate in preservation. Palaios 2010;25:132–40. https://doi.org/10.2110/palo.2008.p08-132r.

Leeuwenhoek's account gave us some idea about how hygiene practices have changed in the developed countries (Box 1.1). Ancient dental, fecal, and environmental samples can show diet and microbes. Neanderthal coprolite samples from a cave in Spain showed many of the same genera as we do in the gut microbiome [34]. Pathogens including viruses, bacteria, fungi, and parasitic worms might have played important roles in shaping both the genome and the microbiome [35,36]. Such historical questions are also up to the readers themselves to further explore.

1.6 Summary

Metagenomics is at its prime for human microbiome studies. Beginning with Antoni van Leeuwenhoek's observation on microorganisms in dental plaques (Box 1.1), this chapter introduces metagenomic studies of the human microbiome in the context of technological development for the advancement of microbiology. Logarithm of the number of microbial cells scales with the biomass of an animal. The current estimate of 3.8×10^{13} microbial cells for an adult human is based on the volume of the colon, where the overwhelming majority of microorganisms reside. A lot of other epithelia, tissues or body fluids have smaller populations of microbes. As detailed in Box 1.2, we will use the term microbiome to include meanings of microbiota or microflora throughout this book.

Besides technology and cost, there is no limit to where we can study for the microbiome. This book is not organized according to the traditional human microbiome body sites, and we expect to accommodate more new discoveries in the years to come. The number of microbial cells, or the microbial biomass in each sample would be an important consideration for metagenomic studies (Chapters 3–5). After covering ecological principles in Chapter 2, this book goes on with more practical knowledge for designing metagenomic studies (Chapters 3 and 4), the taxonomic resolution (Chapter 5), and how we can draw causal conclusions for the role of the microbiome in diseases (Chapter 6). We can then put the knowledge into clinical practice (Chapter 7) and more long-term health management (Chapter 8).

References

[1] Lane N. The unseen World: reflections on Leeuwenhoek (1677) 'Concerning little animals.' Philos Trans R Soc B Biol Sci 2015;370:20140344. https://doi.org/10.1098/rstb.2014.0344.

[2] Cocquyt T, Zhou Z, Plomp J, van Eijck L. Neutron tomography of Van Leeuwenhoek's microscopes. Sci Adv 2021;7. https://doi.org/10.1126/sciadv.abf2402, eabf2402.

[3] van Leeuwenhoek A. The collected letters of Antoni van Leeuwenhoek. Amsterdam: Swets and Zeitlinger; 1952.

[4] Woese CR, Kandler O, Wheelis ML. Towards a natural system of organisms: proposal for the domains Archaea, Bacteria, and Eucarya. Proc Natl Acad Sci U S A 1990;87:4576–9. https://doi.org/10.1073/pnas.87.12.4576.

[5] Handelsman J, Rondon MR, Brady SF, Clardy J, Goodman RM. Molecular biological access to the chemistry of unknown soil microbes: a new frontier for natural products. Chem Biol 1998;5:R245–9.

[6] Goin J. Microbiomes: an origin story. ASM.org, https://asm.org/Articles/2019/March/Microbiomes-An-Origin-Story#; 2019.

[7] Prescott SL. History of medicine: origin of the term microbiome and why it matters. Hum Microbiome J 2017;4:24–5. https://doi.org/10.1016/j.humic.2017.05.004.

[8] Dworkin M. Sergei Winogradsky: a founder of modern microbiology and the first microbial ecologist. FEMS Microbiol Rev 2012;36(2):364–79. https://doi.org/10.1111/j.1574-6976.2011.00299.x.

[9] Lane-Petter W. The provision and use of pathogen-free laboratory animals. Proc R Soc Med 1962;55:253–63. https://doi.org/10.1177/003591576205500402.

[10] Whipp, et al. Fungi in biological control systems. Manchester University Press; 1988.

[11] Kempes CP, Wang L, Amend JP, Doyle J, Hoehler T. Evolutionary tradeoffs in cellular composition across diverse bacteria. ISME J 2016;10:2145–57. https://doi.org/10.1038/ismej.2016.21.

[12] Shi H, Westfall CS, Kao J, Odermatt PD, Anderson SE, Cesar S, et al. Starvation induces shrinkage of the bacterial cytoplasm. Proc Natl Acad Sci U S A 2021;118. https://doi.org/10.1073/pnas.2104686118.

[13] Kieft TL, Simmons KA. Allometry of animal-microbe interactions and global census of animal-associated microbes. Proc R Soc B Biol Sci 2015;282:20150702. https://doi.org/10.1098/rspb.2015.0702.

[14] Sender R, Fuchs S, Milo R. Are we really vastly outnumbered? revisiting the ratio of bacterial to host cells in humans. Cell 2016;164:337–40. https://doi.org/10.1016/j.cell.2016.01.013.

[15] Sender R, Fuchs S, Milo R. Revised estimates for the number of human and bacteria cells in the body. PLoS Biol 2016;14. https://doi.org/10.1371/journal.pbio.1002533, e1002533.

[16] Bouslimani A, Porto C, Rath CM, Wang M, Guo Y, Gonzalez A, et al. Molecular cartography of the human skin surface in 3D. Proc Natl Acad Sci U S A 2015;112:E2120–9. https://doi.org/10.1073/pnas.1424409112.

[17] Byrd AL, Belkaid Y, Segre JA. The human skin microbiome. Nat Rev Microbiol 2018;16:143–55. https://doi.org/10.1038/nrmicro.2017.157.

[18] Hasleton PS. The internal surface area of the adult human lung. J Anat 1972;112:391–400.

[19] Man WH, de Steenhuijsen Piters WAA, Bogaert D. The microbiota of the respiratory tract: gatekeeper to respiratory health. Nat Rev Microbiol 2017;15:259–70. https://doi.org/10.1038/nrmicro.2017.14.

[20] Hou D, Zhou X, Zhong X, Settles ML, Herring J, Wang L, et al. Microbiota of the seminal fluid from healthy and infertile men. Fertil Steril 2013;100:1261–9. https://doi.org/10.1016/j.fertnstert.2013.07.1991.

[21] Chen C, Song X, Wei W, Zhong H, Dai J, Lan Z, et al. The microbiota continuum along the female reproductive tract and its relation to uterine-related diseases. Nat Commun 2017;8:875. https://doi.org/10.1038/s41467-017-00901-0.

[22] Liang G, Bushman FD. The human virome: assembly, composition and host interactions. Nat Rev Microbiol 2021. https://doi.org/10.1038/s41579-021-00536-5.

[23] Hacquard S, Garrido-Oter R, González A, Spaepen S, Ackermann G, Lebeis S, et al. Microbiota and host nutrition across plant and animal kingdoms. Cell Host Microbe 2015;17:603–16. https://doi.org/10.1016/j.chom.2015.04.009.

[24] Tsabouri S, Priftis KN, Chaliasos N, Siamopoulou A. Modulation of gut microbiota downregulates the development of food allergy in infancy. Allergol Immunopathol (Madr) 2014;42(1):69–77. https://doi.org/10.1016/j.aller.2013.03.010.

[25] Matsuura Y, Moriyama M, Łukasik P, Vanderpool D, Tanahashi M, Meng X-Y, et al. Recurrent symbiont recruitment from fungal parasites in cicadas. Proc Natl Acad Sci U S A 2018;115:E5970–9. https://doi.org/10.1073/pnas.1803245115.

[26] Wong AC-N, Wang Q-P, Morimoto J, Senior AM, Lihoreau M, Neely GG, et al. Gut microbiota modifies olfactory-guided microbial preferences and foraging decisions in drosophila. Curr Biol 2017;27:2397–2404.e4. https://doi.org/10.1016/j.cub.2017.07.022.

[27] Reimer LC, Vetcininova A, Carbasse JS, Söhngen C, Gleim D, Ebeling C, et al. BacDive in 2019: bacterial phenotypic data for High-throughput biodiversity analysis. Nucleic Acids Res 2019;47:D631–6. https://doi.org/10.1093/nar/gky879.

[28] Fraune S, Bosch TCG. Why bacteria matter in animal development and evolution. Bioessays 2010;32:571–80. https://doi.org/10.1002/bies.200900192.
[29] McFall-Ngai M, Hadfield MG, Bosch TCG, Carey HV, Domazet-Lošo T, Douglas AE, et al. Animals in a bacterial world, a new imperative for the life sciences. Proc Natl Acad Sci U S A 2013;110:3229–36. https://doi.org/10.1073/pnas.1218525110.
[30] King KC, Brockhurst MA, Vasieva O, Paterson S, Betts A, Ford SA, et al. Rapid evolution of microbe-mediated protection against pathogens in a worm host. ISME J 2016;10:1915–24. https://doi.org/10.1038/ismej.2015.259.
[31] Sherwin E, Bordenstein SR, Quinn JL, Dinan TG, Cryan JF. Microbiota and the social brain. Science 2019;366. https://doi.org/10.1126/science.aar2016, eaar2016.
[32] Martiny JBH, Jones SE, Lennon JT, Martiny AC. Microbiomes in light of traits: a phylogenetic perspective. Science 2015;350:aac9323. https://doi.org/10.1126/science.aac9323.
[33] Jørgensen BB, Marshall IPG. Slow microbial life in the seabed. Ann Rev Mar Sci 2016;8:311–32. https://doi.org/10.1146/annurev-marine-010814-015535.
[34] Rampelli S, Turroni S, Mallol C, Hernandez C, Galván B, Sistiaga A, et al. Components of a Neanderthal gut microbiome recovered from fecal sediments from El salt. Commun Biol 2021;4:169. https://doi.org/10.1038/s42003-021-01689-y.
[35] Enard D, Petrov DA. Evidence that RNA viruses drove adaptive introgression between Neanderthals and modern humans. Cell 2018;175:360–371.e13. https://doi.org/10.1016/j.cell.2018.08.034.
[36] Rasmussen S, Allentoft ME, Nielsen K, Orlando L, Sikora M, Sjögren K-G, et al. Early divergent strains of *Yersinia pestis* in Eurasia 5,000 years ago. Cell 2015;163:571–82. https://doi.org/10.1016/j.cell.2015.10.009.

2 Microbiota

2.1 Trophic levels in macroecology

A classical macroecological community takes in energy from the sun and accumulates biomass in each trophic level (Fig. 2.1). The sun is the only energy source in such a classical system, and this energy is only captured by photosynthetic microorganisms and plants. Each trophic level of animals is added on, as a predator in the food chain. Maintaining overall biodiversity cannot replace the management of particular ecosystem functions [1].

Terms like microbiota, microflora (Box 1.2, Chapter 1), and diversity have been popular ever since the revival of studies on commensal microbes almost two decades ago. We, however, remain humble for how little we understand the ecology of the human microbiome. Besides scavenging on host molecules and undigested food, members of the human microbiome can be complete or partial chemoautotrophs, e.g., to survive with CO_2 fixation, anaerobic nitrate respiration. The exchange of molecules or electrons between microbes and with their host has also largely escaped direct measurements, waiting for further breakthroughs in technology.

2.2 Microbiome stability, diversity, and richness

By analogy to macroecological systems, a more diverse microbiome is likely more stable, which is presumably healthy. α-diversity within each sample, often according to the Shannon index, takes into account species richness and evenness, so that both the number of taxa and their relative abundance distribution are considered. Yet, low-abundance microbes may have a good niche or a steady source, to make us not worried about their extinction.

Investigating Human Diseases with the Microbiome: Metagenomics Bench to Bedside.
https://doi.org/10.1016/B978-0-323-91369-0.00002-9

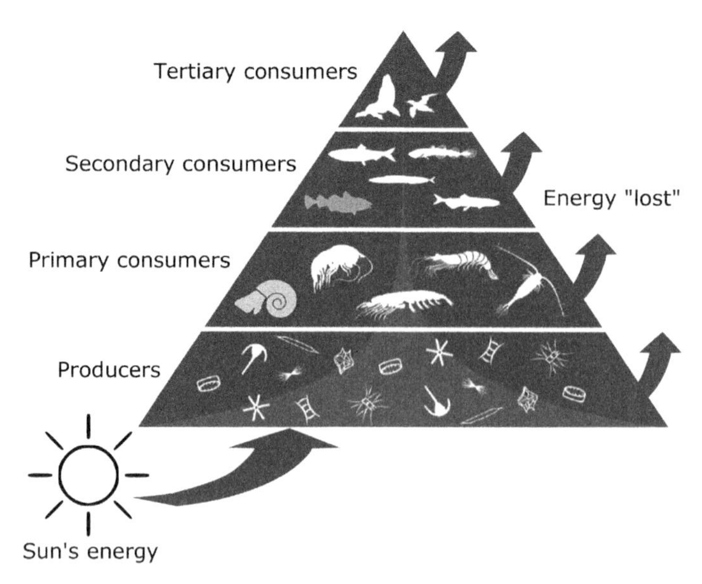

Fig. 2.1 An example of trophic levels in macroecology. Energy from the sun is converted into biomass by producers (plants). Each layer of consumers feeds on the previous layer. Credit: From Fig. 1 of Eddy T.D., Bernhardt J.R., Blanchard J.L., et al. *Energy flow through marine ecosystems: confronting transfer efficiency. Trends Ecol Evol* 2021;36(1):76–86. doi:10.1016/j.tree.2020.09.006. The sun and the arrows have been added to emphasize energy.

For bacteria, much of the functional characteristics are maintained down to the genus level (Chapter 1, Fig. 1.10, more on taxonomy in Chapter 5) [2]. It has been shown with soil and leaf bacteria that the family-level composition would converge under a single nutrient source, regardless of initial concentrations; the variation can be large at the species level, and rare species could co-exist [3]. While cooperation can be efficient, theoretical analyses indicate that competition, rather than cooperation, increases the stability of an ecosystem and is better for the host [4]. Nonspecific cross-feeding networks, instead of pairwise interactions, could help stabilize the competitive ecosystem [3]. Spatial structure that weakens the interaction between species also improves community stability [4]. Bacteriophages could appear to co-exist with their target bacteria due to spatial heterogeneity in the mouse gut [5].

In the human fecal microbiome from developed countries, having a more diverse fecal microbiome usually means having more *Firmicutes* instead of *Bacteroidetes* or *Actinobacteria*, or *Proteobacteria*. Genera in the *Firmicutes* phylum are responsible for many fermentation processes in the gut, and both nonspecific cross-feeding and spatial segregation may be involved. In preagricultural populations such as those in modern-day Amazon, and Hadza hunter-gathers, how *Spirachaetes* (Box 2.1, Fig. 2.2), *Verrumicrobia*, or other taxa contribute to the higher

Box 2.1 *Treponema*, the preagricultural "enterotype?"

Prevotella (*P. copri*) has become known as the gut bacterium that is rarer than *Bacteroides* spp. in people in developed countries, while its relative abundance can be ~ 50% in some people in developing countries [110–112]. *Treponema* spp., however, are likely fundamental in the gut of hunter-gathers. *Treponema* spp. have also been found in the gut of nonhuman primates such as gorillas and baboons [113]. *Treponema* spp. were among the gut bacteria that showed seasonal fluctuations in humans and gorillas [114,115], consistent with increased consumption of dietary fibers. Xylan cellulose-degrading *Treponema* spp. are found in the hindgut of termites, whereas bacteria of the *Bacteroidetes* phylum are mainly responsible for the hydrolysis of xylan in the human gut and in the cow rumen [116]. Cooking followed by cooling slows the digestion of high-amylose starch (i.e., more resistant starch), and high-amylose starch enriches for *Treponema*, concomitant with reduction in the level of *Prevotella* [117,118].

The *Spirochaetes* phylum that contains the *Treponema* genus is present in many animals [113,119]. *Spirochaetes* appeared heritable in metagenomic data from the TwinsUK cohort, but given the fewer than 10 genes detected for this phylum in this modern cohort, we were not confident about the result that points to human genetic variations that help maintain this phylum in the gut [120].

The gut *Treponema* spp. are not the famous periodontal pathogen *T. denticola*, or the sexually transmitted *T. pallidum*. The species *T. succinifaciens* has been detected in multiple rural populations nowadays as well as in Mexico over 1000 years ago [96]. *T. succinifaciens* could not ferment amino acids, but is capable of fixing CO_2 to produce succinate [121]. Acetate is produced through reductive acetogenesis, from H_2 and CO_2. Feeding is a major contributor to the circadian rise in CO_2 level [122]. *Prevotella* spp. and *Bacteroides* spp. produce succinate in addition to propionate, and the succinate could be used for intestinal gluconeogenesis [123,124].

diversity in the gut microbiome remain unclear. Such closely interacting groups of people may have a lower β-diversity among their gut microbiome, i.e., more similar between individuals (Fig. 2.3, Box 2.2), young and old [6,7].

Richness at the gene or genus level only counts the number of genes or genera in each sample. An increase in richness could be due to microbes of low abundance in the sample, with minimum effect on the α-diversity (which takes into account evenness among taxa). And the detection of low abundance microbes depends on a sufficient amount of sequencing (Chapter 3).

For the gut (fecal) microbiome, inflammatory bowel diseases and obesity are known for having a lower α-diversity (Table 2.1). Colorectal cancer, and possibly other diseases with constipation, can show a higher richness in the gut microbiome (Table 2.1), containing low abundance bacteria from the mouth or other body sites. For the teeth and the vagina, a low diversity is considered healthy, in comparison to periodontitis and bacterial vaginosis, respectively (Table 2.1). Low diversity types

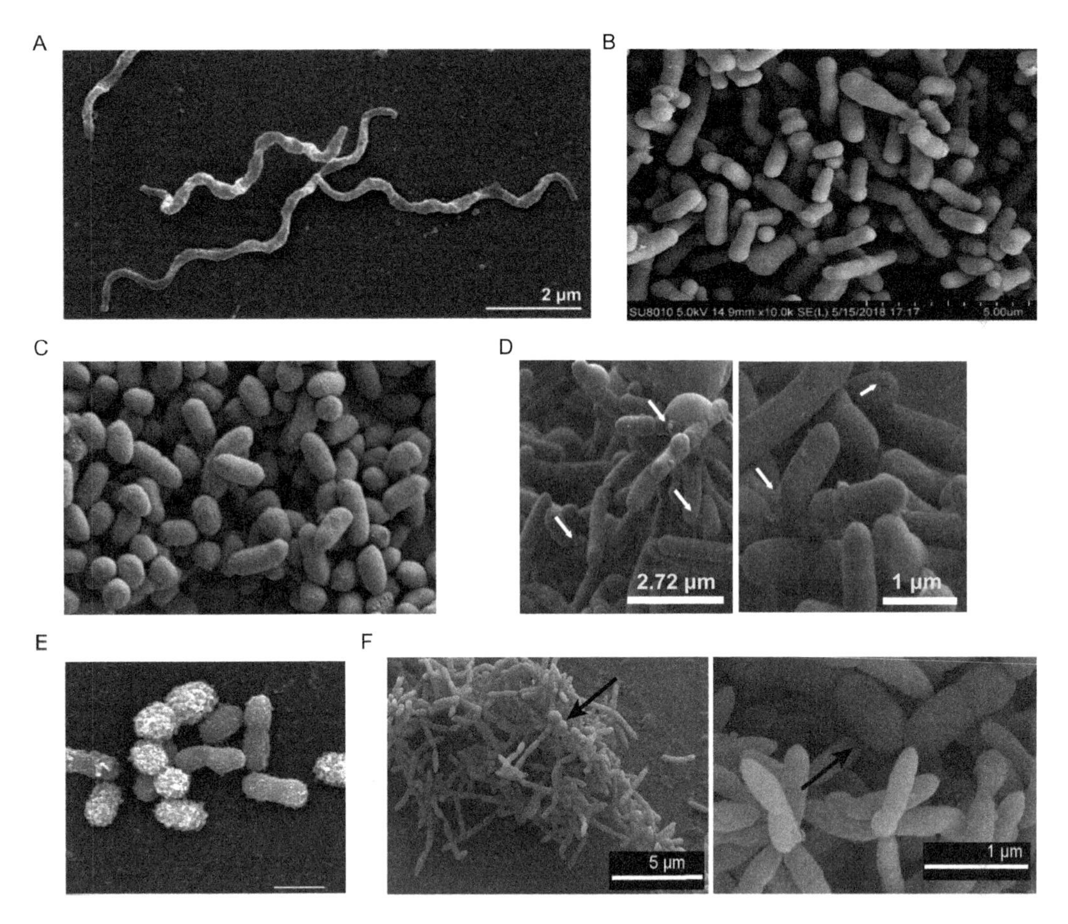

Fig. 2.2 Scanning electron micrographs showing the shape of some human commensal bacteria and archaea. (A) *Treponema succinifaciens.* (B) *Bifidobacterium animalis* subsp. lactis. (C) *Bacteroides thetaiotaomicron.* (D) *Faecalibacterium prausnitzii. Arrows* indicate cell wall extensions, like "swellings." (E) *Akkermansia muciniphila.* (F) Cocultures of *Christensenella minuta* and the archaea *Methanobrevibacter smithii. Arrows* indicate *Methanobrevibacter smithii.* Credit: (A) Han C, Gronow S, Teshima H, Lapidus A, Nolan M, Lucas S, et al. Complete genome sequence of Treponema succinifaciens type strain (6091). Stand Genomic Sci 2011;4:361–70. https://doi.org/10.4056/sigs.1984594. (B) Fig. 3C of Liu W, Chen M, Duo L, Wang J, Guo S, Sun H, et al. Characterization of potentially probiotic lactic acid bacteria and bifidobacteria isolated from human colostrum. J Dairy Sci 2020;103:4013–25. https://doi.org/10.3168/jds.2019-17602. (C) Fig 1A of Stentz R, Horn N, Cross K, et al. Cephalosporinases associated with outer membrane vesicles released by Bacteroides spp. protect gut pathogens and commensals against β-lactam antibiotics. J Antimicrob Chemother. 2015;70(3):701–709. https://doi.org/10.1093/jac/dku466. (D) Miquel S, Martín R, Rossi O, Bermúdez-Humarán L, Chatel J, Sokol H, et al. *Faecalibacterium prausnitzii* and human intestinal health. Curr Opin Microbiol 2013;16:255–61. https://doi.org/10.1016/j.mib.2013.06.003. (E) Derrien M, Belzer C, de Vos WM. Akkermansia muciniphila and its role in regulating host functions. Microb Pathog 2017;106:171–81. https://doi.org/10.1016/j.micpath.2016.02.005. (F) Ruaud A, Esquivel-Elizondo S, de la Cuesta-Zuluaga J, Waters JL, Angenent LT, Youngblut ND, et al. Syntrophy via interspecies H_2 transfer between christensenella and methanobrevibacter underlies their global cooccurrence in the human gut. MBio 2020;11. https://doi.org/10.1128/mBio.03235-19.

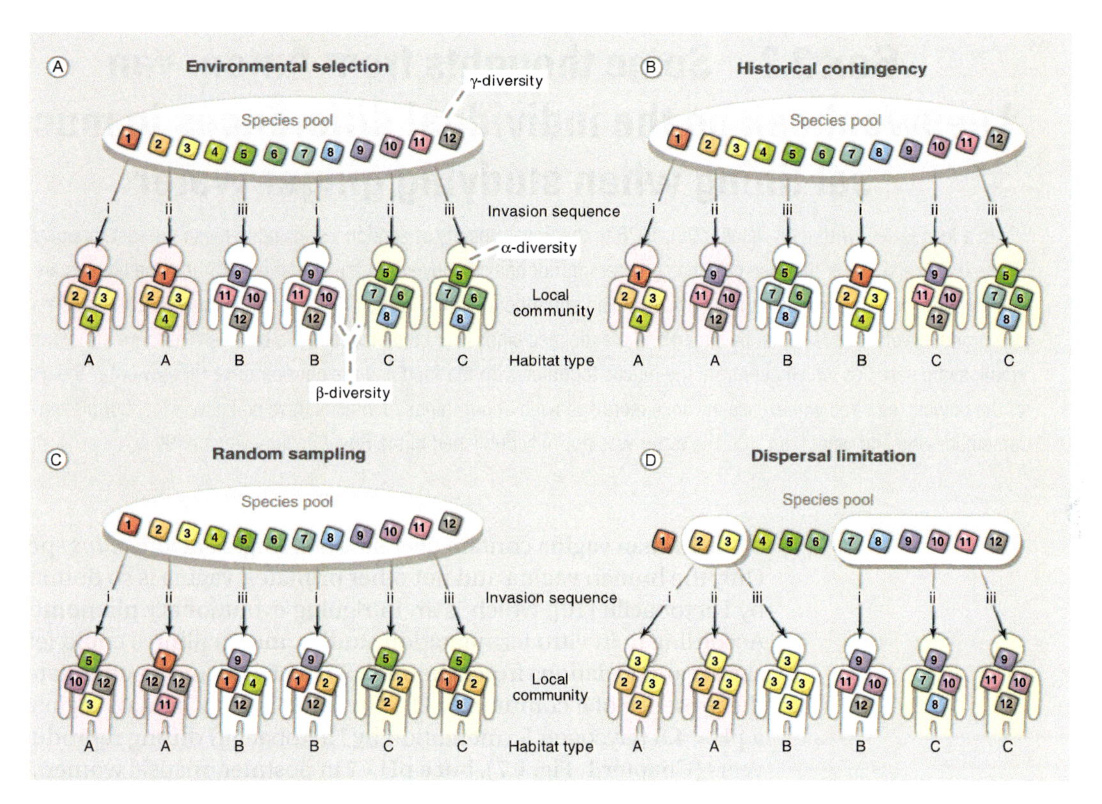

Fig. 2.3 Scenarios for assembly of the human microbiome. Alternative community assembly scenarios could give rise to the compositional variations observed in the human microbiota. Each panel shows the assembly of local communities in different habitat types from a pool of available species. (A–C) Each local community has access to all available colonists, but the order of invasion varies. In (A), local species composition is determined primarily by environmental selection: Regardless of invasion order, habitats with initially similar conditions select for similar assemblages. In (B), the opposite is true: Regardless of initial habitat conditions, historical contingencies (i.e., differences in the timing and order of species invasions) determine assemblage composition. In (C), neither habitat nor history matter: Local communities assemble via random draws from the species pool. (D) Dispersal barriers result in local communities that assemble from different species pools. For each of the pools, local communities may assemble as in (A), (B), or (C). The meaning of three different diversity measures is shown in (A): γ-diversity refers to the "regional" species pool (i.e., the total diversity of the local communities connected via dispersal); β-diversity refers to the differences between local communities (species turnover); and α-diversity refers to the diversity within a local community. Although multiple scenarios are likely to apply to any real-world setting, one may dominate. For example, differences between body habitats may be best explained by environmental selection, differences between siblings for the same habitat may be best explained by historical contingency, differences between monozygotic twins prior to weaning highlight the role of stochasticity, and differences between neonates born by cesarean section versus vaginal delivery are likely to be explained by dispersal limitation. Credit: Fig. 1 of Costello EK, Stagaman K, Dethlefsen L, Bohannan BJM, Relman DA. The application of ecological theory toward an understanding of the human microbiome. Science 2012;336:1255–62. https://doi.org/10.1126/science.1224203. Adapted by Costello, Alm from Chase JM. Oecologia 2003;136:489, and Fukami T. In: Verhoef HA, Morin PJ, editors. Community ecology: processes, models, and applications. Oxford: Oxford Univ. Press; 2010. p. 45–54.

Box 2.2 Some thoughts from Antoni van Leeuwenhoek on the individual differences in mucosal lining when studying ginger water

In a long letter written on October 9th, 1676 to the Royal Society of London, he included these interest thoughts, "Thus it may also occur that one body has greater interior heat or movement than another, or that some bowels are covered with soft thin mucus, others with thick and still others with hard or stiff mucus; consequently one person will be purged by very mild, another by strong and drastic medicines. But I am also ready to believe that there are extremely small, sharp particles which penetrate the mucus together with the food and are not operative till they enter the globules of the bowels, etc. You will excuse my once more speaking about things of which I have no knowledge, but please remember who and what I am" [8]. The letter was not fully published in the Royal Society version [9].

of the human vagina contain over 90% of a single *Lactobacillus* species. Only the human vagina and not other primates' vagina is so dominated by Lactobacilli [10], which is an intriguing evolutionary phenomenon. According to in vitro fermentation studies, monocultures could lead to significant deviations from a neutral pH, which would be attenuated by the presence of a community [3]. The low-diversity human vaginal has a pH < 4.5 (glycogen fermentation by Lactobacilli) during reproductive years (Chapter 1, Fig. 1.7), but a pH ~ 7 in postmenopausal women, who support a smaller number of vaginal bacterial cells [11–13].

2.3 De novo assembly of microbiota and robustness against invasions

Infant receives microbes from the mother through the fecal-oral route and through the reproductive tract, and then through breast milk, skin contact, etc. (Fig. 2.4, more in Chapter 8). Microbes, including spore-forming bacteria, present in the environment from family members and pets could also colonize the infant. They likely contribute to the increasing gut microbiome diversity after cessation of breastfeeding [35], a weaning response together with vitamin A and SCFAs (short-chain fatty acids) that also signal the development of the mucosal immune cells [36]. At least for mice, the introduction of solid foods in a critical time window before weaning reduces the risk of allergy [36,37]. In addition to historical contingency and succession of the microbes themselves (Fig. 2.3), babyhood is a key stage for immune development throughout the body, which has a major impact on which of the existing and newly acquired microbes are going to stay. Recent Metagenome-Genome-wide Association Studies (M-GWAS)

Table 2.1 Diseases and microbiome diversity.

Condition	Sample	Microbiome diversity compared to healthy controls	Microbiome richness compared to healthy controls	Reference
Obesity	Feces	Decreased	Lower in some +	[14–16]
Crohn's disease	Mucosal biopsies; feces	Decreased	Decreased	[17–21]
Ulcerative colitis	Feces	Decreased in severe cases	Decreased in severe cases	[20,22]
Colorectal cancer	Rectal mucosal biopsies; feces	No difference	Increased*	[23,24]
Schizophrenia	Feces	Increased	Increased	[25]
Major depressive disorder	Feces	No difference	NA	[26]
Breast cancer	Feces	Increased in premenopausal cases	Increased in postmenopausal cases	[27]
Indirect‡ breastfeeding	Breast milk	Lower	Lower	[28]
Melanoma	Tumor	Not mentioned	Increased	[29]
Periodontal disease	Dental plaque	Increased	Increased	[30]
Bacterial vaginosis	Vaginal swab	Increased	Increased	[31]
Preterm birth	Vaginal swab	Increased	Increased	[32]
Male infertility	Semen	Increased	NA	[33]

This is not meant to be an exhaustive list, and only serves to show the range of possibilities. Microbiome diversity was compared at the species or genus level, unless otherwise noted. +, Gene richness from metagenomics confirmed by Human intestinal tract chip (HITChip) against ribosomal 16S DNA. *, Lower in [34] potentially due to cleansing for colonoscopy. ‡, i.e., pump instead of a direct manual expression of breast milk; Diversity according to inverse Simpson Index instead of Shannon index.
Credit: Huijue Jia.

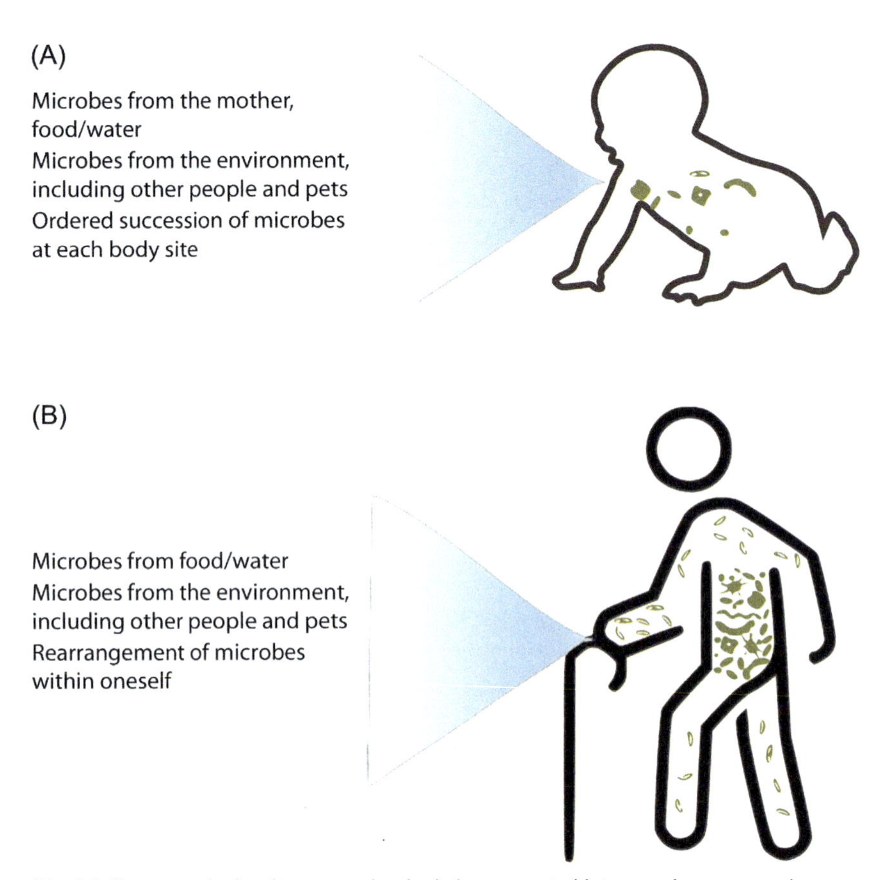

Fig. 2.4 Sources of microbes are selectively incorporated into ones' commensal community. (A) Infant. (B) Old age. Credit: Huijue Jia.

studies have not enrolled infants or children, but it would be a good idea to beware of the genetic influence (Box 2.3).

It appeared to be part of the healthy aging process to have an increasing richness in the gut microbiome, while dominant taxa became not as dominant [38–40]. Yet we do not know whether the numbers change due to a weaker immune system (e.g., fewer antibodies IgA as well as weaker peristalsis to keep the gut bacteria around, Section 2.6), or the microbes are more at different places throughout the aging human body (Fig. 2.4). The vaginal, gut, and oral microbiome show some correlation with hormones [12,41]. Waning of a dominant species could unmask low-abundance species. It will be a difficult decision for the aging immune system whether to try to eliminate an incoming microbe. Some of the fecal microbiome biomarkers for diseases such as colorectal cancer, liver and cardiovascular diseases may be of oral origin, and their abundance in the gut show some associations with human genetic variations [42].

Box 2.3 Prominent human genes associated with the fecal microbiome

LCT

Lactose-resistant individuals could have more *Bifidobacterium* if they ingest lactose, e.g., milk. The GWAS association between *LCT* and *Bifidobacterium* is the strongest signal in multiple European cohorts, while being undetectable in Chinese, and intermediate in a cohort from Israel [42,125]. Other bacteria, such as the Firmicutes *Negativibacillus*, and *Ruminococcus* sp. UBA3855, are also associated with *LCT* [126]. Serum level of the tryptophan metabolite indolepropionate has been reported to associate with nonpersistent *LCT*, possibly through *Bifidobacterium*; Indolepropionate associated with dietary fiber and reduced risk of Type 2 diabetes, and dietary fiber positively associated with Firmicutes such as *Butyrivibrio*, *Ruminococcus*, *Eubacterium*, and the Actinobacteria *Cellulomonas* in the same study [127].

Both historically and in modern days, various animals including camels and horses have served as sources of milk, all with different glycan structures and concentrations [128,129]. According to archeological evidence, consumption of milk products was well documented in Neolithic times, as early as 6000 years ago, 2000 years before the emergence of the lactose-persistent *LCT* allele [130]. The enzyme encoded by *LCT*, lactase phlorizin hydrolase (purified from sheep small intestine), has been shown to hydrolyze flavonoids and isoflavone that are common in plants [131]. Yet, spreading of the lactose-persistent *LCT* alleles is mostly attributed to pastoral populations [128].

ABO

Fecal microbial associations with the ABO loci also reached study-wide significance in European cohorts [126,132]. ABO histo-blood groups and *FUT2* secretor status associated with fecal relative abundances of *Bacteroides* spp. and *Faecalibacterium* spp., with implications for inflammatory bowel diseases and beyond [132]. Population differences in ABO blood group distribution can be rather large. For example, in malaria endemic regions blood group O can take up over 50% of the population [133,134]. Blood group B lowers susceptibility to *Vibrio cholera* infection [135]. The stomach cancer pathogen, *Helicobacter pylori* strains may differentially bind ABO blood groups [133,136].

2.4 Types of habitats for the skin microbiome

Although microbes are also found below the epidermis (Chapter 1, Fig. 1.6), we know too little about these, and are only talking about the skin microbiome on the surface of our body here. The major types of habitats include sebaceous (oily, waxy), moist, foot, and dry, corresponding to very different communities (Fig. 2.5). For both males and females, skin pH becomes less acidic with age [43,44]. High amounts of sebum secretion from adolescence until middle ages create a unique niche for *Cutibacterium acnes* (renamed from *Propionibacterium acnes*). So whenever we see *Cutibacterium acnes* DNA in the upper reproductive tract or in the brain [45–47], we get curious about what

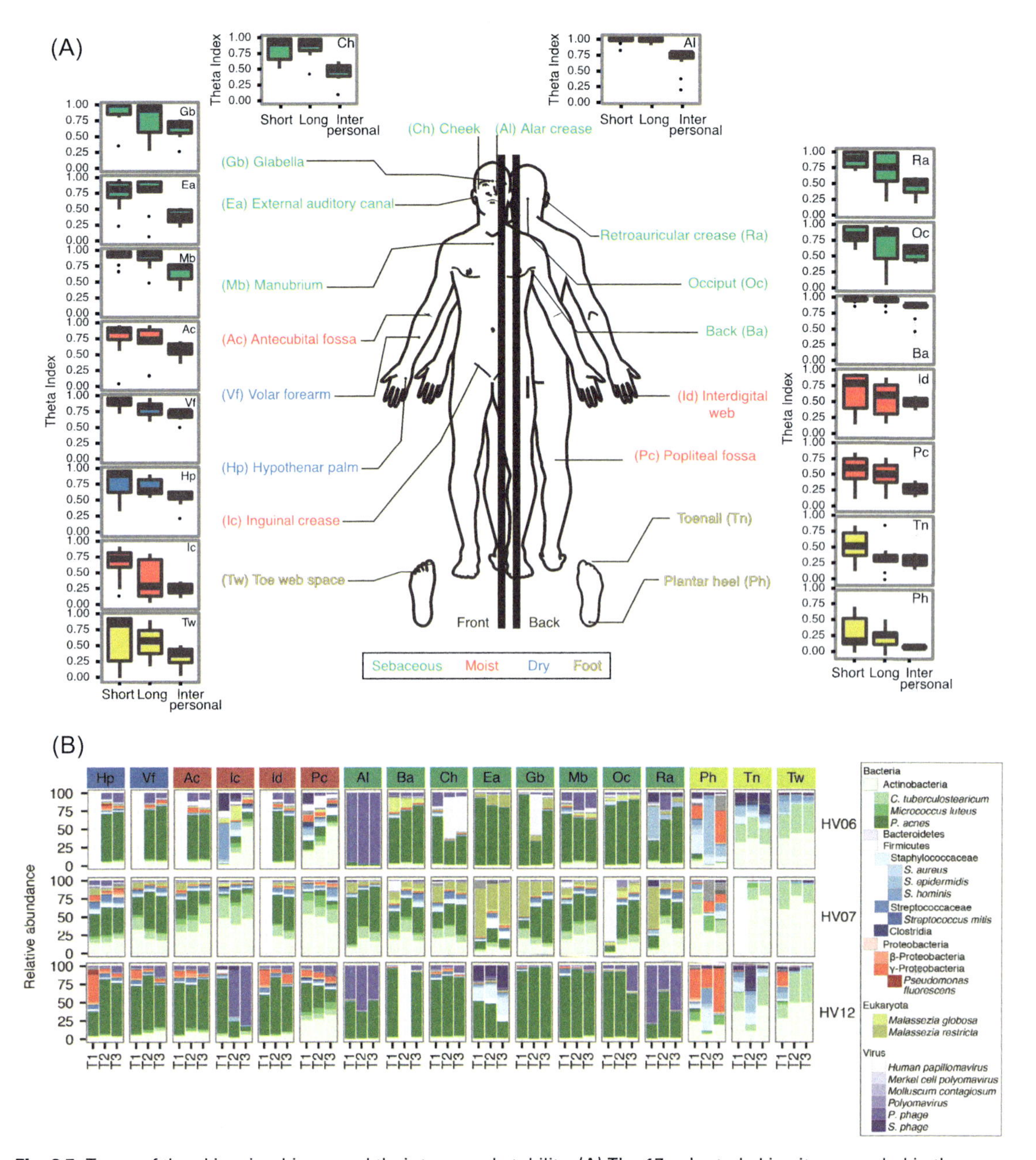

Fig. 2.5 Types of the skin microbiome and their temporal stability. (A) The 17 selected skin sites sampled in the study, and their location on the human body. These sites represent four microenvironments: sebaceous *(green)*, dry *(blue)*, and moist *(red)*, and foot *(black)*. For all sites, boxplots of Yue-Clayton theta indices calculate the similarity between samples in the time series, aggregated by the skin site characteristic. "Long" duration indicates > 1 year between samplings; "Short" duration averages a month. In comparison, "Interpersonal" values show the average distance between individuals. *Black lines* indicate median, *boxes* show first and third quartiles. Panels are color coded by site characteristic. For all intra- versus inter-individual comparisons, $P < .05$ except for the Id ($P = .24$); for all long versus short comparisons, $P > .05$ except for the Ic ($P = .04$). (B) Relative abundances of the most common skin bacteria, fungi, and viruses are shown for three representative individuals. T1, T2, and T3 indicate the order in the time series. Credit: From Fig. S1A and Fig. 2A of Oh J, Byrd AL, Park M, Kong HH, Segre JA. Temporal stability of the human skin microbiome. Cell 2016;165:854–66. https://doi.org/10.1016/j.cell.2016.04.008.

lipids it might be eating there, or maybe it has been killed by its phage. *Staphylococcus epidermis*, and multiple species of *Corynebacterium* could be found in all types of skin sites. Besides phages, polyomavirus such as the potentially oncogenic Merkel cell polyomavirus, and papillomaviruses are also detected rather often [48]. The fungi *Malassezia* can be in hair follicles as well as on the epidermis [48]. Dry sites such as the forearms cannot sustain a high microbial biomass, and the skin microbiome tends to be diverse and changeable [49,50].

The fecal microbiome does not fit Hubbell's neutral theory, but some vaginal, skin, and respiratory microbial communities apparently do [51,52]. Neutral theory, in contrast to the more traditional niche theory, emphasizes random sampling from a source community (Fig. 2.3), followed by stochastic growth and death dynamics in competition for local space.

Chemicals stay along for longer than we think, possibly more so when they are not yet a good food for the microbiome. In a study which asked the two volunteers to avoid showering and application of hygiene or beauty products for 3 days, most of the known metabolites (no more news on whether some of the unknowns are microbial metabolites) annotated to ingredients of personal hygiene or beauty products [53]. For example, the surfactants C12 lauryl ether sulfate and cocoamidopropylbetaine, the sunscreen components avobenzone and octocrylene [53].

Worked sample 2.1

What do you think happen to the skin microbes during perspiration (sweating)?

How might the assembly and maintenance of the skin microbiome be different for different body sites?

According to the taxonomic composition and functional capacity in the skin microbiome, what personal habits we may be able to guess back?

2.5 Forces shaping the oral microbiome

Microbes directly or indirectly adhere to the teeth, and mucosal or keratinized epithelia in the mouth (Fig. 2.6). They are challenged daily with oral hygiene, eating, drinking, and perhaps many other processes (as we heard from Antoni van Leeuwenhoek already in Chapter 1, Box 1.1). Yet, despite daily and monthly (menstrual cycle) fluctuations in microbial biomass, the composition of the oral microbiome is surprisingly stable for each individual [54]. In the absence of fermentable carbohydrates, *Streptococcus* of the *Streptococcus mitis* group bind saliva-coated/protected teeth; *Streptococcus* of the *Streptococcus mutans* group ferments sugar and produce lactic acid, pioneering the

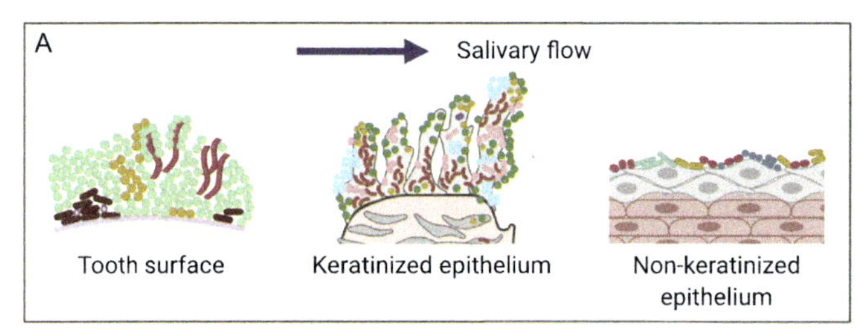

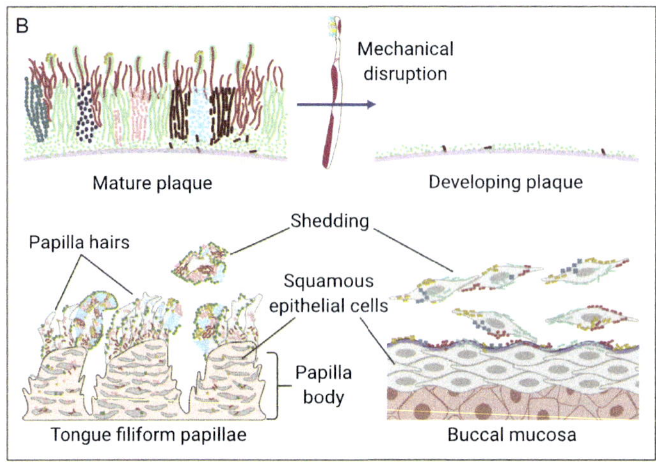

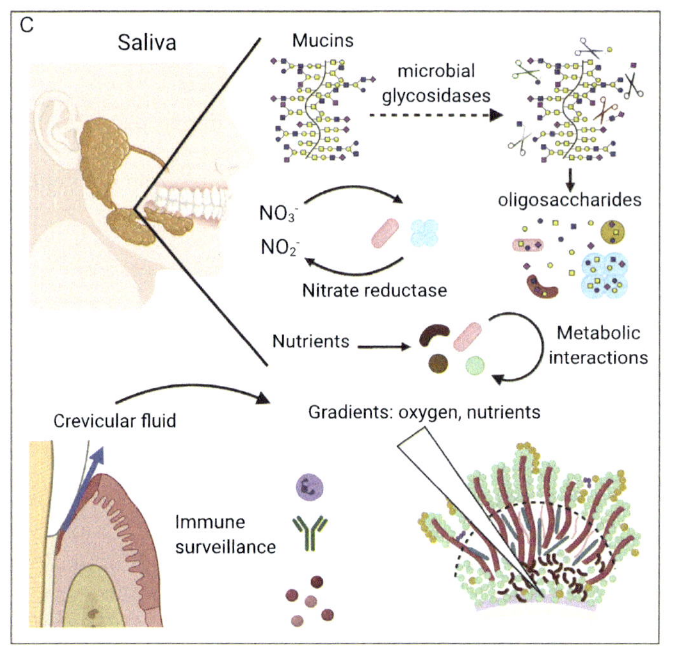

Fig. 2.6 Selective forces for microbes in the oral cavity. (A) Flow and adhesion. (B) Shedding and colonization. (C) Host and microbe. The host and the microbial community exert mutual reciprocal influences on each other through binding interactions, immune surveillance, and gradients of nutrients and solutes. The host secretes salivary mucins, which are complex glycoproteins that support the growth of mixed syntrophic communities of microbes that possess glycosidases capable of releasing oligosaccharides from mucins. Secretion of nitrate and other nutrients into saliva by the host, and release of crevicular fluid from the gingival crevice into the mouth, could also serve to foster the growth of particular microbes, while immune surveillance limits the growth of others. Microbial metabolism, in turn, can generate strongly localized gradients of oxygen and nutrients. The positioning of microbes at favorable locations within these gradients can lead to metabolic interaction and spatial structure within the microbial community. Credit: From Fig. 4 of Mark Welch JL, Ramírez-Puebla ST, Borisy GG. Oral microbiome geography: micron-scale habitat and niche. Cell Host Microbe 2020;28:160–68. https://doi.org/10.1016/j.chom.2020.07.009.

establishment of an acidic environment in the development of dental caries [55] (more on other diseases in Chapter 4). Different salivary glands are responsible for different ingredients in saliva, e.g., parotid saliva contains a large amount of amylase, lesser amounts of lysozyme, and no mucins, while submandibular/sublingual saliva is rich in mucins, cystatins, and lysozyme, but has less amylase [56]. Antibody-producing plasma cells activated in other mucosal sites are recirculated to lymphoid tissues (e.g., tonsils) in the oral cavity, and the IgA (Immunoglobulin A) and IgG (Immunoglobulin G) secreted can bind many oral bacteria. Chewing could induce gingiva-resident T helper 17 (Th17) cells [57]. Other than subgingival plaques from periodontal pockets, the oral microbiome mostly contains aerobes or facultative anaerobes (so it is probably a good idea for the bacteria to go to the lungs when the mouths are shut during sleeping [58]).

Direct visualization of intact structures from the tongue dorsum revealed bacteria domains that expand or shrink relative to their neighbors as they grow more layers (Fig. 2.7) [59]. Differences in growth rates, more shedding into the flow, or attacks by host defense systems could lead to such local dynamics. Remember the number question from Chapter 1, and we hope to have a detailed census for the microbial habitats in the mouth. Other than a major increase or decrease in the ecological niche for a given microbe (Figs. 2.8 and 2.9), scuffles with neighbors would maintain long-term stability in the relative abundances (proportions).

2.6 A stable gut microbiome

2.6.1 Adhesion

The fecal microbiome is not an in situ community, so it takes some imagination and simulations to get a sense of how everyone can stay in the gut (Figs. 2.10 and 2.11) and show up in good proportions in feces every day. Adhesion to the favorite mucin glycan, holding onto glycans or other macromolecules from neighboring microbes, or getting other help from the human host?

Immunoglobulin A (IgA), the major type of mucosal antibodies traditionally known for its ability to remove pathogens by forming large aggregates, often binds weakly to many commensal bacteria, which helps retain them in the gut [60–63], and potentially in the oral cavity if it is not cleaved by *Streptococcus* protease [64]. High-avidity IgA (antibody) against *Salmonella enterica* subspecies enterica serovar Typhimurium prevented separation of daughter cells after division [65]. I would guess that aggregation by IgA depends on a relatively high fraction of dividing bacteria. IgA secretion takes place in the small intestine, but IgA could be precipitated from fecal samples, along with

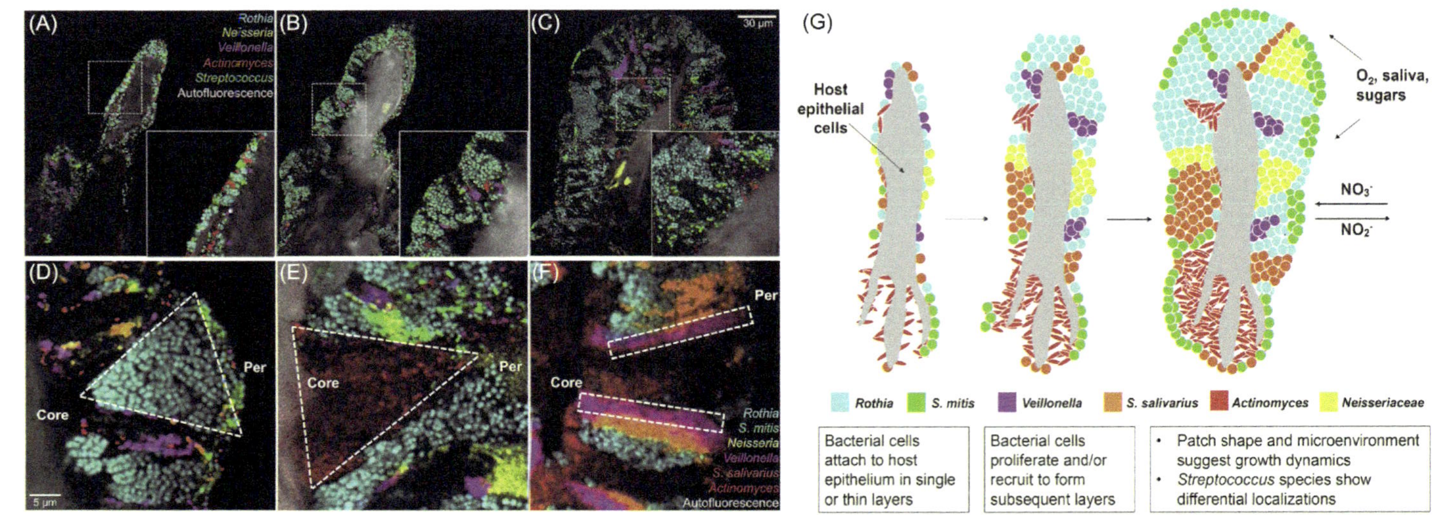

Fig. 2.7 Examples of microbiota dynamics in the tongue dorsum inferred from the difference in width between the inside (core) and the outer (periphery) layers. Host epithelial cells (autofluorescence) are surrounded by a growing number of oral bacterial cells. (A–F) Gradations in biofilm thickness and shape of clonal domains suggest biofilm growth and selective advantage. (A) A thin biofilm is composed of small clusters of cells from each bacterial taxon. (B) A thicker biofilm showing the expansion of the facultative anaerobe *Rothia* and the beginnings of expansion of anaerobes *Veillonella* and *Actinomyces*. (C) A mature structure showing well-defined domains. (D) Increasing width of a clonal domain toward the perimeter suggests a selective advantage toward the periphery. (E) Decreasing width toward the perimeter suggests a disadvantage at the periphery or selective advantage in the interior. (F) Constant width suggests neither selective advantage nor disadvantage with respect to neighboring taxa. (G) Inferred development of the tongue dorsum consortia. The study only focused on this kind of intact shape. The subjects sampled themselves under supervision by gently scraping the dorsal surface of the tongue from back to front using a ridged plastic tongue scraper. In this model, bacterial cells colonize host epithelial cells sparsely. As bacteria proliferate, layers of cells appear in a patch-like structure. Some *Streptococcus mitis* cells form a thin coat on the surface. Domain formation is dependent on neighbors and the microenvironment. Some nutrients may be gained from host epithelial material and other nutrients, O_2, and NO_3^- from the oral cavity by saliva. Credit: From Fig. 6 and Fig. 7 of Wilbert SA, Mark Welch JL, Borisy GG. Spatial ecology of the human tongue dorsum microbiome. Cell Rep 2020;30:4003–15.e3. https://doi.org/10.1016/j.celrep.2020.02.097.

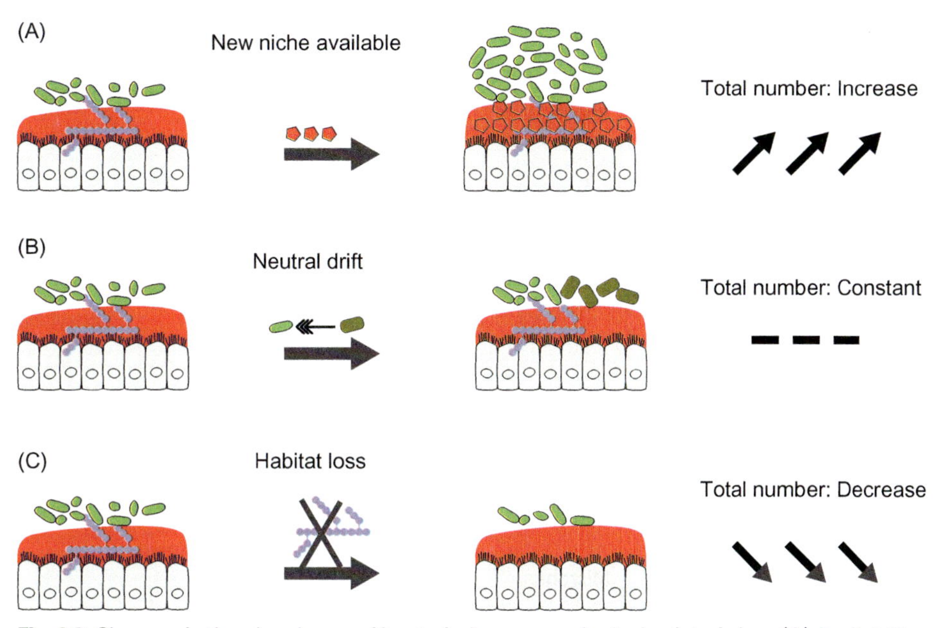

Fig. 2.8 Changes in the abundance of bacteria from an ecological point of view. (A) Availability of previously nonexistent or inaccessible niche. (B) The total capacity does not change, but replacement by other bacteria nearby or from a source community led to dynamics in the number of bacteria. (C) Habitat for bacteria is lost due to remodeling by the host or by other microbes. Competition for food sources or competition for attachment could also be reasons. Credit: Huijue Jia.

intestinal bacteria including species of the *Firmicutes* phylum and *Bifidobacterium* of the *Actinobacteria* phylum [61,63,66]. Bacteria such as *Escherichia coli* are overrepresented in feces from individuals with IgA deficiency, concomitant with the underrepresentation of bacteria such as the *Lachnospiraceae* and *Ruminococcaceae* families [63,67,68]. Lactobacillaceae and Enterobacterobacteriaceae dominate the small intestine [69], and mice experiments have shown small intestinal IgA binding to Enterobacterobacteriaceae and other *Proteobacteria* (e.g., *Acinetobacter*), and *Bifidobacterium* [61].

Note that the IgAs are also a source of complex glycan that can be foraged by bacteria such as *Bacteroides* spp. [70], and the altered glycan might then influence IgA binding to other bacteria.

2.6.2 Peristalsis

Peristaltic mixing not only sets the transit time, but also makes sure that a portion of the microbes stays on the proximal side of the squish (Figs. 2.11 and 2.12). One human peculiarity is that the ascending colon is really vertical as we stand. According to simulation, the ascending colon is the major bioreactor (Fig. 2.12), as remnants of food

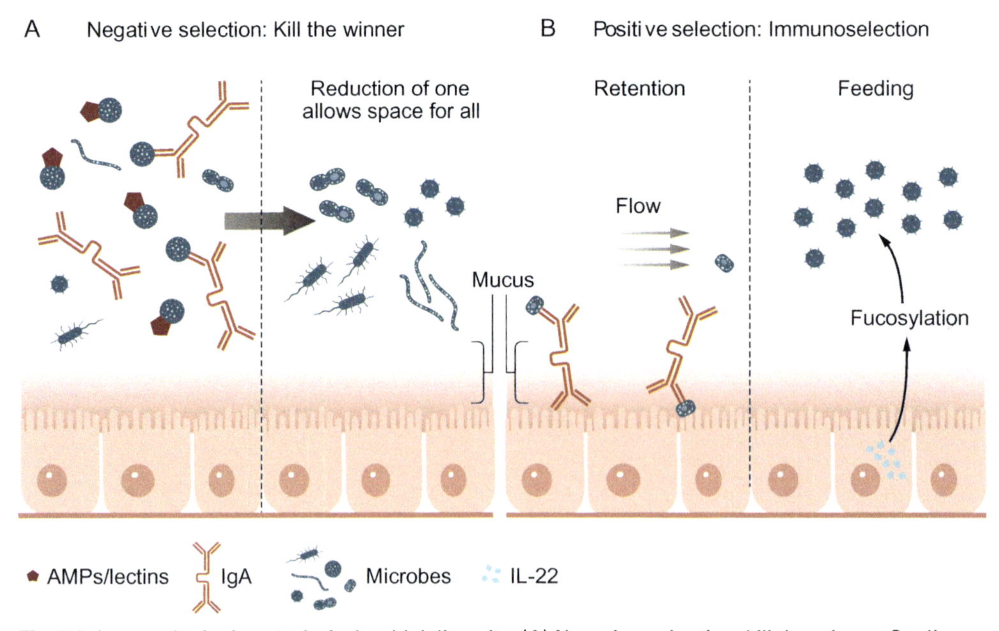

Fig. 2.9 Immunological control of microbial diversity. (A) Negative selection: kill the winner. Studies in macroecosystems demonstrate that predation increases biodiversity by serving to control particularly abundant and well-adapted species. Controlling the population growth of these species can liberate niches and resources that allow other organisms to thrive. The immune system can control gut microbial communities via negative selection through multiple mechanisms. AMPs can mediate direct killing, and IgA can cause aggregation and elimination of specific organisms, as well as downregulation of proteins involved in bacterial motility, invasion, or toxicity, such as flagellin. (B) Positive selection: immunoselection. The immune system might also serve to select particular organisms for residence within the gut. In addition to mediating negative selection, IgA may also help retain specific bacterial taxa by promoting retention of slow-growing species in the mucus or by enabling residence in protected niches, such as the colonic crypt. The immune system can also support the survival and growth of specific taxa by inducing lumenal deposition of specific nutrients; for example, interleukin-22 (IL-22) induces epithelial fucosylation, which nourishes particular beneficial bacterial taxa. Credit: Fig. 3 of Round JL, Palm NW. Causal effects of the microbiota on immune-mediated diseases. Sci Immunol 2018;3:eaao1603. https://doi.org/10.1126/sciimmunol.aao1603.

from the small intestine enter the colon [71]. Many animals, including mice, have a large cecum that keeps microbes. It is speculated that microbes from the human appendix might also help recover the microbiota after major disturbances [72]. Yet, some of the microbes may come from further up in the digestive tract, or be present locally on the colonic mucosa (Fig. 2.11C, Chapter 1, Fig. 1.9), and the lymphatic system might carry some microbes from other body sites. Remnants of food are wrapped with mucin (heavily glycosylated proteins that constitute the mucus hydrogel) in the ascending colon and more mucin toward the end of the colon [73]. There are subtle differences between

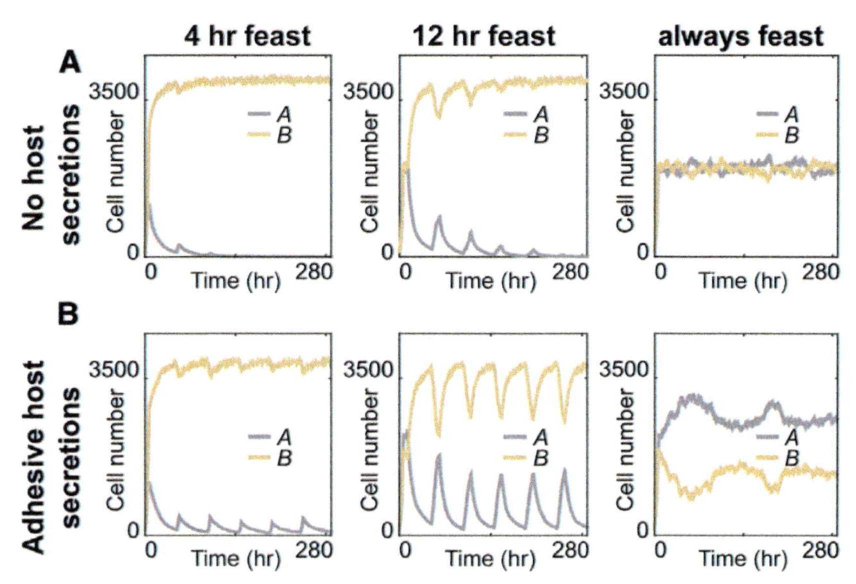

Fig. 2.10 Adhesive host secretions promote microbiota stability in fluctuating environments. Adhesive host secretions prevent the loss of specialist species whose host is only available periodically. McLoughlin et al. simulated two genotypes, a generalist species B that can consume lumen nutrients, which are available at all times. A second nutrient is only available periodically and is the exclusive nutrient source for specialist species A. Simulation results from 4 and 12h feast durations per 48h period are shown; "always feast" is a control where the second nutrient is also available at all times. Without adhesive host secretions, genotype A is lost from the community if "feast" periods are rare and short. When the host secretes an adhesion-promoting factor that creates a horizontal region in which genotype A resists displacement better than genotype B, both genotypes can be maintained even when the environment fluctuates. The variability in the "always feast" condition is due to stochastic fluctuations in population size. Our results are robust to changes in duration and periodicity of feasts. Credit: Fig. 3 of McLoughlin K, Schluter J, Rakoff-Nahoum S, Smith AL, Foster KR. Host selection of microbiota via differential adhesion. Cell Host Microbe 2016:1–10. https://doi.org/10.1016/j.chom.2016.02.021.

the epithelial side and the lumen. *Bacteroides fragilis* has been shown to express more of a sulfatase and glycosyl hydrolase when in mucin or in tissue compared to in the lumen [74]. *Bacteroides thetaiotaomicron* increased N-linked glycosidase transcription in mucus, while *Escherichia coli* showed a difference in the expression of iron acquisition pathway [75].

According to simulation with the standard western diet, microbes would likely spread out in order along the proximal colon, Bacteroidetes (*Bacteroides thetaiotaomicron*) before Firmicutes (*Eubacterium rectale*) [71]. Methane is mainly produced in the distal colon (Box 2.4) [76]. If the diet contains more fiber, the fermentation process could

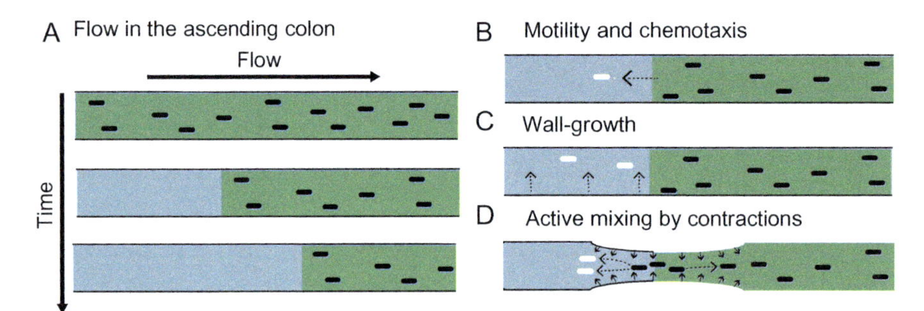

Fig. 2.11 Washout by flow and possible counteracting factors. The movement of luminal content down the colon has a mean velocity of ~ 20 μm/s in the proximal colon of adult humans. (A) Flow alone leads to emptying of channel content over time. Additional factors are required to counteract this washout and help to maintain a stable bacterial density over time. Such factors may include: (B) active motility by bacteria to swim toward the nutrient source; (C) wall growth; and (D) peristaltic mixing, with backflow generated by contractions of colonic walls. Credit: Fig. 1 of Cremer J, Segota I, Yang C, Arnoldini M, Sauls JT, Zhang Z, et al. Effect of flow and peristaltic mixing on bacterial growth in a gut-like channel. Proc Natl Acad Sci U S A 2016;113:11414–419. https://doi.org/10.1073/pnas.1601306113.

go all the way from the ascending colon to the rectum (Chapter 8, Table 8.4) [77]. Interestingly, mice fed with a chow diet have circadian dynamics of *Bacteroidetes*, *Firmicutes*, and *Verrumicrobia*, without limiting the time of food availability [78](Box 2.4). Although microbial growth rates in feces have been studied in metagenomic data of premature infants, *Citrobacter rodentium* infection, inflammatory bowel diseases (IBD), and diabetes patients [79–81], the author tends to believe that in healthy adults, much of the community is not growing, by the time the feces reaches the rectum. The *Firmicutes* (together with the methanogen which consumes H_2) perhaps more readily take advantage of each feeding (Fig. 2.13) [78,82]. The traditional agriculture use of human and animal feces might be a more fulfilling cycle for the microbes, < 100 g/day from a modern human [83] (Box 2.5).

2.7 "Enterotypes" and the Serengeti rules?

The controversial concept of "enterotypes" is essentially unsupervised clustering of fecal microbiome data. The statistically optimal number of clusters depends on the cohort [85]. For urban East Asian populations, and for less studied rural populations such as those in Africa, the *Firmicutes* (including the leanness-associated *Christensenella*) and methanogenic archaea (Box 2.4) (*Methanobrevibacter smithii*, Fig. 2.2) fermentation chain appear not as substantial as for northern European populations [86–88], and the colonic pH is likely not as acidic [69,71] (Fig. 2.12D). *Firmicutes*

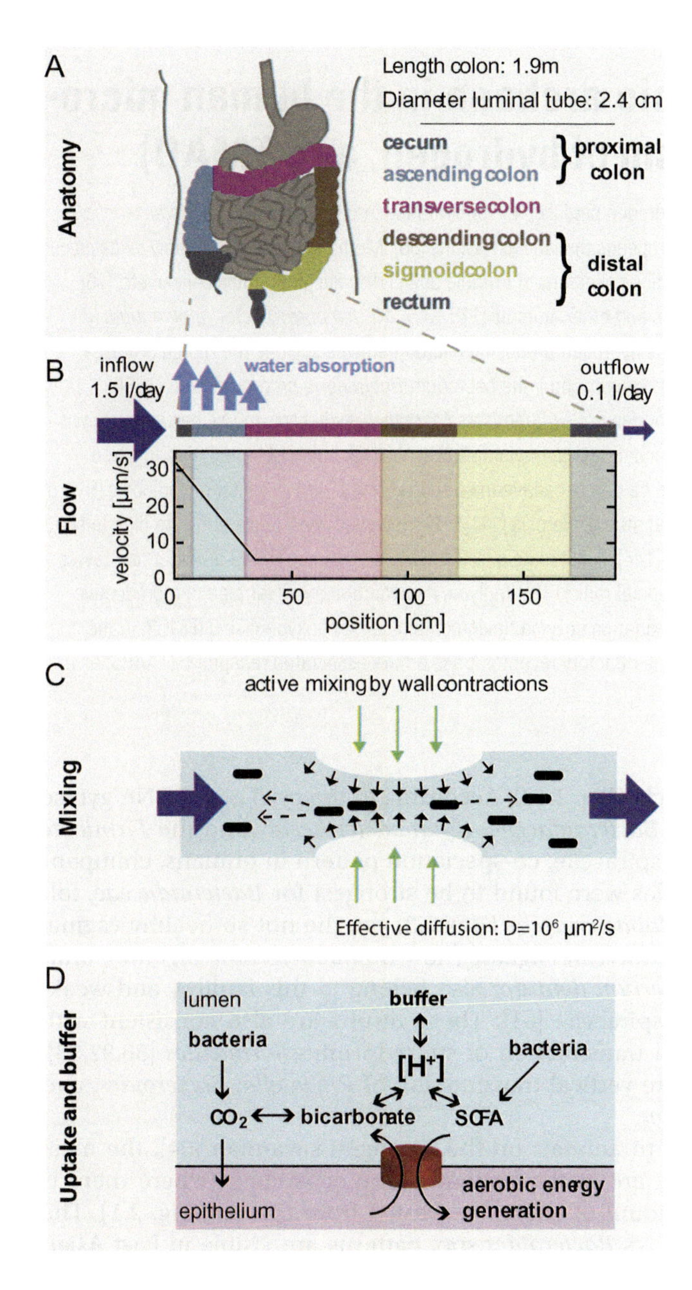

Fig. 2.12 Physiological parameters of the human colon. (A) Anatomical dimensions. Based on measurements of human colonic anatomy during the autopsy, X-ray and CT imaging using contrast media, and magnetic resonance tomography imaging, Cremer et al. derived operational numbers for the lengths, surface areas, and luminal diameters of the different colonic segments. (B) Luminal flow. About 1.5 L of fluid reaches the proximal colon every day. The epithelium absorbs most of this volume, and only 100–200 mL/day exit the colon as feces. This continuous water uptake along the colon leads to a steep gradient in luminal flow rates. We calculated an average flow velocity of about 30 µm/s at the beginning of the colon that drops to about 5 µm/s by the end of the ascending colon. (C) Mixing of luminal contents. Contractions of the intestinal walls can generate local mixing. Based on data on the mixing of radiolabeled dyes in the large intestine, we derive that the measured distributions can be approximated by an effective diffusion constant of $D \sim 10^6 \mu m^2/s$, a value orders of magnitude higher than molecular diffusion. (D) Epithelial SCFA uptake, bicarbonate excretion, and buffer chemistry. Bacterial fermentation leads to the production of SCFAs, which are taken up by the gut epithelium and contribute to the host's energy intake. SCFA uptake is coupled to the excretion of bicarbonate, which, in equilibrium with CO_2 and other luminal components, buffers the luminal acidity ($pH = -\log[H^+]$). All calculations are based on the measured characteristics of epithelial transporters and the buffer capacity of the lumen. Credit: Fig. 2 of Cremer J, Arnoldini M, Hwa T. Effect of water flow and chemical environment on microbiota growth and composition in the human colon. Proc Natl Acad Sci U S A 2017;114:6438–43. https://doi.org/10.1073/pnas.1619598114.

of the Clostridiale order suppress intestinal vitamin A production in mice, possibly finishing up a weaning reaction that induces T regulatory cells (Treg), and fecal Clostridiale species correlated with plasma vitamin A in adult humans [36,41,89,90]. But we are not aware of studies that link population differences in vitamin A levels to gut microbiome composition. The abundance of *Firmicutes* also relates to

Box 2.4 Methanogenic archaea in the human microbiome (metabolism of hydrogen, and TMAO)

Methanobrevibacter smithii reduces enteric hydrogen produced by fermentation into methane (e.g., *Christensenella* in Fig. 2.2). Its abundance is significantly higher in Europeans than in East Asians [86]. Methane production mainly takes place in the distal colon and has been linked to consitipating conditions of irritable bowel disease (IBD), colon cancer, etc. [76].

Methanobrevibacter is also found in ruminants and other animals [119,137]. *Methanobrevibacter ruminantium* M1 from the cow rumen encodes an adhesin that binds hydrogen-producing microorganisms such as the rumen protozoa (including the genera *Epidinium caudatum* and *Entodinium* spp.), the bacterium *Butyrivibrio proteoclasticus* [138].

Methanomassiliicoccus luminyensis, a methanogen not as famous as *Methanobrevibacter smithii*, has been shown to utilize hydrogen to reduce methyl compounds including trimethylamine (TMA) [139]. This TMA-depleting function could potentially decrease the fishy odor (can also be due to cadaverine or putrescine [140]) in bacterial vaginosis (BV), or prevent trimethylamine N-oxide (TMAO)-facilitated atherosclerosis [141,142]. However, TMA and TMAO are high in sea fish [143]; TMA is excreted from urine and sweat [144]; Adult male mice fed with an indirect TMAO inhibitor had fewer victories in social dominance tests regardless of social rank [145], implying its evolutionary advantage in reproductive years. In mice, repression of the trimethylamine oxidation enzyme flavin-containing monooxygenase 3 (FMO3) in the liver is male-specific, and TMA is recognized by the olfactory receptor, trace amine-associated receptor 5 (TAAR5) as a species-specific attraction signal [146,147].

the feeding cycle (Fig. 2.13). According to the *gyrB* gene (DNA gyrase subunit B) in *Bacteroidaceae*, *Bifidobacteriaceae*, and the *Firmicute* family Lachnospiraceae, co-speciation pattern in humans, chimpanzees and gorillas were found to be strongest for *Bacteroidaceae*, followed by *Bifidobacteriaceae* (Box 2.3, yet, the not-so-healthy vaginal bacterium *Gardnerella vaginae* (now *Bifidobacterium vaginae*), oral-gut *Bifidobacterium dentium* also belong to this family), and weakest in Lachnospiraceae [91]. These results are also consistent with interindividual transmission of spore-forming *Firmicutes* [86,92,93], instead of more vertical transmission of *Prevotella*, *Bacteroides*, and *Bifidobacterium*.

In analogy to animals on the Serengeti savannah [94], the most abundant taxa are at the bottom of the ecosystem (where there is the largest amount of energy harvested from the sun, Fig. 2.1). The *Prevotella* spp. vs *Bacteroides* spp. patterns are visible in East Asian samples [95] (Fig. 2.14; Not *Treponema*, Box 2.1). Fecal samples from Mexico both now and over 1000 years ago (coprolites) showed more *P. copri* than samples from the United States and from Europe [96,97]. Many of the well-established vaccines were more effective in *Bacteroides*-dominated than in *Prevotella*-dominated populations [98]; Laboratory mice were also *Bacteroides* and *Firmicutes*-dominated, if a *Prevotella*-dominated model is not chosen intentionally [99,100].

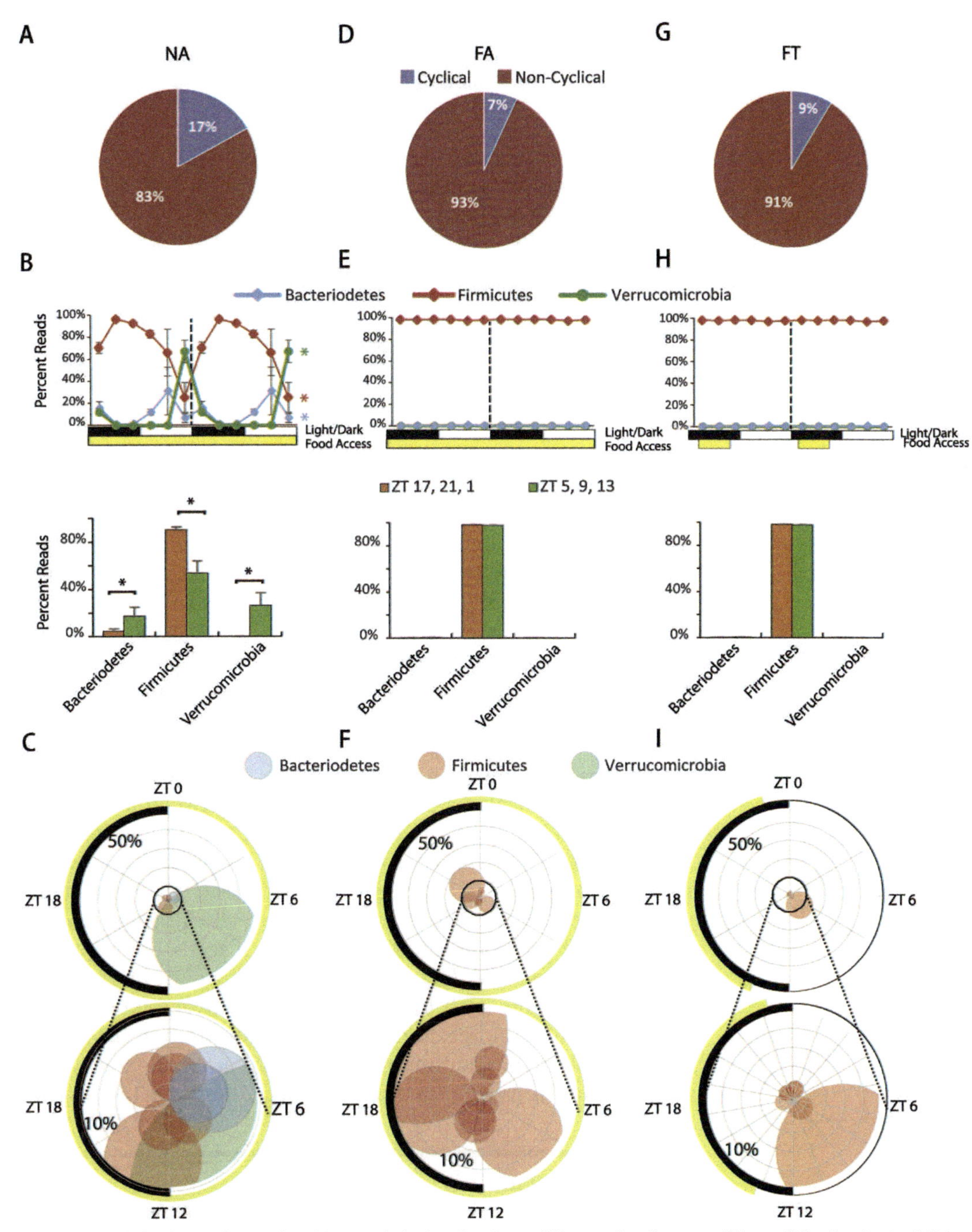

Fig. 2.13 Diurnal rhythms of gut microbiome phyla in mice from different feeding conditions. NA mice had ad libitum access to normal chow. FA mice had ad libitum access to HFD. FT mice had 8-h access (ZT 13–21) to HFD. (A) Pie chart showing the percentage of cycling and noncycling OTUs (across all conditions) in NA mice ($n=18$). (B) Upper double-plot line graph—where the second cycle is a duplicate of the first cycle following the *dashed line*—shows

(Continued)

Fig. 2.13, cont'd the average percent read (± SEM) of the three most predominant phyla at each time point (n=3 per time point). *Black and white* boxes indicate light off and light on, respectively. The *yellow box* shows when mice had access to food. Colored asterisks at the end of lines in the line graph show which phyla were cycling based on JTK analysis (that is ADJ.P < 0.05 and BH.Q. < 0.05). Since it takes > 1 h for a food bolus to reach the cecum [84], lower bar graphs show the average percent reads (± SEM, n=9) for the dark/active feeding phase (ZT 17, 21, and 1), and the light/inactive fasting phase (ZT 5, 9, and 13) *$P<.05$. (C) The top 10 OTUs (based on percent reads) are depicted in a polar plot. The radian indicates the phase of the OTU's peak, the distance from the center is the average percent read across all time points, and the radius of each point indicates the amplitude of cycling. The colors of the circles indicate the phylum of the OTU: Firmicutes *(pink)*, Bacteroidetes *(blue)*, and Verrucomicrobia *(green)*. The *black arc* on the left side of the plot indicates the light/dark cycle. The *yellow arc* depicts access to food. The bottom polar plot shows a magnified view of the inner ring (10%) of the top polar plot. These descriptions also apply to panels for FA mice (D–F) and FT mice (G–I). Credit: From Fig. 2 of Zarrinpar A, Chaix A, Yooseph S, Panda S. Diet and feeding pattern affect the diurnal dynamics of the gut microbiome. Cell Metab 2014;20:1006–17. https://doi.org/10.1016/j.cmet.2014.11.008.

People with higher relative abundances of *Bacteroides* tend to be free of worms and protists [101–103]. Human genetics also contributes to abundance differences in *Prevotella* spp. vs *Bacteroides* spp. [42,104]. Instead of harvesting energy from the sun and build up the trophic levels, the human microbiome, if not just living there and being a chemoautotroph, harvest energy from host molecules or from feeds such as food, drink, and medication. *Bacteroides* spp., especially *Bacteroides thetaiotaomicron*, are known for extracellular digestion of complex glycans from mucin, IgA, and from the diet [70,105–108]. Before these chopped-up glycans are transported for further metabolism intracellularly, there could be opportunities for scavenging by other bacteria, including many *Firmicutes*. *Bacteroides* spp. stayed together in a 15-species mix gavaged into gnotobiotic mice, and image analyses indicated a positive correlation between the numbers of *Bacteroides cellulolyticus* and *Bacteroides vulgatus*, whereas the numbers of *Bacteroides cellulolyticus* and *Ruminococcus torques* showed a negative correlation in 19 × 19 μm grid squares [109]. Extracellular digestion of inulin has been shown to increase the fitness of *Bacteroides ovatus* owing to reciprocal benefits when it feeds other gut species such as *Bacteroides vulgatus* [108]. We are currently limited by technology to examine such local cooperation between members of the human gut microbiome. Simple communities from insects provide a nice example of the genomic and metabolic complementarity in shaping the spatial organization of a microbial community (Fig. 2.15).

Worked sample 2.2

When do you think we will be able to visualize all the microbes in a well-preserved sample, along with their complete genomic or transcriptomic information?

What questions do you think can be addressed with such information?

Box 2.5 Circadian rhythm in the gut microbiome and their products

Lack of microbiota (i.e., germfree mice) induces a prediabetic state due to overproduction of corticosterone in the ileum [148]. The nuclear receptors RORα activator and RevErbα repressor together generate a circadian rhythm in the intestinal epithelial expression of TLRs (Toll-like receptors). Microbes signal through TLRs to induce rhythmic JNK and IKKβ activities, which prevents RevErbα activation by PPARα. The arhythmic gut microbiome has recently been shown in T2D and obese patients [149].

If the *Verrucomicrobia* in Fig. 2.13 is indeed the mucin-degrading *Akkermansia*, a probable explanation for its rise toward the end of the day (remember that mice are active at night) for mice on a chow diet would be that, after consuming everything from food, the gut microbiome sustains on host mucin. The mucin instead of diet would not be so favorable for maintaining a large population of *Firmicutes* before feeding time [78,149,150]. Butyrate peaked immediately after sleeping [151]. The acetate or propionate-producing *Akkermansia muciniphila* increases thermogenesis but decreased with colder temperature [152,153]. The bacterium is well adapted to the loose outer mucus layer in the gut lining and can utilize nanomolar concentrations of oxygen in the presence of carbon dioxide [154]. The oxygen level in the body is also circadian [122,155]. *A. muciniphila* was more abundant in people with conditions such as Alzheimer's disease, atherosclerotic cardiovascular disease and schizophrenia [25,156–158].

Fecal samples from volunteers are typically taken in the morning, without or with breakfast. A sampling of the gut microbiome many times a day in human volunteers has been technically challenging [159]. The above-mentioned T2D study carefully recorded the time of defecation, which explained some variations in the gut microbiome [149]. Fecal bacteria that showed altered circadian oscillations in the T2D patients included *Akkermansia*, *Roseburia*, *Bifidobacterium longum*, *Faecalibacterium prausnitzii*, etc.

Histone deacetylase 3 (HDAC3) is rhythmically expressed in epithelial cells of the small intestine [160]. HDAC3 drives the expression of genes in nutrient transport and lipid metabolism, and activates estrogen-related receptor α, which promotes lipid absorption. Histone methylation in intestinal epithelial cells also showed rhythmicity [78].

Polyamines (putrescine, spermidine, and spermine) contribute to the normal circadian clock; the cycle lengthens with age with declining polyamine levels [161]. A diet deficient in polyamines did impact the diurnal oscillation of the liver transcriptome [78].

Besides, mice fed with a high-fat diet showed hydrogen sulfite (H_2S) production from dsrAB (dissimilatory sulfite reductase) after feeding, which would be inflammatory in the gut [151].

The lung microbiome may also be circadian, with more microbial migration from the mouth overnight, and slowed immigration and enhanced elimination during the day [58].

Independent of a host animal, some microbes may encode their own clock genes. Some *Cyanobacteria* have the complete set of *KaiA*, *KaiB*, *KaiC* genes for circadian oscillations; Clock systems with only *KaiB* and *KaiC*, or only *KaiC* also have time-keeping functions and have been found in some *Proteobacteria* and *Cyanobacteria* [162]. The *Proteobacteria* species *Klebsiella aerogenes* (used to be *Enterobacter aerogenes*) has a circadian pattern of swarming and motility in vitro, which showed enhanced robustness after the addition of the hormone melatonin [163,164]. Besides, the rhythm was impacted by temperature. This illustrates likely modes of receiving circadian clues from the host animal. *Bacillus subtilis* does not encode KaiABC, but has photoreceptors and Per-Arnt-Sim (PAS) domains; *Bacillus subtilis* shows a 24-h clock controlled by light or temperature [165].

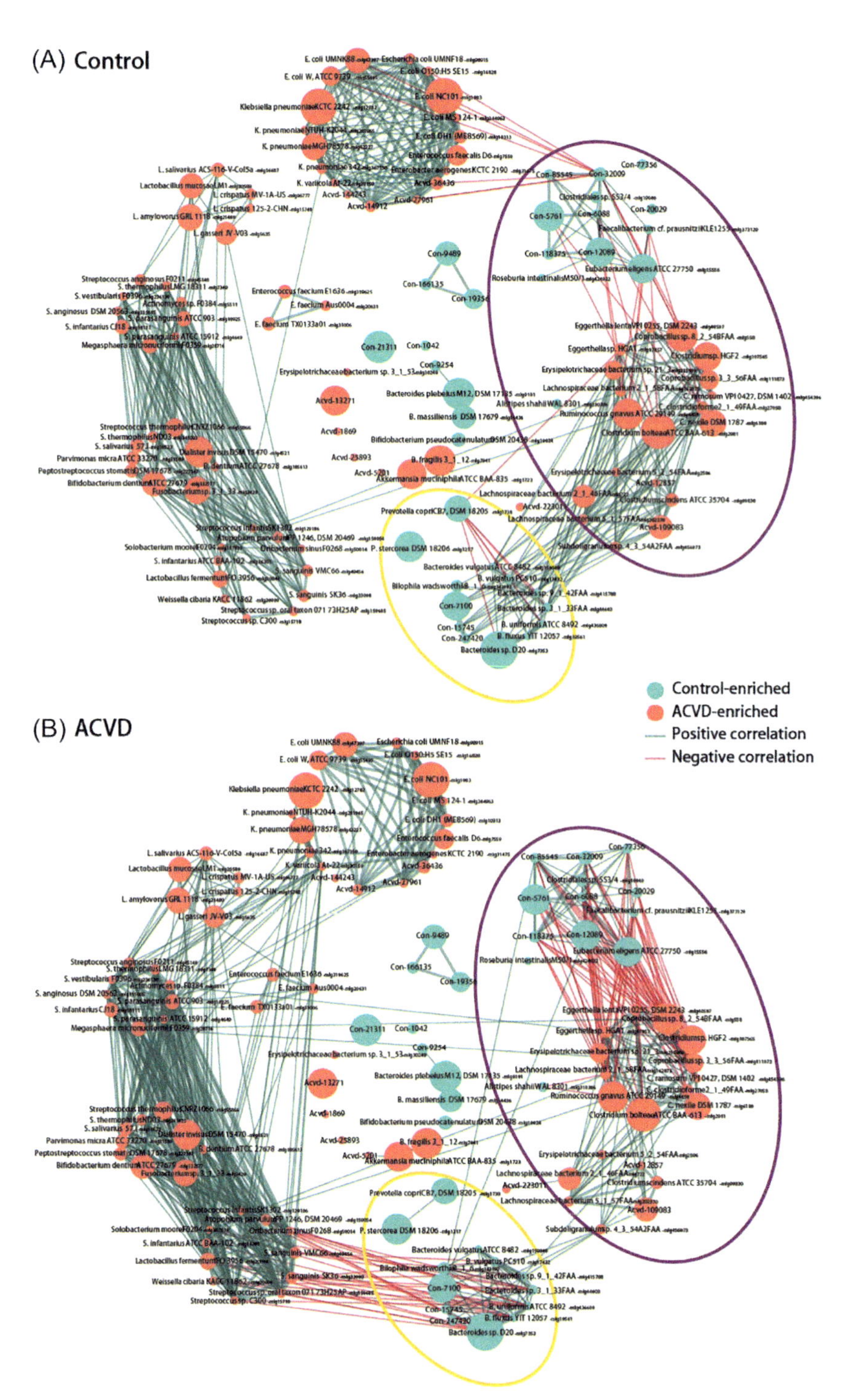

Fig. 2.14 See legend in next page

Fig. 2.14, Cont'd Fecal microbial species show patterns of abundance covariations in healthy individuals and in patients such as those with atheroschlerotic cardiovascular diseases (ACVD). The *yellow circle* highlights *Bacteroides* spp. and *Prevotella* spp. More on compositional data in Chapter 5. The *purple circle* highlights members of the Firmicutes phylum that likely scavenge metabolites including glycans from the *Bacteroides* spp. (A) Correlations between species in 187 healthy controls. The *circles* are colored according to Wilcoxon rank-sum test $q < 0.05$ between their relative abundance in controls and in ACVD patients (*cyan circles*, controlled enriched; *red*, ACVD enriched). *Green lines*, Spearman's correlation coefficient > 0.3; *Red lines*, Spearman's correlation coefficient < -0.3. (B) Correlations between species in 218 ACVD patients. Two hundred and five of the patients had stable angina, 8 had unstable angina, and 5 had an acute myocardial infarction (AMI). Credit: The *yellow and purple circles* are added onto Fig. 2 of Jie Z, Xia H, Zhong S-L, Feng Q, Li S, Liang S, et al. The gut microbiome in atherosclerotic cardiovascular disease. Nat Commun 2017;8:845. https://doi.org/10.1038/s41467-017-00900-1.

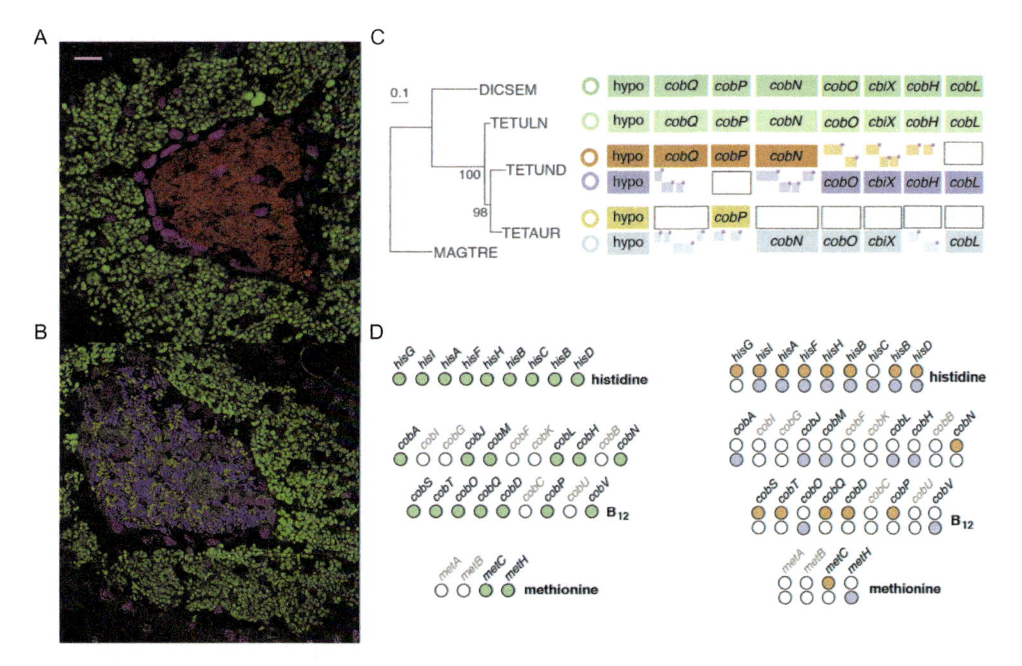

Fig. 2.15 Symbiosis in situ. Having information on the spatial organization and the genomes (metabolic potential) of microbial communities allowed researchers to discover interdependence between TETUND1 and TETUND2 (*Tettigades undata* chromosome 1 and 2) at the metabolic level, both are intracellular Candidatus Hodgkinia cicadicola bacteria that live inside cicada insects from the genus Tettigades. (A) FISH against rRNA to distinguish between the endosymbionts Candidatus Sulcia muelleri *(green)* from Hodgkinia *(red)*. (B) FISH against genomes to distinguish among the three bacteria, Sulcia *(green)*, TETUND1 *(yellow)* and TETUND2 *(blue)*. (C and D) Genomic divergence of TETUND1 *(orange)* and TETUND2 *(blue)* to have complementary functions. The cicada *Diceroprocta semicincta* (DICSEM) only have Sulcia and one Hodgkinia *(green)*. Credit: Cropped from Fig. 4, Fig. 1, Fig. 2 of Van Leuven JT, Meister RC, Simon C, McCutcheon JP. Sympatric speciation in a bacterial endosymbiont results in two genomes with the functionality of one. Cell 2014;158:1270–80. https://doi.org/10.1016/j.cell.2014.07.047.

2.8 Summary

This chapter brings in concepts from macroecology regarding diversity and trophic levels. Microbial diversity or richness could appear higher in some diseased conditions, especially when going beyond the gut microbiome. Single species dominance in the vaginal microbiome becomes more diverse in the majority of postmenopausal women (metagenomic applications in Chapter 8), as the pH is no longer maintained as acidic (attributed to glycogen fermentation into lactic acid by Lactobacilli). Lipids, microelements, and other ingredients feed a stable skin microbiome, which also tends to be acidic, before sebum secretion wanes with age. Spatial segregation and nonspecific cross-feeding are key to microbiome stability. Mucins, immunoglobulins, or even glycans expressed as blood group antigens, help microbes with localization and food. For the gut microbiome, the primary bioreactor is the ascending colon, while the microbes might spread out along the length of the colon, fermenting what they can ferment. In infants, the microbiome at multiple body sites recruit more members to reach new states; whereas in the elderly, ineffective containment of microbes likely lead to their spatial spread and shift in composition. Human genetics, historical contingency, and circadian rhythms (Box 2.5) all contribute to what we currently see in the microbiome, together with more studied factors such as nutrients and immune responses. Trophic levels in a human microbiome site may need to consider different scenarios of physiological states, during which different molecules are available, and different functions may be evolutionarily prioritized.

Worked sample 2.3

Try to draw a few scenarios for tropic levels in the human microbiome. For example, the gut microbiome on a fiber-rich diet, similar to the cow rumen [119]; the gut microbiome when starved overnight, grazing on mucin and potentially other host molecules.

References

[1] Meyer ST, Ptacnik R, Hillebrand H, Bessler H, Buchmann N, Ebeling A, et al. Biodiversity-multifunctionality relationships depend on identity and number of measured functions. Nat Ecol Evol 2018;2:44–9. https://doi.org/10.1038/s41559-017-0391-4.

[2] Martiny JBH, Jones SE, Lennon JT, Martiny AC. Microbiomes in light of traits: a phylogenetic perspective. Science 2015;350:aac9323. https://doi.org/10.1126/science.aac9323.

[3] Goldford JE, Lu N, Bajić D, Estrela S, Tikhonov M, Sanchez-Gorostiaga A, et al. Emergent simplicity in microbial community assembly. Science 2018;361:469–74. https://doi.org/10.1126/science.aat1168.

[4] Coyte KZ, Schluter J, Foster KR. The ecology of the microbiome: networks, competition, and stability. Science 2015;350:663–6. https://doi.org/10.1126/science.aad2602.
[5] Lourenço M, Chaffringeon L, Lamy-Besnier Q, Pédron T, Campagne P, Eberl C, et al. The spatial heterogeneity of the gut limits predation and fosters coexistence of bacteria and bacteriophages. Cell Host Microbe 2020;28:390–401.e5. https://doi.org/10.1016/j.chom.2020.06.002.
[6] Costello EK, Stagaman K, Dethlefsen L, Bohannan BJM, Relman DA. The application of ecological theory toward an understanding of the human microbiome. Science 2012;336:1255–62. https://doi.org/10.1126/science.1224203.
[7] Song SJ, Lauber C, Costello EK, Lozupone CA, Humphrey G, Berg-Lyons D, et al. Cohabiting family members share microbiota with one another and with their dogs. Elife 2013;2. https://doi.org/10.7554/eLife.00458, e00458.
[8] van Leeuwenhoek A. The collected letters of Antoni van Leeuwenhoek. Amsterdam: Swets and Zeitlinger; 1952.
[9] van Leeuwenhoek A. Observations, communicated to the publisher by Mr. Antony van Leewenhoeck, in a dutch letter of the 9th Octob. 1676. here English'd: concerning little animals by him observed in rain-well-sea- and snow water; as also in water wherein pepper had lain infus. Philos Trans R Soc London 1677;12:821–31. https://doi.org/10.1098/rstl.1677.0003.
[10] Yildirim S, Yeoman CJ, Janga SC, Thomas SM, Ho M, Leigh SR, et al. Primate vaginal microbiomes exhibit species specificity without universal Lactobacillus dominance. ISME J 2014;8:2431–44. https://doi.org/10.1038/ismej.2014.90.
[11] Gliniewicz K, Schneider GM, Ridenhour BJ, Williams CJ, Song Y, Farage MA, et al. Comparison of the vaginal microbiomes of premenopausal and postmenopausal women. Front Microbiol 2019;10:193. https://doi.org/10.3389/fmicb.2019.00193.
[12] Jie Z, Chen C, Hao L, Li F, Song L, Zhang X, et al. Life history recorded in the vagino-cervical microbiome along with multi-omics. Genomics Proteomics Bioinformatics 2021. https://doi.org/10.1016/j.gpb.2021.01.005.
[13] Qi W, Li H, Wang C, Li H, Fan A, Han C, et al. The effect of pathophysiological changes in the vaginal milieu on the signs and symptoms of genitourinary syndrome of menopause (GSM). Menopause 2021;28:102–8. https://doi.org/10.1097/GME.0000000000001644.
[14] Turnbaugh PJ, Hamady M, Yatsunenko T, Cantarel BL, Duncan A, Ley RE, et al. A core gut microbiome in obese and lean twins. Nature 2009;457:480–4. https://doi.org/10.1038/nature07540.
[15] Cotillard A, Kennedy SP, Kong LC, Prifti E, Pons N, Le Chatelier E, et al. Dietary intervention impact on gut microbial gene richness. Nature 2013;500:585–8. https://doi.org/10.1038/nature12480.
[16] Le Chatelier E, Nielsen T, Qin J, Prifti E, Hildebrand F, Falony G, et al. Richness of human gut microbiome correlates with metabolic markers. Nature 2013;500:541–6. https://doi.org/10.1038/nature12506.
[17] He Q, Gao Y, Jie Z, Yu X, Laursen JMJM, Xiao L, et al. Two distinct metacommunities characterize the gut microbiota in Crohn's disease patients. Gigascience 2017;6:1–11. https://doi.org/10.1093/gigascience/gix050.
[18] Ricanek P, Lothe SM, Frye SA, Rydning A, Vatn MH, Tønjum T. Gut bacterial profile in patients newly diagnosed with treatment-naïve Crohn's disease. Clin Exp Gastroenterol 2012;5:173–86. https://doi.org/10.2147/CEG.S33858.
[19] Gevers D, Kugathasan S, Denson LA, Vázquez-Baeza Y, Van Treuren W, Ren B, et al. The treatment-naive microbiome in new-onset Crohn's disease. Cell Host Microbe 2014;15:382–92. https://doi.org/10.1016/j.chom.2014.02.005.
[20] Imhann F, Vich Vila A, Bonder MJ, Fu J, Gevers D, Visschedijk MC, et al. Interplay of host genetics and gut microbiota underlying the onset and clinical

presentation of inflammatory bowel disease. Gut 2018;67:108–19. https://doi.org/10.1136/gutjnl-2016-312135.

[21] Manichanh C, Rigottier-Gois L, Bonnaud E, Gloux K, Pelletier E, Frangeul L, et al. Reduced diversity of faecal microbiota in Crohn's disease revealed by a metagenomic approach. Gut 2006;55:205–11. https://doi.org/10.1136/gut.2005.073817.

[22] Maldonado-Arriaga B, Sandoval-Jiménez S, Rodríguez-Silverio J, Lizeth Alcaráz-Estrada S, Cortés-Espinosa T, Pérez-Cabeza de Vaca R, et al. Gut dysbiosis and clinical phases of pancolitis in patients with ulcerative colitis. MicrobiologyOpen 2021;10. https://doi.org/10.1002/mbo3.1181, e1181.

[23] Feng Q, Liang S, Jia H, Stadlmayr A, Tang L, Lan Z, et al. Gut microbiome development along the colorectal adenoma–carcinoma sequence. Nat Commun 2015;6:6528. https://doi.org/10.1038/ncomms7528.

[24] Sanapareddy N, Legge RM, Jovov B, McCoy A, Burcal L, Araujo-Perez F, et al. Increased rectal microbial richness is associated with the presence of colorectal adenomas in humans. ISME J 2012;6:1858–68. https://doi.org/10.1038/ismej.2012.43.

[25] Zhu F, Ju Y, Wang W, Wang Q, Guo R, Ma Q, et al. Metagenome-wide association of gut microbiome features for schizophrenia. Nat Commun 2020;11:1612. https://doi.org/10.1038/s41467-020-15457-9.

[26] Yang J, Zheng P, Li Y, Wu J, Tan X, Zhou J, et al. Landscapes of bacterial and metabolic signatures and their interaction in major depressive disorders. Sci Adv 2020;6. https://doi.org/10.1126/sciadv.aba8555, eaba8555.

[27] Zhu J, Liao M, Yao Z, Liang W, Li Q, Liu J, et al. Breast cancer in postmenopausal women is associated with an altered gut metagenome. Microbiome 2018;6:136. https://doi.org/10.1186/s40168-018-0515-3.

[28] Moossavi S, Sepehri S, Robertson B, Bode L, Goruk S, Field CJ, et al. Composition and variation of the human milk microbiota are influenced by maternal and early-life factors. Cell Host Microbe 2019;25:324–335.e4. https://doi.org/10.1016/j.chom.2019.01.011.

[29] Kalaora S, Nagler A, Nejman D, Alon M, Barbolin C, Barnea E, et al. Identification of bacteria-derived HLA-bound peptides in melanoma. Nature 2021;592:138–43. https://doi.org/10.1038/s41586-021-03368-8.

[30] Lamont RJ, Koo H, Hajishengallis G. The oral microbiota: dynamic communities and host interactions. Nat Rev Microbiol 2018;16:745–59. https://doi.org/10.1038/s41579-018-0089-x.

[31] Ravel J, Gajer P, Abdo Z, Schneider GM, Koenig SSK, Mcculle SL, et al. Vaginal microbiome of reproductive-age women. Proc Natl Acad Sci U S A 2010;108:4680–7. http://www.pnas.org/cgi/doi/10.1073/pnas.1002611107.

[32] DiGiulio DB, Callahan BJ, McMurdie PJ, Costello EK, Lyell DJ, Robaczewska A, et al. Temporal and spatial variation of the human microbiota during pregnancy. Proc Natl Acad Sci U S A 2015;112:11060–5. https://doi.org/10.1073/pnas.1502875112.

[33] Lundy SD, Sangwan N, Parekh NV, Selvam MKP, Gupta S, McCaffrey P, et al. Functional and taxonomic dysbiosis of the gut, urine, and semen microbiomes in male infertility. Eur Urol 2021;79:826–36. https://doi.org/10.1016/j.eururo.2021.01.014.

[34] Yu J, Feng Q, Wong SHSH, Zhang D, Liang QY, Qin Y, et al. Metagenomic analysis of faecal microbiome as a tool towards targeted non-invasive biomarkers for colorectal cancer. Gut 2017;66:70–8. https://doi.org/10.1136/gutjnl-2015-309800.

[35] Bäckhed F, Roswall J, Peng Y, Feng Q, Jia H, Kovatcheva-Datchary P, et al. Dynamics and stabilization of the human gut microbiome during the first year of life. Cell Host Microbe 2015;17:690–703. https://doi.org/10.1016/j.chom.2015.04.004.

[36] Al Nabhani Z, Dulauroy S, Marques R, Cousu C, Al Bounny S, Déjardin F, et al. A weaning reaction to microbiota is required for resistance to immunopathologies in the adult. Immunity 2019;50:1276–1288.e5. https://doi.org/10.1016/j.immuni.2019.02.014.
[37] Knoop KA, Gustafsson JK, McDonald KG, Kulkarni DH, Coughlin PE, McCrate S, et al. Microbial antigen encounter during a preweaning interval is critical for tolerance to gut bacteria. Sci Immunol 2017;2. https://doi.org/10.1126/sciimmunol.aao1314, eaao1314.
[38] Wilmanski T, Diener C, Rappaport N, Patwardhan S, Wiedrick J, Lapidus J, et al. Gut microbiome pattern reflects healthy ageing and predicts survival in humans. Nat Metab 2021;3:274–86. https://doi.org/10.1038/s42255-021-00348-0.
[39] Biagi E, Franceschi C, Rampelli S, Severgnini M, Ostan R, Turroni S, et al. Gut microbiota and extreme longevity. Curr Biol 2016;26:1480–5. https://doi.org/10.1016/j.cub.2016.04.016.
[40] Zhang X, Zhong H, Li Y, Shi Z, Ren H, Zhang Z, et al. Sex- and age-related trajectories of the adult human gut microbiota shared across populations of different ethnicities. Nat Aging 2021;1:87–100. https://doi.org/10.1038/s43587-020-00014-2.
[41] Jie Z, Liang S, Ding Q, Tang S, Wang D, Zhong H, et al. A multi-omic cohort as a reference point for promoting a healthy gut microbiome. Med Microecol 2021. https://doi.org/10.1101/585893.
[42] Liu X, Tang S, Zhong H, Tong X, Jie Z, Ding Q, et al. A genome-wide association study for gut metagenome in Chinese adults illuminates complex diseases. Cell Discov 2021;7:9. https://doi.org/10.1038/s41421-020-00239-w.
[43] Man MQ, Xin SJ, Song SP, Cho SY, Zhang XJ, Tu CX, et al. Variation of skin surface pH, sebum content and stratum corneum hydration with age and gender in a large Chinese population. Skin Pharmacol Physiol 2009;22:190–9. https://doi.org/10.1159/000231524.
[44] Luebberding S, Krueger N, Kerscher M. Age-related changes in skin barrier function - quantitative evaluation of 150 female subjects. Int J Cosmet Sci 2013;35:183–90. https://doi.org/10.1111/ics.12024.
[45] Chen C, Song X, Wei W, Zhong H, Dai J, Lan Z, et al. The microbiota continuum along the female reproductive tract and its relation to uterine-related diseases. Nat Commun 2017;8:875. https://doi.org/10.1038/s41467-017-00901-0.
[46] Leheste JR, Ruvolo KE, Chrostowski JE, Rivera K, Husko C, Miceli A, et al. P acnes-driven disease pathology: current knowledge and future directions. Front Cell Infect Microbiol 2017;7:81. https://doi.org/10.3389/fcimb.2017.00081.
[47] Javurek AB, Spollen WG, Ali AMM, Johnson SA, Lubahn DB, Bivens NJ, et al. Discovery of a novel seminal fluid microbiome and influence of estrogen receptor alpha genetic status. Sci Rep 2016;6:23027. https://doi.org/10.1038/srep23027.
[48] Byrd AL, Belkaid Y, Segre JA. The human skin microbiome. Nat Rev Microbiol 2018;16:143–55. https://doi.org/10.1038/nrmicro.2017.157.
[49] The Human Microbiome Project Consortium. Structure, function and diversity of the healthy human microbiome. Nature 2012;486:207–14. https://doi.org/10.1038/nature11234.
[50] Oh J, Byrd AL, Park M, Kong HH, Segre JA. Temporal stability of the human skin microbiome. Cell 2016;165:854–66. https://doi.org/10.1016/j.cell.2016.04.008.
[51] Sloan WT, Lunn M, Woodcock S, Head IM, Nee S, Curtis TP. Quantifying the roles of immigration and chance in shaping prokaryote community structure. Environ Microbiol 2006;8:732–40. https://doi.org/10.1111/j.1462-2920.2005.00956.x.
[52] Li L, Ma Z. Testing the neutral theory of biodiversity with human microbiome datasets. Sci Rep 2016;6:31448. https://doi.org/10.1038/srep31448.
[53] Bouslimani A, Porto C, Rath CM, Wang M, Guo Y, Gonzalez A, et al. Molecular cartography of the human skin surface in 3D. Proc Natl Acad Sci U S A 2015;112:E2120–9. https://doi.org/10.1073/pnas.1424409112.

[54] Ding T, Schloss PD. Dynamics and associations of microbial community types across the human body. Nature 2014;509:357–60. https://doi.org/10.1038/nature13178.

[55] Valm AM. The structure of dental plaque microbial communities in the transition from health to dental caries and periodontal disease. J Mol Biol 2019;431:2957–69. https://doi.org/10.1016/j.jmb.2019.05.016.

[56] Rudney JD. Does variability in salivary protein concentrations influence oral microbial ecology and oral health? Crit Rev Oral Biol Med 1995;6:343–67. https://doi.org/10.1177/10454411950060040501.

[57] Dutzan N, Abusleme L, Bridgeman H, Greenwell-Wild T, Zangerle-Murray T, Fife ME, et al. On-going mechanical damage from mastication drives homeostatic Th17 cell responses at the oral barrier. Immunity 2017;46:133–47. https://doi.org/10.1016/j.immuni.2016.12.010.

[58] Dickson RP, Erb-Downward JR, Martinez FJ, Huffnagle GB. The microbiome and the respiratory tract. Annu Rev Physiol 2016;78:481–504. https://doi.org/10.1146/annurev-physiol-021115-105238.

[59] Wilbert SA, Mark Welch JL, Borisy GG. Spatial ecology of the human tongue dorsum microbiome. Cell Rep 2020;30:4003–4015.e3. https://doi.org/10.1016/j.celrep.2020.02.097.

[60] McLoughlin K, Schluter J, Rakoff-Nahoum S, Smith AL, Foster KR. Host selection of microbiota via differential adhesion. Cell Host Microbe 2016;1–10. https://doi.org/10.1016/j.chom.2016.02.021.

[61] Bunker JJ, Erickson SA, Flynn TM, Henry C, Koval JC, Meisel M, et al. Natural polyreactive IgA antibodies coat the intestinal microbiota. Science 2017;358. https://doi.org/10.1126/science.aan6619, eaan6619.

[62] Donaldson GP, Ladinsky MS, Yu KB, Sanders JG, Yoo BB, Chou W-C, et al. Gut microbiota utilize immunoglobulin A for mucosal colonization. Science 2018;360:795–800. https://doi.org/10.1126/science.aaq0926.

[63] Fadlallah J, El Kafsi H, Sterlin D, Juste C, Parizot C, Dorgham K, et al. Microbial ecology perturbation in human IgA deficiency. Sci Transl Med 2018;10. https://doi.org/10.1126/scitranslmed.aan1217, eaan1217.

[64] Weiser JN, Ferreira DM, Paton JC. Streptococcus pneumoniae: transmission, colonization and invasion. Nat Rev Microbiol 2018;1–13. https://doi.org/10.1038/s41579-018-0001-8.

[65] Moor K, Diard M, Sellin ME, Felmy B, Wotzka SY, Toska A, et al. High-avidity IgA protects the intestine by enchaining growing bacteria. Nature 2017;544:498–502. https://doi.org/10.1038/nature22058.

[66] Palm NW, de Zoete MR, Cullen TW, Barry NA, Stefanowski J, Hao L, et al. Immunoglobulin a coating identifies colitogenic bacteria in inflammatory bowel disease. Cell 2014;158:1000–10. https://doi.org/10.1016/j.cell.2014.08.006.

[67] Moll JM, Myers PN, Zhang C, Eriksen C, Wolf J, Appelberg KS, et al. Gut microbiota perturbation in IgA deficiency is influenced by IgA-autoantibody status. Gastroenterology 2021. https://doi.org/10.1053/j.gastro.2021.02.053.

[68] Jørgensen SF, Holm K, Macpherson ME, Storm-Larsen C, Kummen M, Fevang B, et al. Selective IgA deficiency in humans is associated with reduced gut microbial diversity. J Allergy Clin Immunol 2019;143:1969–1971.e11. https://doi.org/10.1016/j.jaci.2019.01.019.

[69] Donaldson GP, Lee SM, Mazmanian SK. Gut biogeography of the bacterial microbiota. Nat Rev Microbiol 2015;14:20–32. https://doi.org/10.1038/nrmicro3552.

[70] Briliūtė J, Urbanowicz PA, Luis AS, Baslé A, Paterson N, Rebello O, et al. Complex N-glycan breakdown by gut Bacteroides involves an extensive enzymatic apparatus encoded by multiple co-regulated genetic loci. Nat Microbiol 2019. https://doi.org/10.1038/s41564-019-0466-x.

[71] Cremer J, Arnoldini M, Hwa T. Effect of water flow and chemical environment on microbiota growth and composition in the human colon. Proc Natl Acad Sci U S A 2017;114:6438–43. https://doi.org/10.1073/pnas.1619598114.
[72] Randal Bollinger R, Barbas AS, Bush EL, Lin SS, Parker W. Biofilms in the large bowel suggest an apparent function of the human vermiform appendix. J Theor Biol 2007;249:826–31. https://doi.org/10.1016/j.jtbi.2007.08.032.
[73] Bergstrom K, Shan X, Casero D, Batushansky A, Lagishetty V, Jacobs JP, et al. Proximal colon–derived O-glycosylated mucus encapsulates and modulates the microbiota. Science 2020;370:467–72. https://doi.org/10.1126/science.aay7367.
[74] Donaldson GP, Chou W-C, Manson AL, Rogov P, Abeel T, Bochicchio J, et al. Spatially distinct physiology of *Bacteroides fragilis* within the proximal colon of gnotobiotic mice. Nat Microbiol 2020. https://doi.org/10.1038/s41564-020-0683-3.
[75] Li H, Limenitakis JP, Fuhrer T, Geuking MB, Lawson MA, Wyss M, et al. The outer mucus layer hosts a distinct intestinal microbial niche. Nat Commun 2015;6:8292. https://doi.org/10.1038/ncomms9292.
[76] Sahakian AB, Jee SR, Pimentel M. Methane and the gastrointestinal tract. Dig Dis Sci 2010;55:2135–43. https://doi.org/10.1007/s10620-009-1012-0.
[77] Eswaran S, Muir J, Chey WD. Fiber and functional gastrointestinal disorders. Am J Gastroenterol 2013;108:718–27. https://doi.org/10.1038/ajg.2013.63.
[78] Zarrinpar A, Chaix A, Yooseph S, Panda S. Diet and feeding pattern affect the diurnal dynamics of the gut microbiome. Cell Metab 2014;20:1006–17. https://doi.org/10.1016/j.cmet.2014.11.008.
[79] Korem T, Zeevi D, Suez J, Weinberger A, Avnit-Sagi T, Pompan-Lotan M, et al. Growth dynamics of gut microbiota in health and disease inferred from single metagenomic samples. Science 2015;349:1101–6. https://doi.org/10.1126/science.aac4812.
[80] Gao Y, Li H. Quantifying and comparing bacterial growth dynamics in multiple metagenomic samples. Nat Methods 2018;15:1041–4. https://doi.org/10.1038/s41592-018-0182-0.
[81] Brown CT, Olm MR, Thomas BC, Banfield JF. Measurement of bacterial replication rates in microbial communities. Nat Biotechnol 2016;34:1256–63. https://doi.org/10.1038/nbt.3704.
[82] von Schwartzenberg RJ, Bisanz JE, Lyalina S, Spanogiannopoulos P, Ang QY, Cai J, et al. Caloric restriction disrupts the microbiota and colonization resistance. Nature 2021;1–6. https://doi.org/10.1038/s41586-021-03663-4.
[83] Stephen AM, Cummings JH. The microbial contribution to human faecal mass. J Med Microbiol 1980;13:45–56. https://doi.org/10.1099/00222615-13-1-45.
[84] Padmanabhan P, Grosse J, Asad Abu Bakar Md Ali, Radda GK, Golay X. Gastrointestinal transit measurements in mice with 99mTc-DTPA-labeled activated charcoal using NanoSPECT-CT. EJNMMI Res 2013;3(1):60. https://doi.org/10.1186/2191-219X-3-60.
[85] Costea PI, Hildebrand F, Arumugam M, Bäckhed F, Blaser MJ, Bushman FD, et al. Enterotypes in the landscape of gut microbial community composition. Nat Microbiol 2018;3:8–16. https://doi.org/10.1038/s41564-017-0072-8.
[86] Li J, Jia H, Cai X, Zhong H, Feng Q, Sunagawa S, et al. An integrated catalog of reference genes in the human gut microbiome. Nat Biotechnol 2014;32:834–41. https://doi.org/10.1038/nbt.2942.
[87] Ayeni FA, Biagi E, Rampelli S, Fiori J, Soverini M, Audu HJ, et al. Infant and adult gut microbiome and metabolome in rural Bassa and urban settlers from Nigeria. Cell Rep 2018;23:3056–67. https://doi.org/10.1016/j.celrep.2018.05.018.
[88] Schnorr SL, Candela M, Rampelli S, Centanni M, Consolandi C, Basaglia G, et al. Gut microbiome of the Hadza hunter-gatherers. Nat Commun 2014;5:3654. https://doi.org/10.1038/ncomms4654.

[89] Grizotte-Lake M, Zhong G, Duncan K, Kirkwood J, Iyer N, Smolenski I, et al. Commensals suppress intestinal epithelial cell retinoic acid synthesis to regulate interleukin-22 activity and prevent microbial dysbiosis. Immunity 2018;49:1103–1115.e6. https://doi.org/10.1016/j.immuni.2018.11.018.
[90] Atarashi K, Tanoue T, Oshima K, Suda W, Nagano Y, Nishikawa H, et al. Treg induction by a rationally selected mixture of Clostridia strains from the human microbiota. Nature 2013;500:232–6. https://doi.org/10.1038/nature12331.
[91] Moeller AH, Caro-Quintero A, Mjungu D, Georgiev AV, Lonsdorf EV, Muller MN, et al. Cospeciation of gut microbiota with hominids. Science 2016;353:380–2. https://doi.org/10.1126/science.aaf3951.
[92] Browne HP, Forster SC, Anonye BO, Kumar N, Neville BA, Stares MD, et al. Culturing of 'unculturable' human microbiota reveals novel taxa and extensive sporulation. Nature 2016;533:543–6. https://doi.org/10.1038/nature17645.
[93] Hildebrand F, Gossmann TI, Frioux C, Özkurt E, Myers PN, Ferretti P, et al. Dispersal strategies shape persistence and evolution of human gut bacteria. Cell Host Microbe 2021;29:1167–1176.e9. https://doi.org/10.1016/j.chom.2021.05.008.
[94] Caroll SB. The serengeti rules. Princeton University Press; 2016.
[95] Tett A, Pasolli E, Masetti G, Ercolini D, Segata N. Prevotella diversity, niches and interactions with the human host. Nat Rev Microbiol 2021. https://doi.org/10.1038/s41579-021-00559-y.
[96] Wibowo MC, Yang Z, Borry M, Hübner A, Huang KD, Tierney BT, et al. Reconstruction of ancient microbial genomes from the human gut. Nature 2021. https://doi.org/10.1038/s41586-021-03532-0.
[97] Tett A, Huang KD, Asnicar F, Fehlner-Peach H, Pasolli E, Karcher N, et al. The *Prevotella copri* complex comprises four distinct clades underrepresented in westernized populations. Cell Host Microbe 2019. https://doi.org/10.1016/j.chom.2019.08.018.
[98] Lynn DJ, Benson SC, Lynn MA, Pulendran B. Modulation of immune responses to vaccination by the microbiota: implications and potential mechanisms. Nat Rev Immunol 2021. https://doi.org/10.1038/s41577-021-00554-7.
[99] Xiao L, Feng Q, Liang S, Sonne SB, Xia Z, Qiu X, et al. A catalog of the mouse gut metagenome. Nat Biotechnol 2015;33:1103–8. https://doi.org/10.1038/nbt.3353.
[100] Gálvez EJCC, Iljazovic A, Amend L, Lesker TR, Renault T, Thiemann S, et al. Distinct polysaccharide utilization determines interspecies competition between intestinal Prevotella spp. Cell Host Microbe 2020;28:838–852.e6. https://doi.org/10.1016/j.chom.2020.09.012.
[101] Ramanan D, Bowcutt R, Lee SC, Tang MS, Kurtz ZD, Ding Y, et al. Helminth infection promotes colonization resistance via type 2 immunity. Science 2016;352:608–12. https://doi.org/10.1126/science.aaf3229.
[102] Chabé M, Lokmer A, Ségurel L. Gut protozoa: friends or foes of the human gut microbiota? Trends Parasitol 2017. https://doi.org/10.1016/j.pt.2017.08.005.
[103] Gabrielli S, Furzi F, Fontanelli Sulekova L, Taliani G, Mattiucci S. Occurrence of Blastocystis-subtypes in patients from Italy revealed association of ST3 with a healthy gut microbiota. Parasite Epidemiol Control 2020;9. https://doi.org/10.1016/j.parepi.2020.e00134, e00134.
[104] Li J, Fu R, Yang Y, Horz H-P, Guan Y, Lu Y, et al. A metagenomic approach to dissect the genetic composition of enterotypes in Han Chinese and two Muslim groups. Syst Appl Microbiol 2017. https://doi.org/10.1016/j.syapm.2017.09.006.
[105] Sonnenburg JL, Xu J, Leip DD, Chen C-H, Westover BP, Weatherford J, et al. Glycan foraging in vivo by an intestine-adapted bacterial symbiont. Science 2005;307:1955–9. https://doi.org/10.1126/science.1109051.

[106] Cuskin F, Lowe EC, Temple MJ, Zhu Y, Cameron EA, Pudlo NA, et al. Human gut Bacteroidetes can utilize yeast mannan through a selfish mechanism. Nature 2015;517:165–9. https://doi.org/10.1038/nature13995.

[107] Ndeh D, Rogowski A, Cartmell A, Luis AS, Baslé A, Gray J, et al. Complex pectin metabolism by gut bacteria reveals novel catalytic functions. Nature 2017. https://doi.org/10.1038/nature21725.

[108] Rakoff-Nahoum S, Foster KR, Comstock LE. The evolution of cooperation within the gut microbiota. Nature 2016;533:255–9. https://doi.org/10.1038/nature17626.

[109] Mark Welch JL, Hasegawa Y, McNulty NP, Gordon JI, Borisy GG. Spatial organization of a model 15-member human gut microbiota established in gnotobiotic mice. Proc Natl Acad Sci U S A 2017;114:E9105–14. https://doi.org/10.1073/pnas.1711596114.

[110] Raman AS, Gehrig JL, Venkatesh S, Chang H-W, Hibberd MC, Subramanian S, et al. A sparse covarying unit that describes healthy and impaired human gut microbiota development. Science 2019;365. https://doi.org/10.1126/science.aau4735, eaau4735.

[111] Vangay P, Johnson AJ, Ward TL, Al-Ghalith GA, Shields-Cutler RR, Hillmann BM, et al. US immigration westernizes the human gut microbiome. Cell 2018;175:962–972.e10. https://doi.org/10.1016/j.cell.2018.10.029.

[112] Wexler AG, Goodman AL. An insider's perspective: Bacteroides as a window into the microbiome. Nat Microbiol 2017;2:17026. https://doi.org/10.1038/nmicrobiol.2017.26.

[113] Angelakis E, Bachar D, Yasir M, Musso D, Djossou F, Gaborit B, et al. Treponema species enrich the gut microbiota of traditional rural populations but are absent from urban individuals. New Microbes New Infect 2019;27:14–21. https://doi.org/10.1016/j.nmni.2018.10.009.

[114] Smits SA, Leach J, Sonnenburg ED, Gonzalez CG, Lichtman JS, Reid G, et al. Seasonal cycling in the gut microbiome of the Hadza hunter-gatherers of Tanzania. Science 2017;357:802–6. https://doi.org/10.1126/science.aan4834.

[115] Gomez A, Rothman JM, Petrzelkova K, Yeoman CJ, Vlckova K, Umaña JD, et al. Temporal variation selects for diet-microbe co-metabolic traits in the gut of Gorilla spp. ISME J 2016;10:514–26. https://doi.org/10.1038/ismej.2015.146.

[116] Tokuda G, Mikaelyan A, Fukui C, Matsuura Y, Watanabe H, Fujishima M, et al. Fiber-associated spirochetes are major agents of hemicellulose degradation in the hindgut of wood-feeding higher termites. Proc Natl Acad Sci U S A 2018;115:E11996–2004. https://doi.org/10.1073/pnas.1810550115.

[117] Bui AT, Williams BA, Hoedt EC, Morrison M, Mikkelsen D, Gidley MJ. High amylose wheat starch structures display unique fermentability characteristics, microbial community shifts and enzyme degradation profiles. Food Funct 2020;11:5635–46. https://doi.org/10.1039/d0fo00198h.

[118] Li H, Gidley MJ, Dhital S. High-amylose starches to bridge the "Fiber gap": development, structure, and nutritional functionality. Compr Rev Food Sci Food Saf 2019;18:362–79. https://doi.org/10.1111/1541-4337.12416.

[119] Mizrahi I, Wallace RJ, Moraïs S. The rumen microbiome: balancing food security and environmental impacts. Nat Rev Microbiol 2021. https://doi.org/10.1038/s41579-021-00543-6, 0123456789.

[120] Xie H, Guo R, Zhong H, Feng Q, Lan Z, Qin B, et al. Shotgun metagenomics of 250 adult twins reveals genetic and environmental impacts on the gut microbiome. Cell Syst 2016;3:572–584.e3. https://doi.org/10.1016/j.cels.2016.10.004.

[121] Cwyk WM, Canale-Parola E. Treponema succinifaciens sp. nov., an anaerobic spirochete from the swine intestine. Arch Microbiol 1979;122:231–9. https://doi.org/10.1007/BF00411285.

[122] Adamovich Y, Ladeuix B, Sobel J, Manella G, Neufeld-Cohen A, Assadi MH, et al. Oxygen and carbon dioxide rhythms are circadian clock controlled and differentially directed by behavioral signals. Cell Metab 2019;29:1092–1103.e3. https://doi.org/10.1016/j.cmet.2019.01.007.
[123] De Vadder F, Kovatcheva-Datchary P, Zitoun C, Duchampt A, Bäckhed F, Mithieux G. Microbiota-produced succinate improves glucose homeostasis via intestinal gluconeogenesis. Cell Metab 2016;24:151–7. https://doi.org/10.1016/j.cmet.2016.06.013.
[124] Koh A, De Vadder F, Kovatcheva-Datchary P, Bäckhed F. From dietary fiber to host physiology: short-chain fatty acids as key bacterial metabolites. Cell 2016;165:1332–45. https://doi.org/10.1016/j.cell.2016.05.041.
[125] Kurilshikov A, Medina-Gomez C, Bacigalupe R, Radjabzadeh D, Wang J, Demirkan A, et al. Large-scale association analyses identify host factors influencing human gut microbiome composition. Nat Genet 2021;53:156–65. https://doi.org/10.1038/s41588-020-00763-1.
[126] Qin Y, Havulinna AS, Liu Y, Jousilahti P, Ritchie SC, Tokolyi A, et al. Combined effects of host genetics and diet on human gut microbiota and incident disease in a single population cohort. MedRxiv 2020. https://doi.org/10.1101/2020.09.12.20193045. 2020.09.12.20193045.
[127] Qi Q, Li J, Yu B, Moon J-Y, Chai JC, Merino J, et al. Host and gut microbial tryptophan metabolism and type 2 diabetes: an integrative analysis of host genetics, diet, gut microbiome and circulating metabolites in cohort studies. Gut 2021. https://doi.org/10.1136/gutjnl-2021-324053.
[128] Ségurel L, Bon C. On the evolution of lactase persistence in humans. Annu Rev Genomics Hum Genet 2017;18:297–319. https://doi.org/10.1146/annurev-genom-091416-035340.
[129] McKeen S, Young W, Fraser K, Roy NC, McNabb WC. Glycan utilisation and function in the microbiome of weaning infants. Microorganisms 2019;7. https://doi.org/10.3390/microorganisms7070190.
[130] Charlton S, Ramsøe A, Collins M, Craig OE, Fischer R, Alexander M, et al. New insights into Neolithic milk consumption through proteomic analysis of dental calculus. Archaeol Anthropol Sci 2019;11:6183–96. https://doi.org/10.1007/s12520-019-00911-7.
[131] Day AJ, Cañada FJ, Díaz JC, Kroon PA, Mclauchlan R, Faulds CB, et al. Dietary flavonoid and isoflavone glycosides are hydrolysed by the lactase site of lactase phlorizin hydrolase. FEBS Lett 2000;468:166–70. https://doi.org/10.1016/s0014-5793(00)01211-4.
[132] Rühlemann MC, Hermes BM, Bang C, Doms S, Moitinho-Silva L, Thingholm LB, et al. Genome-wide association study in 8,956 German individuals identifies influence of ABO histo-blood groups on gut microbiome. Nat Genet 2021. https://doi.org/10.1038/s41588-020-00747-1.
[133] Ewald DR, Sumner SCJ. Blood type biochemistry and human disease. Wiley Interdiscip Rev Syst Biol Med 2016;8:517–35. https://doi.org/10.1002/wsbm.1355.
[134] Yamamoto F, Cid E, Yamamoto M, Blancher A. ABO research in the modern era of genomics. Transfus Med Rev 2012;26:103–18. https://doi.org/10.1016/j.tmrv.2011.08.002.
[135] Glass RI, Holmgren J, Haley CE, Khan MR, Svennerholm AM, Stoll BJ, et al. Predisposition for cholera of individuals with O blood group. Possible evolutionary significance. Am J Epidemiol 1985;121:791–6. https://doi.org/10.1093/oxfordjournals.aje.a114050.
[136] Aspholm-Hurtig M, Dailide G, Lahmann M, Kalia A, Ilver D, Roche N, et al. Functional adaptation of BabA, the H. pylori ABO blood group antigen binding adhesin. Science 2004;305:519–22. https://doi.org/10.1126/science.1098801.

[137] Moissl-Eichinger C, Pausan M, Taffner J, Berg G, Bang C, Schmitz RA. Archaea are interactive components of complex microbiomes. Trends Microbiol 2017. https://doi.org/10.1016/j.tim.2017.07.004.
[138] Ng F, Kittelmann S, Patchett ML, Attwood GT, Janssen PH, Rakonjac J, et al. An adhesin from hydrogen-utilizing rumen methanogen *Methanobrevibacter ruminantium* M1 binds a broad range of hydrogen-producing microorganisms. Environ Microbiol 2016;18:3010–21. https://doi.org/10.1111/1462-2920.13155.
[139] Brugère J-F, Borrel G, Gaci N, Tottey W, O'Toole PW, Malpuech-Brugère C. Archaebiotics: proposed therapeutic use of archaea to prevent trimethylaminuria and cardiovascular disease. Gut Microbes 2014;5:5–10. https://doi.org/10.4161/gmic.26749.
[140] Yeoman CJ, Thomas SM, Miller MEB, Ulanov AV, Torralba M, Lucas S, et al. A multi-omic systems-based approach reveals metabolic markers of bacterial vaginosis and insight into the disease. PLoS One 2013;8. https://doi.org/10.1371/journal.pone.0056111, e56111.
[141] Wang Z, Roberts AB, Buffa JA, Levison BS, Zhu W, Org E, et al. Non-lethal inhibition of gut microbial trimethylamine production for the treatment of atherosclerosis. Cell 2015;163:1585–95. https://doi.org/10.1016/j.cell.2015.11.055.
[142] Zhu W, Gregory JC, Org E, Buffa JA, Gupta N, Wang Z, et al. Gut microbial metabolite TMAO enhances platelet hyperreactivity and thrombosis risk. Cell 2016;165:111–24. https://doi.org/10.1016/j.cell.2016.02.011.
[143] Cho CE, Taesuwan S, Malysheva OV, Bender E, Tulchinsky NF, Yan J, et al. Trimethylamine-N-oxide (TMAO) response to animal source foods varies among healthy young men and is influenced by their gut microbiota composition: a randomized controlled trial. Mol Nutr Food Res 2017;61:1–12. https://doi.org/10.1002/mnfr.201600324.
[144] Chhibber-Goel J, Gaur A, Singhal V, Parakh N, Bhargava B, Sharma A. The complex metabolism of trimethylamine in humans: endogenous and exogenous sources. Expert Rev Mol Med 2016;18. https://doi.org/10.1017/erm.2016.6, e8.
[145] Mao J, Zhao P, Wang Q, Chen A, Li X, Li X, et al. Repeated 3,3-dimethyl-1-butanol exposure alters social dominance in adult mice. Neurosci Lett 2021;758:136006. https://doi.org/10.1016/j.neulet.2021.136006.
[146] Li Q, Korzan WJ, Ferrero DM, Chang RB, Roy DS, Buchi M, et al. Synchronous evolution of an odor biosynthesis pathway and behavioral response. Curr Biol 2013;23:11–20. https://doi.org/10.1016/j.cub.2012.10.047.
[147] Apps PJ, Weldon PJ, Kramer M. Chemical signals in terrestrial vertebrates: search for design features. Nat Prod Rep 2015;32:1131–53. https://doi.org/10.1039/c5np00029g.
[148] Mukherji A, Kobiita A, Ye T, Chambon P. Homeostasis in intestinal epithelium is orchestrated by the circadian clock and microbiota cues transduced by TLRs. Cell 2013;153:812–27. https://doi.org/10.1016/j.cell.2013.04.020.
[149] Reitmeier S, Kiessling S, Clavel T, List M, Almeida EL, Ghosh TS, et al. Arrhythmic gut microbiome signatures predict risk of type 2 diabetes. Cell Host Microbe 2020;28:258–272.e6. https://doi.org/10.1016/j.chom.2020.06.004.
[150] Thaiss CA, Levy M, Korem T, Dohnalová L, Shapiro H, Jaitin DA, et al. Microbiota diurnal rhythmicity programs host transcriptome oscillations. Cell 2016;167:1495–1510.e12. https://doi.org/10.1016/j.cell.2016.11.003.
[151] Leone V, Gibbons SM, Martinez K, Hutchison AL, Huang EY, Cham CM, et al. Effects of diurnal variation of gut microbes and high-fat feeding on host circadian clock function and metabolism. Cell Host Microbe 2015;17:681–9. https://doi.org/10.1016/j.chom.2015.03.006.
[152] Yoon HS, Cho CH, Yun MS, Jang SJ, You HJ, Hyeong KJ, et al. *Akkermansia muciniphila* secretes a glucagon-like peptide-1-inducing protein that improves glucose homeostasis and ameliorates metabolic disease in mice. Nat Microbiol 2021;1–11. https://doi.org/10.1038/s41564-021-00880-5.

[153] Chevalier C, Stojanović O, Colin DJ, Suarez-Zamorano N, Tarallo V, Veyrat-Durebex C, et al. Gut microbiota orchestrates energy homeostasis during cold. Cell 2015;163:1360–74. https://doi.org/10.1016/j.cell.2015.11.004.
[154] Ouwerkerk JP, van der Ark KCH, Davids M, Claassens NJ, Finestra TR, de Vos WM, et al. Adaptation of *Akkermansia muciniphila* to the oxic-anoxic interface of the mucus layer. Appl Environ Microbiol 2016;82:6983–93. https://doi.org/10.1128/AEM.01641-16.
[155] Adamovich Y, Ladeuix B, Golik M, Koeners MP, Asher G. Rhythmic oxygen levels reset circadian clocks through HIF1α. Cell Metab 2017;25:93–101. https://doi.org/10.1016/j.cmet.2016.09.014.
[156] Li B, He Y, Ma J, Huang P, Du J, Cao L, et al. Mild cognitive impairment has similar alterations as Alzheimer's disease in gut microbiota. Alzheimers Dement 2019;1–10. https://doi.org/10.1016/j.jalz.2019.07.002.
[157] Wu H, Esteve E, Tremaroli V, Khan MT, Caesar R, Mannerås-Holm L, et al. Metformin alters the gut microbiome of individuals with treatment-naive type 2 diabetes, contributing to the therapeutic effects of the drug. Nat Med 2017;23:850–8. https://doi.org/10.1038/nm.4345.
[158] Jie Z, Xia H, Zhong S-L, Feng Q, Li S, Liang S, et al. The gut microbiome in atherosclerotic cardiovascular disease. Nat Commun 2017;8:845. https://doi.org/10.1038/s41467-017-00900-1.
[159] Thaiss CA, Zeevi D, Levy M, Zilberman-Schapira G, Suez J, Tengeler AC, et al. Transkingdom control of microbiota diurnal oscillations promotes metabolic homeostasis. Cell 2014;159:514–29. https://doi.org/10.1016/j.cell.2014.09.048.
[160] Kuang Z, Wang Y, Li Y, Ye C, Ruhn KA, Behrendt CL, et al. The intestinal microbiota programs diurnal rhythms in host metabolism through histone deacetylase 3. Science 2019;365:1428–34. https://doi.org/10.1126/science.aaw3134.
[161] Zwighaft Z, Aviram R, Shalev M, Rousso-Noori L, Kraut-Cohen J, Golik M, et al. Circadian clock control by polyamine levels through a mechanism that declines with age. Cell Metab 2015;22:874–85. https://doi.org/10.1016/j.cmet.2015.09.011.
[162] Johnson CH, Zhao C, Xu Y, Mori T. Timing the day: what makes bacterial clocks tick? Nat Rev Microbiol 2017;15:232–42. https://doi.org/10.1038/nrmicro.2016.196.
[163] Paulose JK, Wright JM, Patel AG, Cassone VM. Human gut bacteria are sensitive to melatonin and express endogenous circadian rhythmicity. PLoS One 2016;11. https://doi.org/10.1371/journal.pone.0146643, e0146643.
[164] Paulose JK, Cassone CV, Graniczkowska KB, Cassone VM. Entrainment of the circadian clock of the enteric bacterium *Klebsiella aerogenes* by temperature cycles. IScience 2019;19:1202–13. https://doi.org/10.1016/j.isci.2019.09.007.
[165] Eelderink-Chen Z, Bosman J, Sartor F, Dodd AN, Kovács ÁT, Merrow M. A circadian clock in a nonphotosynthetic prokaryote. Sci Adv 2021;7. https://doi.org/10.1126/sciadv.abe2086, eabe2086.

3

Collecting samples for metagenomics

3.1 Nonmicrobial components in the sample that could influence DNA extraction and sequencing amount

The Bristol's stool score (BSS), whereby fecal samples are graded according to shape (Fig. 3.1A), is a useful approximation for water content, gut transit time, as well the number of bacterial cells per gram of feces [1–3]. With illustrations in questionnaires, volunteers can fill out the BSS as they collect their own feces. Such self-reported BSS showed correlations with enzymes detected in the fecal metagenome that metabolize secondary bile acids [3], which makes mechanistic sense.

A mountable toilet system can automatically record values such as fecal BSS, volume, and flow rate of urine, but the published version does not collect samples yet [4].

The number of microbial cells per gram of feces could be affected by the presence of food debris, which becomes substantial when the diet is as fibrous as that of a panda (Fig. 3.1A). A study of fecal samples from healthy adults on a British diet (1980), with vigorous agitation and detergent treatment of the feces, reported the fecal mass to be 55% bacteria, 17% fiber, and 24% soluble material. Proportion of the dry mass can actually be different with different microbes, but we are not too worried about that.

Nylon swabs with detachable ends can be used for fecal and mucosal samples (Fig. 3.1B). Swabs for skin and nasal samples are typically wetted with physiological saline or buffer solutions before sampling, but it is not clear how such transient exposure to minerals might change the microbial community. Larger volumes may be necessary for some low-biomass samples. Make sure to screw the lids tightly, to avoid liquid leaks or drying during transport. If the samples are going to be collected by volunteers themselves, clear illustrations or videos

Investigating Human Diseases with the Microbiome: Metagenomics Bench to Bedside.
https://doi.org/10.1016/B978-0-323-91369-0.00001-7

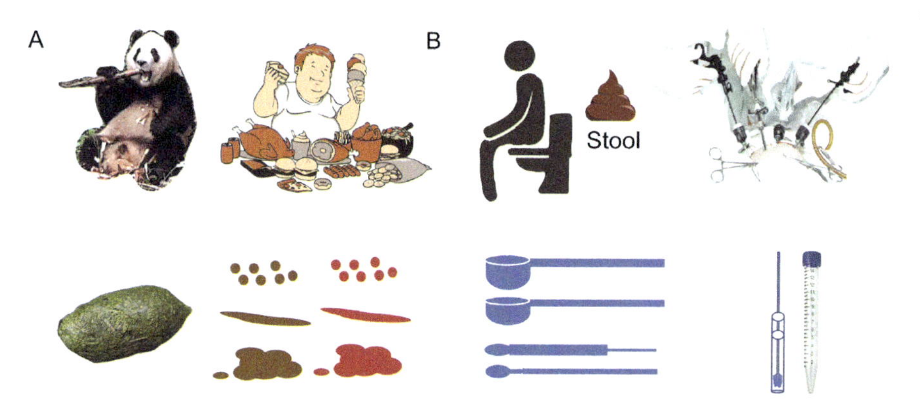

Fig. 3.1 Getting enough microbes when collecting samples. (A) Metagenomic samples may contain materials other than microbes. The human feces are illustrated on a simplified Bristol's Stool Scale (BSS), the hardest has a BSS of 1, a BSS of 4 in the middle, and the watery form has a BSS of 7. (B) Scoops or swabs for solid samples or surfaces, and tubes for liquids. Laparoscopes or other new technologies that make a small cut are presumably less prone to contamination for the collection of microbiome samples in the operation room. Skin and other surfaces are decontaminated. Swabs or brushes can be in protective tubing before reaching a sampling site. Credit: Huijue Jia, Xin Tong of BGI-Shenzhen.

for the procedure would be a good idea, as well as photo records for the sample. Compliance rates could vary between cohorts. Quality checks should be performed early.

Fecal samples can become bloody with inflammatory bowel diseases, hemorrhoids, etc., making the proportion of human reads in a metagenomic sample to go much higher than 1% (Table 3.1). Human microbiome samples other than feces and supragingival plaques all have a higher percentage of the human genomic sequence, going over 99% in some tissue samples (Table 3.1). With informed consent, the low-depth human genome sequence can be analyzed together with the microbiome, while beware of tissue differences in comparison to blood.

Removal of host cells by low-speed centrifugation has been shown to lead to loss of bacterial species in bronchoalveolar lavage samples (BAL) [17]. Experimental removal of host sequences using molecular biology or chemical ways also affects the microbiome composition [15,18] but may be optimized in the future. So we currently recommend bioinformatic removal after unbiased sequencing.

The combination of technologies used depends on the questions we would like to address (e.g., bacteria in pancreatic cancer, Chapter 1, Fig. 1.8). For a good sample, it would be a pity not to fully investigate its potential, just because the sequencing amount appears intimidating. After all, a lot of the microbial evolution and interactions would be local.

Table 3.1 Varying percentage of human sequences in shotgun metagenomic data of samples from different body sites.

Body region	Body site	% human sequences	Reference
Gut	Feces	1%	[5,6]
Gut	Feces (Crohn's disease)	20% or more in some samples	[7]
Oral	Buccal mucosa	82%–90%	[5,8,9]
Oral	Supragingival plaque	40%; 5.55%	[5,10]
Oral	Subgingival plaque	79%	[5]
Oral	Tongue dorsum	30%	[9,11]
Oral	Saliva	77%–91%	[5,11]
Skin	Dry (e.g., volar forearm)	36%	[12]
Skin	Moist (e.g., antecubital fossa)	44%	[12]
Skin	Sebaceous (e.g., retroauricular crease)	59%–73%	[5,12]
Urogenital	Vagina (including the posterior fornix)	90%–98%	[5,13]
Urogenital	Cervical orifice	98%	[14]
Urogenital	Peritoneal fluid	99.8%	[15]
Urogenital	Placenta	> 99%	[16]
Airway	Anterior nares	96%	[5]

Approximate average values are summarized from multiple studies, and if different, separated by ";". This is not meant to be an exhaustive list. It should serve the purpose for a general understanding, when researchers and clinicians decide on the sequencing amount for metagenomic shotgun sequencing (e.g., 10 Gb of paired-end 100 bp reads), or opt for amplicon sequencing, in situ hybridization, etc.
Credit: Huijue Jia.

3.2 Beware of contamination in each step, from stools to low-biomass samples

As briefly noted in Chapter 1, claims for the presence of a microbiome in samples with a low microbial biomass, e.g., placenta samples (Chapter 1, Fig. 1.7), are greeted with skepticism, but were more readily accepted in tumor samples (e.g., Chapter 1, Fig. 1.8) [19–24]. Studying these low biomass samples helps us reach a clear understanding of things to take care of for all kinds of metagenomic samples (Fig. 3.2). On the other hand, it may turn out to be more difficult to prove the absence of microbes at a site, than the presence. What kind of extreme environment can be free of any bacteria, archaea, and fungi?

For fecal, oral (Chapter 2, Fig. 2.6), and vaginal samples, there are typically over 10^{10} microbial cells (cfu, colony forming units on a culture plate) in a swab sample. This high number of microbial genomes relieves the caution for contamination during sampling, DNA extraction, library construction, and sequencing. For low biomass samples (Figs. 3.2 and 3.3), the reagents used for sampling and

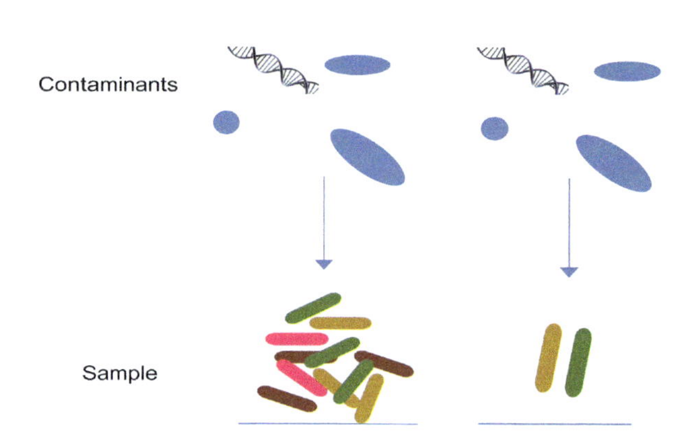

Fig. 3.2 Contaminating DNA or microbes can be introduced at any step from sampling to sequencing. It is just that samples with low biomass are more easily overwhelmed by the presence of contaminants. Credit: Huijue Jia, Xin Tong of BGI-Shenzhen.

subsequent steps, and the facility should all be routinely checked for the presence of live or dead microbes [25], e.g., by using amplicon sequencing (Fig. 3.4). The most cited study on contamination performed serial dilution of *Salmonella bongori*, and other taxa dominated the 16S rRNA gene amplicon results at the 5th serial dilution, corresponding to about 10^3 *Salmonella bongori* cells [26]. While the focus of the study was on reagents contamination [26], the author could not find information regarding the pipettes for dilution (the PCR amplification was performed in a hood using autoclaved microcentrifuge tubes and filtered pipette tips) and the protection from human contamination, as is well known in the ancient DNA field. More caution against batch effects should also be taken with amplicon sequencing (e.g., [27]).

Hospital rooms, or even space station rooms, do accumulate microbes after being used [28–32], with potential microbiome contribution from both patients and staff. Air and ventilation systems in a hospital room may show up positive with bacteria or fungi (Fig. 3.3). As long as they are different from the samples of interest (e.g., Worked sample 3.3), one can still tentatively believe that there is a microbiome in the samples and look for more evidence. The negative controls also apply to live culture on plates, and quantitative polymerase chain reactions (qPCR).

In samples dominated by human sequences, however, amplicon sequencing may pick up lots of human sequences that were nonspecifically amplified (e.g., in kidney stones [33]). Besides optimizing the PCR condition, such samples may need to be purified according to fragment length, or subject to targeted sequencing, while controlling for contamination.

Having multiple samples from the same individual, including samples from different body sites, could also lend some support that the microbes detected are not random contamination, or systematic contamination in reagents, but show patterns that are specific for each individual [15,34,35]. When studying the female reproductive tract in nonpregnant women without inflammation, our collaborating

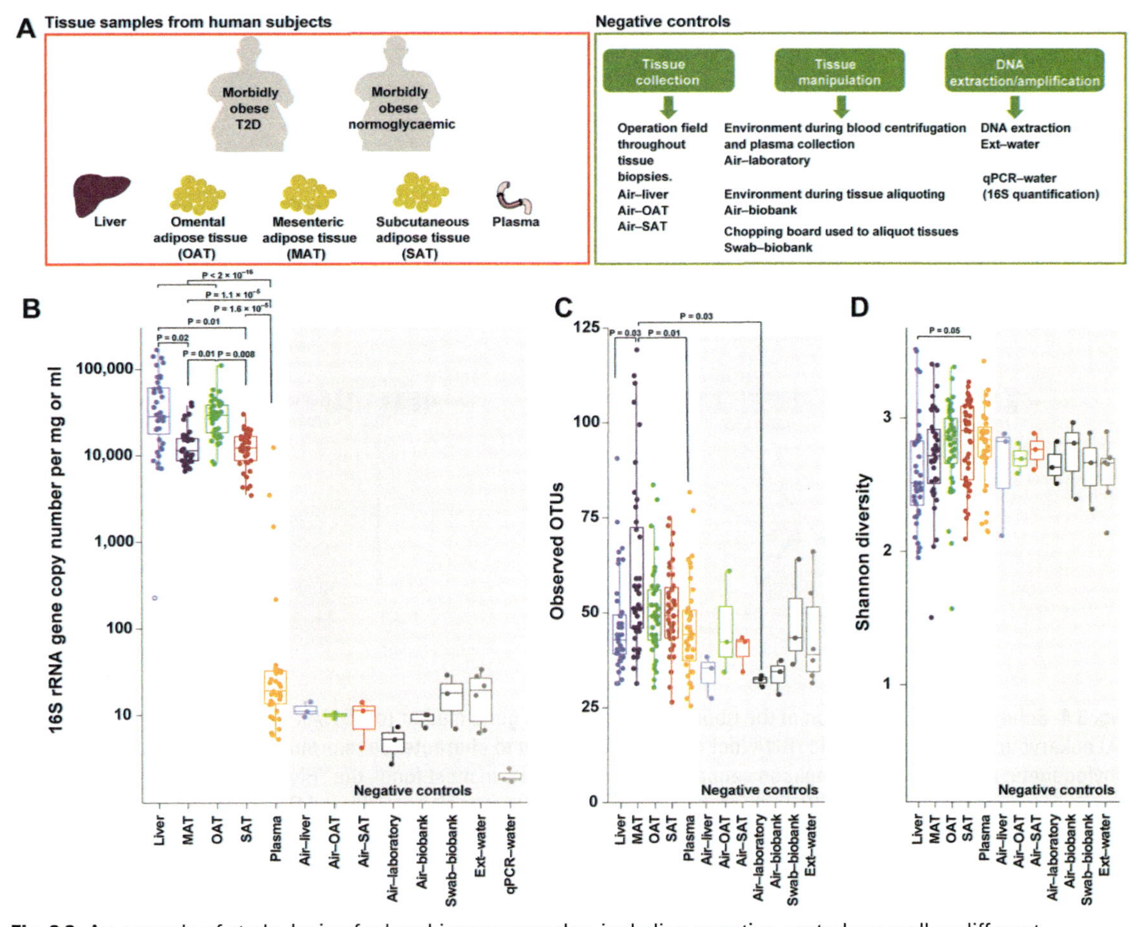

Fig. 3.3 An example of study design for low-biomass samples, including negative controls as well as different physiological conditions for association study. (A) Liver, three different adipose tissue depots (OAT, MAT, and SAT), and plasma samples were collected from individuals with morbid obesity who had T2D ($n=20$) and from those who had normoglycaemia ($n=20$). DNA extraction and amplification procedures were carried out using optimized conditions for bacterial DNA detection in blood plasma and tissues. A comprehensive set of negative controls was tested to control for environmental sample contamination at major steps in the analysis: tissue collection, tissue manipulation, and DNA extraction and amplification. During tissue collection, tubes were kept open next to the operation field throughout the entire procedure (air-liver, air-OAT, and air-SAT). Contamination coming from tissue manipulation was controlled by another set of tubes that were kept open next to the operator throughout blood centrifugation and plasma collection (air-laboratory) as well as during tissue aliquoting (air-biobank). The chopping board used to aliquot tissues was sampled prior to tissue manipulation (swab-biobank). Water samples were used to control for labware, reagent, and/or environmental contamination during DNA extraction (ext-water) and amplification steps for tissue 16S rRNA quantification by quantitative PCR (qPCR-water). After thorough validation of negative controls on a case-by-case basis, 16S quantification and sequencing data were used in the discovery of tissue-specific bacterial signatures linked to T2D. (B–D) Number of bacteria across body sites. (B) 16S rRNA gene counts. (C) Observed OTUs. (D) Shannon index in the liver, three different adipose tissue depots (OAT, MAT, and SAT), and plasma of participants with obesity. Negative controls were tested to control for environmental sample contamination at major steps in the analysis: tissue collection (air-liver, air-OAT, air-SAT), tissue manipulation (air-laboratory, air-biobank, and swab-biobank), and DNA extraction or amplification (ext-water, qPCR-water). In panels (B–D), groups were compared using a Kruskal-Wallis one-way ANOVA followed by Dunn's test for pairwise comparison. Box plots depict the first and the third quartile with the median represented by a vertical line within the box; the whiskers extend from the first and third quartiles to the highest and lowest observation, respectively, not exceeding 1.5×IQR. Credit: Cropped from Fig. 1, Fig. 2 of Anhê FF, Jensen BAH, Varin TV, Servant F, Van Blerk S, Richard D, et al. Type 2 diabetes influences bacterial tissue compartmentalisation in human obesity. Nat Metab 2020;2:233–42. https://doi.org/10.1038/s42255-020-0178-9.

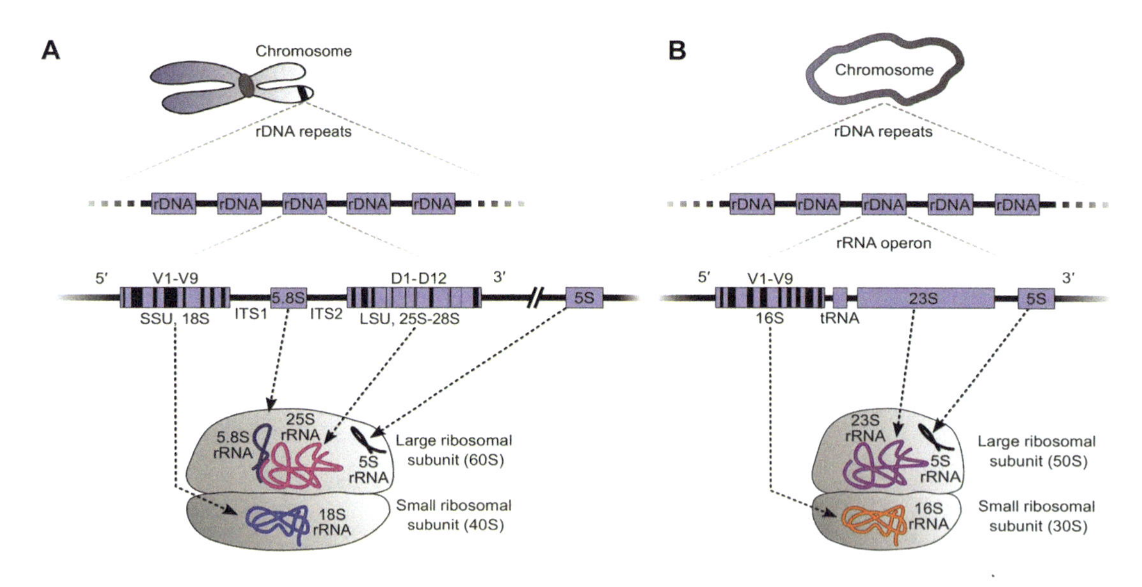

Fig. 3.4 Schematic representation of the ribosomal RNA (rRNA) gene cluster (or rDNA). The variable regions of (A) eukaryotic and (B) prokaryotic rRNA loci are commonly used to characterize microbial taxa and resolve their phylogenetic relationships by amplicon sequencing and analyses. In most fungi, the rRNA gene cluster includes the small ribosomal subunit (SSU, 18S), with internal transcribed spacer regions (ITS1 and ITS2) flanking the 5.8S, and large ribosomal subunit (LSU, 25–28S) regions. In bacteria, the rRNA operon comprises the SSU (16S), LSU (23S), and 5S loci. *Black vertical lines* in serial order illustrate the variable regions in SSU (V1–V9) and LSU (D1–D12), best suited for biodiversity assessments through microbial communities profiling. Credit: From Fig. 1 of Lavrinienko A, Jernfors T, Koskimäki JJ, Pirttilä AM, Watts PC. Does intraspecific variation in rDNA copy number affect analysis of microbial communities? Trends Microbiol 2021;29:19–27. https://doi.org/10.1016/j.tim.2020.05.019.

gynecologist Dr. Ruifang Wu made sure that after the subcentimeter cut on the skin, her team started with the peritoneal fluid from the pouch of Douglas, before the laparoscope moved on to the fallopian tubes (e.g., for those with an obstruction there), and then the endometrium (fibroids, endometriosis, or adenomyosis) that we suspected to have a denser microbial community. The vaginal and cervical samples were taken on the day of the initial visit to the clinic, with protective tubing for the cervical mucus sampling before it reached into the cervix. Despite the overdominance of Lactobacilli, the vaginal and cervical samples nicely matched upper reproductive tract samples that were collected a few days later in the operation room for each volunteer [34,35]. The peritoneal fluid, with a more neutral pH, might also be a source community (Chapter 2) of low biomass but diverse microbes [34,35].

The placenta microbiome has remained controversial (Chapter 1, Fig. 1.7), as it is pushing the detection limit of existing methods. According to fluorescent in situ hybridization (FISH) against conserved regions in the 16S rRNA, Dr. Kjersti M. Aagaard reported that small clusters of bacteria were mostly localized to the villous parenchyma or syncytiotrophoblast, and less commonly in the chorion and the maternal intervillous space [36]. A recent study collected samples from the placental terminal villi, required bacteria to be detected by both 16S rRNA gene amplicon sequencing (V1–V2 region, Fig. 3.4) and metagenomic shotgun sequencing, and discredited results identified in controls including vaginal samples [25]. Only the neonatal pathogen *Streptococcus agalactiae* (Group B *Streptococcus*) remained as a placental microbe [25], which we could detect in the cervix [14].

Contamination from blood has always been a concern (sampled in the adipose tissue microbiome study, Fig. 3.3). The sequencing amount in the placenta study was so low (26.5 million reads on average, > 99% human genome (Table 3.1); e.g., if 99.9% human, there would be 26.5 thousand nonhuman reads per sample on average, less than 1 × coverage for the genome of a single bacterium) that broad functional capacity seen from KEGG level 2 pathways fluctuated among samples [16]. For cervical orifice samples, we observed a correlation between the dominant bacteria species and the percentage of human sequences [9,14]. However, some of the taxa identified in the original study may still be genuinely present in the placenta. The placenta microbiome in this metagenomic study showed a high relative abundance of *Escherichia coli* [16], which is known in the meconium (newborn feces) [37–41], along with potentially gut or oral bacteria such as *Bacteroides, Streptococcus parasanguinis, Prevotella melaninogenica,* reproductive tract bacteria such as *Cutibacterium acnes* (renamed from *Propionibacterium acnes*), *Lactobacillus iners, L. crispatus,* which were more or less recapitulated in the abovementioned recent study with plenty of controls [25]. Bacteria such as *Cutibacterium acnes* and *Streptococcus parasanguinis* are also part of the infant oral and gut microbiome [16,38,40,42]. And if the species assignment for *Neisseria lactamica* holds true despite the low sequencing coverage [16] (more on taxonomy in Chapter 5), it will provide a potential explanation for carriage of the bacteria in the nasopharynx of young children. More taxa have been detected by 16S rRNA gene amplicon sequencing [36], which was not overloaded with the human sequences.

According to interesting studies on bacteria in adipose tissues and their dysbiosis in T2D patients, apparent bacterial diversity in plasma and in negative controls was not lower than in adipose tissue samples, but the copy numbers of the total 16S rRNA gene were orders of magnitude higher than the negative controls (Fig. 3.3) [43,44]. The per mg (milligram) tissue unit [44] looks better than the per μg DNA unit [43], because the latter includes the human genome (2 copies of 3.2 GB). The diameter of white adipocytes varies from less than 30 μm to over 300 μm [45]. Try some calculations in light of Chapter 1. If these are all very small adipocytes with a diameter of 20 μm, the bacteria cell (~ 16S rRNA gene copy number in Fig. 3.3, per unit tissue weight or volume [44]) to human cell ratio would be roughly 1:10. With a more mediocre adipocyte diameter of 100 μm, however, the bacteria cell to human cell ratio would be flip to about 10:1.

As we move on to study the brain, the lungs (Figs. 3.5 and 3.6), or other tissues, are we collecting all the related samples, and in a good order? For bronchoalveolar lavage (BAL) samples, the good news is that there appeared to be no difference between samples taken via the mouth versus samples taken via the nose (with protective tubing) [46],

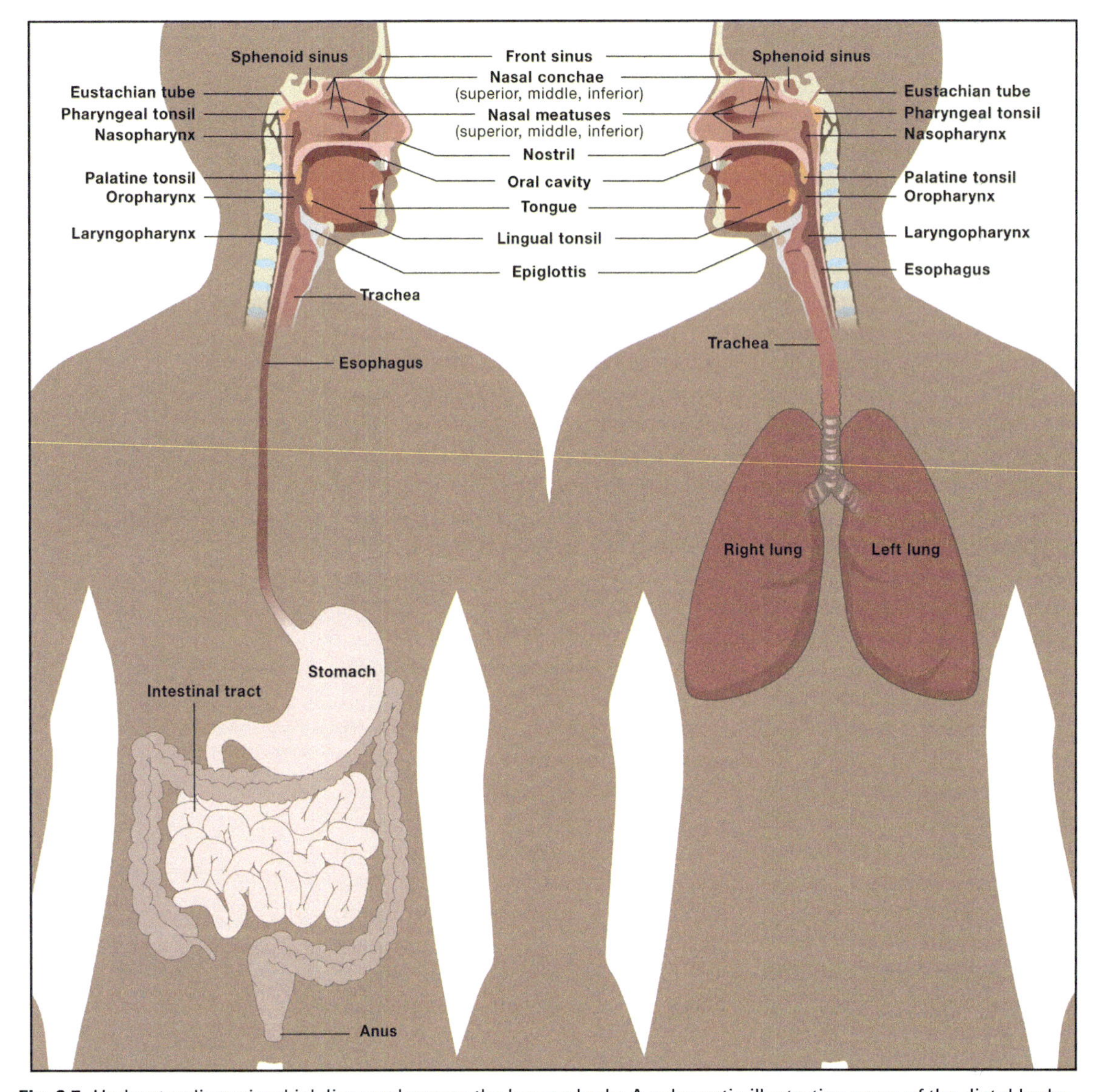

Fig. 3.5 Understanding microbial dispersal across the human body. A schematic illustrating some of the distal body sites to which microbes may disperse from the nose, mouth, or throat, as well as some interconnections among sites. For example, the pharyngeal tonsils, also known as the adenoids (a major site of lymphoid tissue in the oral/nasal pharyngeal), may serve as a reservoir for middle ear infections as a result of dispersal via the eustachian tube. Credit: Fig. 3 of Proctor DM, Relman DA. The landscape ecology and microbiota of the human nose, mouth, and throat. Cell Host Microbe 2017;21:421–32. https://doi.org/10.1016/j.chom.2017.03.011.

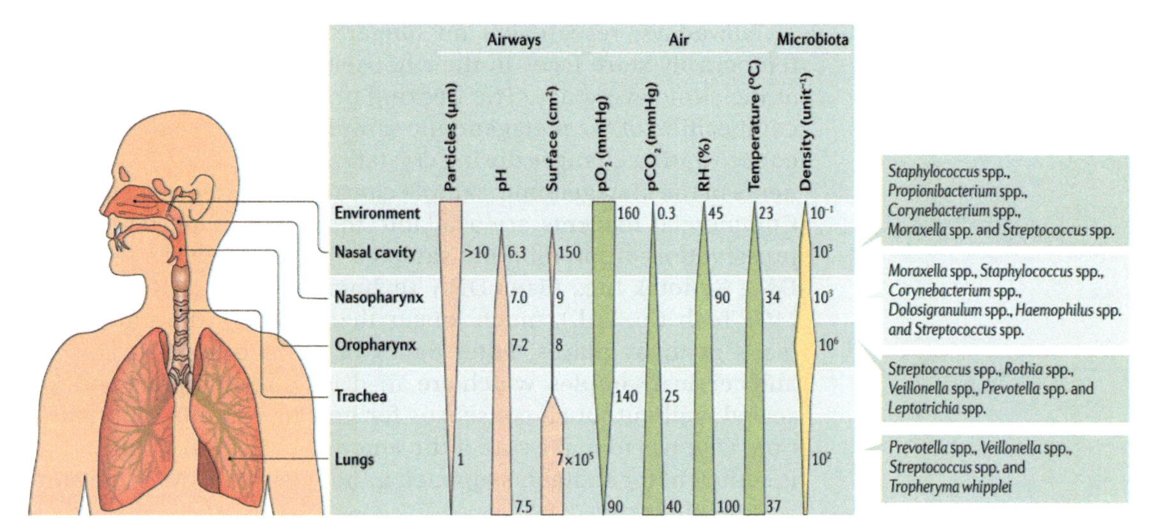

Fig. 3.6 Physiological and microbial gradients along the respiratory tract. Physiological and microbial gradients exist along the nasal cavity, nasopharynx, oropharynx, trachea, and the lungs. The pH gradually increases along the respiratory tract, whereas most of the increases in relative humidity (RH) and temperature occur in the nasal cavity Furthermore, the partial pressures of oxygen (pO_2) and carbon dioxide (pCO_2) have opposing gradients that are determined by environmental air conditions and gas exchange at the surface of the lungs. Inhalation results in the deposition of particles from the environment into the respiratory tract; inhaled particles that are more than 10 μm in diameter are deposited in the upper respiratory tract, whereas particles less than 1 μm in diameter can reach the lungs. These particles include bacteria-containing and virus-containing particles, which are typically larger than 0.4 μm in diameter. These physiological parameters determine the niche-specific selective growth conditions that ultimately shape the microbial communities along the respiratory tract. The unit by which bacterial density is measured varies per niche; the density in the environment is depicted as bacteria per cm^3 (indoor) air, density measures in the nasal cavity and nasopharynx are shown as an estimated number of bacteria per nasal swab, and the densities in the oropharynx and the lungs represent the estimated number of bacteria per ml of oral wash or bronchoalveolar lavage (BAL), respectively. Credit: Fig. 1 of Man WH, de Steenhuijsen Piters WAA, Bogaert D. The microbiota of the respiratory tract: gatekeeper to respiratory health. Nat Rev Microbiol 2017;15:259–70. https://doi.org/10.1038/nrmicro.2017.14.

although the related samples should be compared pairwise for each individual instead of shown in a crude PCA (Principle Component Analysis). How many microbial cells (Chapter 1) are we going to have in the samples? For the viruses and fungi, would we need other information to better understand the local habitat and morphology?

3.3 Reagents that prevent microbial growth after sampling

In the early days of metagenomic studies, fecal samples with a relative abundance of the facultative anaerobe *E. coli* that exceeded ~ 30% were often suspected of prolonged exposure to room temperature, but they can reflect genuine conditions such as colorectal cancer, Crohn's disease, IgA deficiency, Type 2 diabetes [7,47–50].

Nowadays, researchers no longer have to ask volunteers to temporarily store feces in their household fridge, or have dry ice at the clinic every day. The freezing procedure can also affect the composition of the metagenomic sample, e.g., due to pH and other concentration changes during crystallization of water, and components in the metagenomic sample can affect the freezing efficiency. Commercial reagents are available that allow room-temperature preservation of microbiome samples for 2 or 4 weeks (e.g., from DNA Genotek Inc., Mawi DNA Technologies LLC., MGIEasy from MGI Tech Co. Ltd.), much longer than delivery time for courier mails in many places. Filter paper has also been used for fecal and cervical samples, which are air-dried after sampling and then sealed, without much consensus for how to minimize contamination. One has to make sure if the amount of DNA from filter paper is sufficient for shotgun sequencing, not just 16S rRNA gene amplicon sequencing.

There are not enough studies published for metatranscriptomics, which in addition to high-quality preservation of RNA, also require (however incomplete) removal of ribosomal RNAs before sequencing [51,52]. Metaproteomics is also on the rise, and we have only tried fresh or frozen samples.

The stabilizing reagents only stop bacterial growth and decay, and do not kill bacteria. So at least some of the microbes could still be cultured on a plate. Similarly, analyzing the metabolome of the same samples using mass spectrometry typically requires the swabs to be in a reagent (e.g., 50:50 ethanol:water for skin swabs [53]) different from the one used for microbiome storage, but there are commercial products that try to work for both purposes.

Worked sample 3.1

Please think of a kind of microbiome sample you are interested in.

What is the current estimate for the number of microbial cells and the number of species at this body site? Do you expect the numbers to change for the disease in question?

How much DNA would you need for constructing a metagenomic library (e.g., 0.5 µg)? Would you use a swab or some other plasticware?

Would you be able to process or freeze the samples immediately (e.g., save some for other omics in the future)? If a commercial reagent would be used to preserve the samples for a few weeks at room temperature and possibly through courier mail, do you think it might affect some microbes more than the others in this particular microbial community?

Would you have a standard mix and real samples to compare the fresh, frozen, and reagent-preserved community?

3.4 DNA extraction for metagenomic samples

Even in the absence of too many plant fibers or protists (Fig. 3.1), DNA extraction for metagenomic samples is more complex than DNA extraction for mammalian cells or single microbial species, because of the myriad of cell walls that we all try to break (Fig. 3.7) [54–56]. Physical and chemical steps can all be added, and control samples could be used in parallel or as a quantitative spike-in, if sufficiently different in sequence. Nanopore sequencing that requires the DNA fragments to be more than 20 kb long would be a different story.

For bead beating, the harder tetragonal Zirconium polycrystals are superior to glass beads. Size of the beads also needs to be considered for the bacteria and fungi in the sample. Smaller beads make more contact points within a given time but may not have enough momentum to break some fungi.

Unlike amplicon sequencing, metagenomic shotgun sequencing workflows that do not involve any PCR steps can better tolerate impurities in DNA. For example, DNA from the fecal sample may still come out a little yellowish.

Automated platforms (e.g., 96 samples per plate) can achieve a more uniform quality of sample processing, and reduce the chance of random contamination, in addition to saving time. Such automation could also better protect the staff from whatever is in the clinical samples.

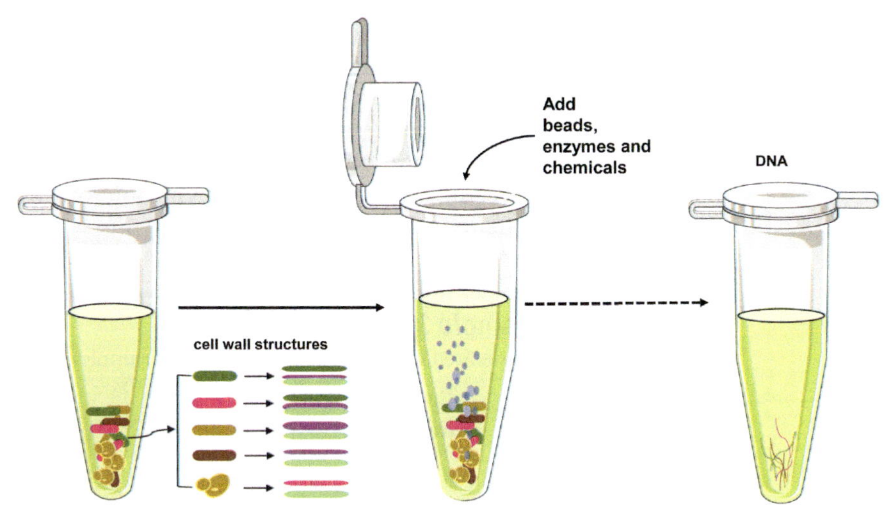

Fig. 3.7 DNA extraction for a complex community of microorganisms. The thick lines represent different cell wall structures for gram-negative, gram-positive bacteria, as well as fungi. Credit: Xin Tong of BGI-Shenzhen.

For constructing sequencing libraries, fragmentation of the extracted DNA used to be performed by sonication. The throughput can be much higher using enzymatic reactions (e.g., Tn5 transposase), which is compatible with automation.

3.5 Sequencing amount

In theory, the unique alignment of a single sequencing read is sufficient to detect a microbe (more on taxonomy in Chapter 5). Metagenomic shotgun sequencing that has no PCR amplification step has a negligible error rate [57]. For a metagenomic sample sequenced with 100 million reads (e.g., paired-end 100 bp reads, making $100 \times 10^6 \times 100 = 10$ Gb data), the lowest possible relative abundance that is directly detected for a taxa or a gene is 10^{-8}. For a sample containing 10^{11} microbial cells (< 1 g from the total biomass of ~ 200 g [58,59], Chapter 1), the detection limit should be single-cell, which could be verified with serially diluted spike-ins [22,60]. For low-abundance taxa, there is still a large gap in sequencing amount. It may be wise to look for samples in the same individual or in other people where this taxon is more abundant or try to culture it first.

Due to the physics of polymer (DNA) bending, bridge-PCR-based sequencing platforms tend to oversequence high-GC regions (e.g., Bifidobacterium genomes have a GC content of ~ 60%), and numerical adjustment may be needed for the abundances [57,61,62].

16S rRNA gene amplicon sequencing for bacteria and ITS (internal transcribed spacer regions) amplicon sequencing for fungi (Fig. 3.4) could detect microbes even when the concentration of DNA in a sample is below detection using ultraviolet light (e.g., in some urine samples [35]). The different hypervariable regions (e.g., V4–V5, V1–V2) have different taxonomic resolutions. Full-length amplicon sequencing spanning the entire region of 16S for bacteria, and 18S-ITS for fungi, is a promising way of reliably classifying microbial species using amplicon sequencing (e.g., [63]).

Worked sample 3.2

Shown below are DNA extraction results from 3 samples:

Sample	Concentration (ng/μL)	Volume (μL)	Total DNA (μg)
A001	8.56	80	0.68
A002	1.32	80	0.11
A003	24.4	80	1.95

If the average genome size in the above fecal samples is 5 Mb, about how many bacterial cells do we have in each sample?
If these are vaginal samples with an average bacterial genome size of 2.3 Mb, with 96% human sequences in the metagenome, about how many bacterial cells do we have in each sample?
If these vaginal samples are dominated by *L. iners*, whose genome size is only 1.3 Mb, about how many bacterial cells do we have in each sample?
If these vaginal samples are dominated by the fungus *Candida albicans*, how many copies of the fungal genome do we have in each sample?

3.6 Taxonomic and functional profiles, absolute abundance

Taxonomic and functional profiles can then be derived from the sequencing data, typically according to genes (More in Chapter 5). Some of the retroviral sequences in the human genome are understudied and may show up as RNA viruses in the taxonomic assignment.

When the total relative abundance is normalized to 1, please beware of the unclassified portion, which can be high in some samples.

The Kyoto Encyclopedia of Genes and Genomes (KEGG) is a classic database for functional annotation [64], with curated new add-ons such as the gut-brain modules [65,66]. Many KOs (KEGG Orthology groups) are present in multiple modules, and a cutoff such as 60% completeness is often imposed when reporting the presence of a module in a sample. The exact function of an enzyme may lie in a single amino acid difference from the database output (e.g., production of imidazole propionate [67]). There is a lot of basic research to catch on here.

If one takes pains to quantitatively perform each of the above-mentioned steps from sampling to sequencing, in comparison to standards, one can estimate the absolute number of particular microbes in a sample. A recent fecal microbiome study on preterm infants used spike-in cells and reported dynamics in absolute abundance, e.g., the fungi *Candida albicans* appeared to inhibit many bacteria [68], or rather it boomed when not effectively outcompeted by bacteria. The bacterium *Salinibacter ruber* DSM 13855, archaea *Haloarcula hispanica* ATCC 33960 and fungus *Trichoderma reesei* ATCC 13631 chosen as spike-ins were all absent from the samples, and were only detected in the samples because they were added to weighed aliquots of feces.

Worked sample 3.3

Find the taxonomic profile (16S rRNA gene amplicon sequencing) from the following study, and try to plot the differences between the breast milk samples (some pumped, some not) and the negative controls. Besides a table, what do you think would be a good plot, PCoA (Principal Coordinate Analysis) using Bray-Curtis distance? Box plot or violin plot for each taxon?

Core milk microbiota[a] among 393 mothers in the CHILD (Canadian Healthy Infant Longitudinal Development) cohort in comparison to the negative controls (Table S3 of [69])

		Samples (n=393)			Negative controls (n=15)
Lineage	**Genus**	**Mean ± SD**	**Maximum**	**Prevalence**	**Prevalence**
Proteobacteria—Burkholderiales	Unclassified	5.86 ± 3.43	12.56	100%	13%
Firmicutes—*Staphylococcaceae*	*Staphylococcus*	4.86 ± 11.5	87.5	100%	20%
Proteobacteria—*Oxalobacteraceae*	*Ralstonia*	4.79 ± 2.76	9.41	100%	7%
Proteobacteria—*Comamonadaceae*	Unclassified	4.42 ± 2.58	9.75	100%	7%
Proteobacteria—*Comamonadaceae*	*Acidovorax*	3.95 ± 2.34	13.33	100%	20%
Proteobacteria—*Oxalobacteraceae*	*Massilia*	2.37 ± 1.40	6.47	100%	13%
Proteobacteria—Uncl. Alteromonadales	*Rheinheimera*	1.89 ± 1.15	4.74	100%	0
Proteobacteria—*Rhizobiaceae*	*Agrobacterium*	1.85 ± 1.08	4.51	100%	7%
Proteobacteria—*Rhodospirillaceae*	Unclassified	1.61 ± 1.07	5.24	100%	7%
Proteobacteria—*Neisseriaceae*	*Vogesella*	1.23 ± 0.74	3.04	100%	0
Actinobacteria—*Nocardioidaceae*	*Nocardioides*	1.09 ± 0.65	2.61	100%	13%
Proteobacteria—Burkholderiales	Unclassified	1.07 ± 0.64	2.63	100%	0

[a] Defined as amplicon sequence variants (ASVs) present in at least 95% of samples with a mean relative abundance of more than 1% after removing potential reagent contaminants. Uncl, unclassified. The unit for the "Mean ± SD" and the "Maximum" columns is probably % relative abundance.

3.7 Sample size for metagenome-wide association studies

Microbiome data are notoriously heterogeneous, fat-tailed (high probability of extreme values, in contrast to normal distributions), and zero-inflated (a taxon can have zero value in many samples). Therefore, all the commonly used statistical tests for metagenomic data are nonparametric, and do not depend on a distribution. For example, Wilcoxon rank-sum test, also known as Mann-Whitney *U* test, is commonly used to look for differences between two groups, and significant results do not need to have a large effect.

For sample size estimation and power analyses, we have a modified version of [70] which removed the limitation on the initial statistical sampling space. Ideally, one needs to start with about 40 samples to get the pseudo-R^2 to then accurately calculate the required sample size for a power of 0.8 or higher. But that would have already exceeded the sample size needed for comparing groups of samples with a strong and relatively homogenous difference, such as dental plaques from rheumatoid arthritis patients and healthy controls, fecal samples from IgA deficient people and their normal spouse. Including more samples could be adding heterogeneity (Fig. 3.8), and would not necessarily follow the same statistical distribution, e.g., different people might get the same disease in a different season. So another dataset that differs in some way should rather be an independent validation, and could also allow the discovery of some different biomarkers.

Current knowledge of a disease can help us decide on the most important body sites and the number of samples to collect for metagenome-wide association studies (MWAS), along with other metadata that needs to be recorded or controlled for. For example, gingival samples for rheumatoid arthritis, fecal samples for Crohn's disease or IgA-deficient individuals show a strong dysbiosis with only a few samples [7,10,48,71–73]. Fecal microbiome markers for colorectal cancer converged nicely with a few dozens of samples from each country (more in Chapter 7), while for metabolic diseases such as obesity and type 2 diabetes (T2D) (more in Chapter 6), one either goes for extreme phenotypes or larger sample sizes [74,75]. Treatment-naïve samples come from population-wide disease screening efforts (Chapter 7), or from remote regions without access to modern treatments. For most doctors, checking on previous patients before a relapse would be a more realistic and clinically meaningful study design. Patients who relapsed after being without medication for at least 3 months could show the same disease markers, despite other differences in their microbiome with longer disease duration and previous medication [10,66]. Otherwise, medication information is recorded and statistically adjusted for in analyses.

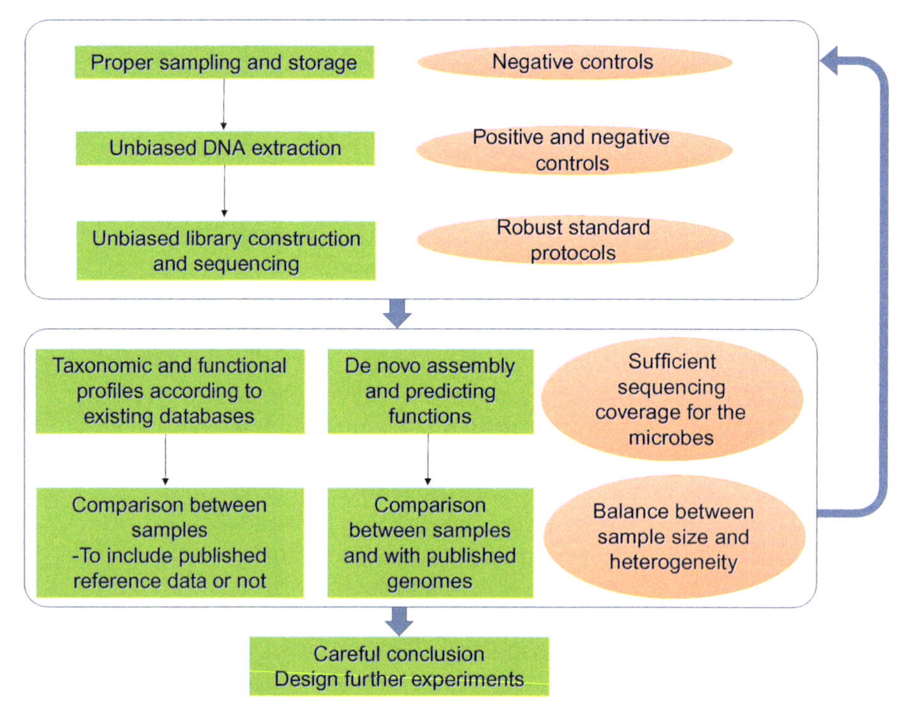

Fig. 3.8 A simplified step-by-step guide for an ideal metagenomics study from samples to data analyses and conclusions. After about two dozen samples, it would be a good idea to check the phenotype distributions and see whether the recruitment of volunteers is working well to address the questions one intends to study. Credit: Huijue Jia.

Permutational Analyses of Variances (PERMANOVA) [76,77] is a nonparametric statistical test that is suitable for analyzing potential influence from phenotypes and questionnaire entries on metagenomic data. Although it is a multivariant test, a typical start is to analyze each phenotype by itself. Questionnaires and other omics data, if well designed for a particular body site, e.g., pregnancy history and hormones for the cervical microbiome, oral hygiene and immune features for the oral microbiome, can explain notable portions of the microbiome variances [14,78], more than the single-digit percentiles typically seen for fecal studies of adult cohorts [3,79–81].

For body sites that are known to have clearly different community types, e.g., fecal microbiome, vaginal microbiome, the samples may need to be stratified according to the community types during analyses, at the cost of apparent sample size [9,82,83]. Stratifying the Human Microbiome Project (HMP) fecal data according to enterotypes was reported to increase statistical power [84]. Sex differences in microbiome composition and immune responses also mean that stratifying a microbiome study according to sex could lead to interesting discoveries that differ from the overall analyses [11,83].

Researchers need to know the important factors in a cohort, so that whenever we see some differences between groups, we can then make sure that the difference is due to the disease (or other condition) we are comparing, instead of a difference in the other factors between groups (e.g. genetic and lifestyle differences behind ethnic or geographical groups [50,85,86]). Brute force statistical adjustments would always lead to loss of true signals, before we can rationally model the microbiome. In a study of colorectal cancer, levels of the iron-binding protein ferritin and weekly intake of red meat significantly differed among the control, adenoma, and carcinoma groups, which are relevant for colorectal cancer and needs no adjustment [47] (more on causality in Chapter 6). Days since the last menstrual cycle (in numbers instead of divided into two or three categories of menstrual phase) associated with many bacteria in the upper and the lower reproductive tract (from the peritoneal fluid, uterus, cervix, to mid-vagina), so we simply cannot be sure about some of the bacteria's associations with hysteromyoma (uterine fibroids) and adenomyosis [34]. These are not to be discarded because of statistics, but to be further studied in more conclusive ways.

In addition to healthy controls, other conditions that share some feature of the disease should also be considered for sampling. For example, hyposalivation due to other reasons are included, to better understand the oral microbiome in Sjögren's syndrome [87]. A study of the respiratory microbiome in people with obstructive sleep apnea probably needs to consider differences in body fat and cardiovascular health in the volunteers. Again, if published data are included as the related conditions for comparison, considerable heterogeneity may be introduced [52,88]. One way to circumvent the metadata and sample handling differences is to always compare the disease samples with control samples from the same study [6,89] before better standards are worked out for specific fields of study [90].

3.8 Summary

Low biomass samples, i.e., where the microbial populations are small (Chapter 1) and not much above current detection limits, are more sensitive to contamination during sample collection, DNA extraction, etc. Multiple methods in addition to metagenomic sequencing would be needed to estimate the number of microbes in low biomass samples and to link the microbes to potential functions. It will always be a good idea to collect related samples from the same

Worked sample 3.4

Seeing the PERMANOVA results below from the fecal microbiome of monozygotic and dizygotic twins in the TwinsUK cohort (Table S2 of [50]; smaller sample size than the amplicon studies on the cohort [91,92]), which are the phenotypes you would be cautious about when making conclusions?

Try to have a visual sense of the distribution of phenotypes, by making plots. BMI, for example, is mostly lean in some East Asian cohorts (e.g., Ref. [3]), and would not show up as strongly as for the old ladies here; Waist-to-hip ratio, muscle mass, and fat mass estimated from conductance or scans, could better reflect the belly fat than BMI.

In what ways do you think "Year of birth" might be different from "Age of metagenomics sample?"

According to current knowledge and technologies, how do you think the potential influence of physical exercises on the gut microbiome could be better investigated? (also for Chapter 8)

Phenotypes	Number of twins	Groups	Sample size	Degree of freedom	Sums of squares	Mean square	F model	Pseudo-R^2	$P(>F)$	P adjusted (BH)
Twin pair number	246	NA	246	122	43.28	0.355	1.196	0.543	0.000	0.002
BMI	249	NA	249	1	0.511	0.511	1.571	0.006	0.002	0.017
Drugs (diabetic tablets)	230	Y	5	1	0.463	0.463	1.428	0.006	0.015	0.083
		N	225							
Has a doctor ever diagnosed or treated you for any of the following conditions?/diabetes	222	Y	10	1	0.429	0.429	1.319	0.006	0.028	0.091
		N	212							
Year of birth	250	NA	250	1	0.424	0.424	1.303	0.005	0.035	0.091
Current location (Geo-clusters)	247	Cluster 1	10	3	1.133	0.378	1.161	0.014	0.037	0.091
		Cluster 2	134							
		Cluster 3	52							
		Cluster 4	51							
Vegetarian or vegan	198	N	179	1	0.420	0.420	1.291	0.007	0.041	0.091
		Y	19							
Age at metagenomic sample	250	NA	250	1	0.417	0.417	1.280	0.005	0.043	0.091
Number of units of alcohol drunk per week	241	1–5 units	88	1	0.389	0.389	1.193	0.005	0.100	0.173
		6–10 units	21							
		11–15 units	56							
		16–20 units	7							
		21–40 units	34							
		40 + units	4							
		None	31							

Phenotypes	Number of twins	Groups	Sample size	Degree of freedom	Sums of squares	Mean square	F model	Pseudo-R^2	P(> F)	P adjusted (BH)
Menopausal status	240	Postmenopausal	174	2	0.734	0.367	1.129	0.009	0.102	0.173
		Premenopausal	38							
		Going through menopause	28							
Smoking status	249	Smoker	92	1	0.367	0.367	1.126	0.005	0.160	0.247
		Never smoked	157							
Currently, how many minutes per week do you spend walking briskly/gardening vigorously?	196	0	22	1	0.354	0.354	1.096	0.006	0.214	0.303
		≥ 1	174							
Currently, how many minutes per week do you spend in nonweight bearing activity? e.g., swimming, cycling, yoga, aqua aerobics etc.	188	0	102	1	0.323	0.323	0.998	0.005	0.447	0.559
		≥ 1	86							
Drugs (Insulin)	230	Y	4	1	0.321	0.321	0.988	0.004	0.460	0.559
		N	226							
Currently, how many minutes per week do you spend on weight-bearing activity? E.g., aerobics, running, dance, football, basketball, racquet sports, etc. (do not include walking or gardening)	191	0	103	1	0.309	0.309	0.953	0.005	0.589	0.667
		≥ 1	88							
Outdoor sports	108	Y	56	1	0.29	0.29	0.885	0.008	0.798	0.848
		N	52							

PERMANOVA for the influence of each phenotype on the gut microbial gene profile (11.4 million genes of the fecal microbiome [50]). 9999 permutations, Bray-Curtis distance. As one PERMANOVA test was performed for each phenotype, multiple testing was controlled using the Benjamini-Hochberg procedure. The phenotypes are not analyzed in combination. Y, yes; N, no.

individual. As complex communities, caution needs to be taken in storage, DNA extraction, and sequencing to avoid underrepresentation of some of the taxa in a sample. We will see in Chapter 5 that metagenomic assembly would require a higher sequencing coverage than metagenomic detection. Questionnaires and other information need to be optimized for each body site, in order not to miss important information during statistical analyses, and to consistently identify biomarkers from MWAS on multiple cohorts.

References

[1] Vandeputte D, Falony G, Vieira-Silva S, Tito RY, Joossens M, Raes J. Stool consistency is strongly associated with gut microbiota richness and composition, enterotypes and bacterial growth rates. Gut 2016;65:57–62. https://doi.org/10.1136/gutjnl-2015-309618.

[2] Vandeputte D, Kathagen G, D'hoe K, Vieira-Silva S, Valles-Colomer M, Sabino J, et al. Quantitative microbiome profiling links gut community variation to microbial load. Nature 2017;551:507–11. https://doi.org/10.1038/nature24460.

[3] Jie Z, Liang S, Ding Q, Li F, Tang S, Wang D, et al. A transomic cohort as a reference point for promoting a healthy gut microbiome. Med Microecol 2021;8:100039. https://doi.org/10.1016/j.medmic.2021.100039.

[4] Park S, Won DD, Lee BJ, Escobedo D, Esteva A, Aalipour A, et al. A mountable toilet system for personalized health monitoring via the analysis of excreta. Nat Biomed Eng 2020;4:624–35. https://doi.org/10.1038/s41551-020-0534-9.

[5] Methé BA, Nelson KE, Pop M, Creasy HH, Giglio MG, Huttenhower C, et al. A framework for human microbiome research. Nature 2012;486:215–21. https://doi.org/10.1038/nature11209.

[6] Jie Z, Liang S, Ding Q, Li F, Tang S, Sun X, et al. Disease trends in a young Chinese cohort according to fecal metagenome and plasma metabolites. Med Microecol 2021. https://doi.org/10.1016/j.medmic.2021.100037.

[7] He Q, Gao Y, Jie Z, Yu X, Laursen JM, Xiao L, et al. Two distinct metacommunities characterize the gut microbiota in Crohn's disease patients. Gigascience 2017;6:1–11. https://doi.org/10.1093/gigascience/gix050.

[8] Lloyd-Price J, Mahurkar A, Rahnavard G, Crabtree J, Orvis J, Hall AB, et al. Strains, functions and dynamics in the expanded human microbiome project. Nature 2017. https://doi.org/10.1038/nature23889.

[9] Chen C, Hao L, Zhang Z, Tian L, Song L, Zhang X, et al. Dynamics in the vaginocervical microbiome after oral probiotics. J Genet Genomics 2021. https://doi.org/10.1101/2020.06.16.155929.

[10] Zhang X, Zhang D, Jia H, Feng Q, Wang D, Di Liang D, et al. The oral and gut microbiomes are perturbed in rheumatoid arthritis and partly normalized after treatment. Nat Med 2015;21:895–905. https://doi.org/10.1038/nm.3914.

[11] Zhu J, Tian L, Chen P, Han M, Song L, Tong X, et al. Over 50,000 metagenomically assembled draft genomes for the human oral microbiome reveal new taxa. Genomics Proteomics Bioinformatics 2021. https://doi.org/10.1016/j.gpb.2021.05.001.

[12] Oh J, Byrd AL, Park M, Kong HH, Segre JA. Temporal stability of the human skin microbiome. Cell 2016;165:854–66. https://doi.org/10.1016/j.cell.2016.04.008.

[13] Fettweis JM, Serrano MG, Brooks JP, Edwards DJ, Girerd PH, Parikh HI, et al. The vaginal microbiome and preterm birth. Nat Med 2019;25:1012–21. https://doi.org/10.1038/s41591-019-0450-2.

[14] Jie Z, Chen C, Hao L, Li F, Song L, Zhang X, et al. Life history recorded in the vagino-cervical microbiome along with multi-omics. Genomics Proteomics Bioinformatics 2021. https://doi.org/10.1016/j.gpb.2021.01.005.

[15] Li F, Chen C, Wei W, Wang Z, Dai J, Hao L, et al. The metagenome of the female upper reproductive tract. Gigascience 2018;7. https://doi.org/10.1093/gigascience/giy107.

[16] Aagaard K, Ma J, Antony KM, Ganu R, Petrosino J, Versalovic J. The placenta harbors a unique microbiome. Sci Transl Med 2014;6:237ra65. https://doi.org/10.1126/scitranslmed.3008599.

[17] Dickson RP, Erb-Downward JR, Prescott HC, Martinez FJ, Curtis JL, Lama VN, et al. Cell-associated bacteria in the human lung microbiome. Microbiome 2014;2:28. https://doi.org/10.1186/2049-2618-2-28.

[18] Marotz CA, Sanders JG, Zuniga C, Zaramela LS, Knight R, Zengler K. Improving saliva shotgun metagenomics by chemical host DNA depletion. Microbiome 2018;6:42. https://doi.org/10.1186/s40168-018-0426-3.

[19] Castellarin M, Warren RL, Freeman JD, Dreolini L, Krzywinski M, Strauss J, et al. *Fusobacterium nucleatum* infection is prevalent in human colorectal carcinoma. Genome Res 2012;22:299–306. https://doi.org/10.1101/gr.126516.111.

[20] Kostic AD, Gevers D, Pedamallu CS, Michaud M, Duke F, Earl AM, et al. Genomic analysis identifies association of Fusobacterium with colorectal carcinoma. Genome Res 2012;22:292–8. https://doi.org/10.1101/gr.126573.111.

[21] Riley DR, Sieber KB, Robinson KM, White JR, Ganesan A, Nourbakhsh S, et al. Bacteria-human somatic cell lateral gene transfer is enriched in cancer samples. PLoS Comput Biol 2013;9. https://doi.org/10.1371/journal.pcbi.1003107, e1003107.

[22] Geller LT, Barzily-Rokni M, Danino T, Jonas OH, Shental N, Nejman D, et al. Potential role of intratumor bacteria in mediating tumor resistance to the chemotherapeutic drug gemcitabine. Science 2017;357:1156–60. https://doi.org/10.1126/science.aah5043.

[23] Poore GD, Kopylova E, Zhu Q, Carpenter C, Fraraccio S, Wandro S, et al. Microbiome analyses of blood and tissues suggest cancer diagnostic approach. Nature 2020;579:567–74. https://doi.org/10.1038/s41586-020-2095-1.

[24] Nejman D, Livyatan I, Fuks G, Gavert N, Zwang Y, Geller LT, et al. The human tumor microbiome is composed of tumor type–specific intracellular bacteria. Science 2020;368:973–80. https://doi.org/10.1126/science.aay9189.

[25] de Goffau MC, Lager S, Sovio U, Gaccioli F, Cook E, Peacock SJ, et al. Human placenta has no microbiome but can contain potential pathogens. Nature 2019;572:329–34. https://doi.org/10.1038/s41586-019-1451-5.

[26] Salter SJ, Cox MJ, Turek EM, Calus ST, Cookson WO, Moffatt MF, et al. Reagent and laboratory contamination can critically impact sequence-based microbiome analyses. BMC Biol 2014;12:87. https://doi.org/10.1186/s12915-014-0087-z.

[27] Sun X, Hu Y-H, Wang J, Fang C, Li J, Han M, et al. Efficient and stable metabarcoding sequencing data using a DNBSEQ-G400 sequencer validated by comprehensive community analyses. GigaByte 2021. https://doi.org/10.46471/gigabyte.16.

[28] Lax S, Sangwan N, Smith D, Larsen P, Handley KM, Richardson M, et al. Bacterial colonization and succession in a newly opened hospital. Sci Transl Med 2017;9:1–11.

[29] Pidot SJ, Gao W, Buultjens AH, Monk IR, Guerillot R, Carter GP, et al. Increasing tolerance of hospital *Enterococcus faecium* to handwash alcohols. Sci Transl Med 2018;10:eaar6115. https://doi.org/10.1126/scitranslmed.aar6115.

[30] Mora M, Wink L, Kögler I, Mahnert A, Rettberg P, Schwendner P, et al. Space station conditions are selective but do not alter microbial characteristics relevant to human health. Nat Commun 2019;10:3990. https://doi.org/10.1038/s41467-019-11682-z.

[31] Checinska A, Probst AJ, Vaishampayan P, White JR, Kumar D, Stepanov VG, et al. Microbiomes of the dust particles collected from the international space station and spacecraft assembly facilities. Microbiome 2015;3:50. https://doi.org/10.1186/s40168-015-0116-3.
[32] Lee MD, O'Rourke A, Lorenzi H, Bebout BM, Dupont CL, Everroad RC. Reference-guided metagenomics reveals genome-level evidence of potential microbial transmission from the ISS environment to an astronaut's microbiome. IScience 2021;24:102114. https://doi.org/10.1016/j.isci.2021.102114.
[33] Saw JJ, Sivaguru M, Wilson EM, Dong Y, Sanford RA, Fields CJ, et al. In vivo entombment of bacteria and fungi during calcium oxalate, brushite, and struvite urolithiasis. Kidney360 2021;2:298–311. https://doi.org/10.34067/kid.0006942020.
[34] Chen C, Song X, Wei W, Zhong H, Dai J, Lan Z, et al. The microbiota continuum along the female reproductive tract and its relation to uterine-related diseases. Nat Commun 2017;8:875. https://doi.org/10.1038/s41467-017-00901-0.
[35] Chen C, Hao L, Wei W, Li F, Song L, Zhang X, et al. The female urinary microbiota in relation to the reproductive tract microbiota. Gigabyte 2020;2020:1–9. https://doi.org/10.46471/gigabyte.9.
[36] Seferovic MD, Pace RM, Carroll M, Belfort B, Major AM, Chu DM, et al. Visualization of microbes by 16S in situ hybridization in term and preterm placentas without intraamniotic infection. Am J Obstet Gynecol 2019;221:146.e1–146.e23. https://doi.org/10.1016/j.ajog.2019.04.036.
[37] Gosalbes MJ, Llop S, Vallès Y, Moya A, Ballester F, Francino MP. Meconium microbiota types dominated by lactic acid or enteric bacteria are differentially associated with maternal eczema and respiratory problems in infants. Clin Exp Allergy 2013;43:198–211. https://doi.org/10.1111/cea.12063.
[38] Bäckhed F, Roswall J, Peng Y, Feng Q, Jia H, Kovatcheva-Datchary P, et al. Dynamics and stabilization of the human gut microbiome during the first year of life. Cell Host Microbe 2015;17:690–703. https://doi.org/10.1016/j.chom.2015.04.004.
[39] Collado MC, Rautava S, Aakko J, Isolauri E, Salminen S. Human gut colonisation may be initiated in utero by distinct microbial communities in the placenta and amniotic fluid. Sci Rep 2016;6:23129. https://doi.org/10.1038/srep23129.
[40] Wang J, Zheng J, Shi W, Du N, Xu X, Zhang Y, et al. Dysbiosis of maternal and neonatal microbiota associated with gestational diabetes mellitus. Gut 2018. https://doi.org/10.1136/gutjnl-2018-315988. gutjnl-2018-315988.
[41] He Q, Kwok L-Y, Xi X, Zhong Z, Ma T, Xu H, et al. The meconium microbiota shares more features with the amniotic fluid microbiota than the maternal fecal and vaginal microbiota. Gut Microbes 2020;12:1794266. https://doi.org/10.1080/19490976.2020.1794266.
[42] Ferretti P, Pasolli E, Tett A, Asnicar F, Gorfer V, Fedi S, et al. Mother-to-infant microbial transmission from different body sites shapes the developing infant gut microbiome. Cell Host Microbe 2018;24:133–145.e5. https://doi.org/10.1016/j.chom.2018.06.005.
[43] Massier L, Chakaroun R, Tabei S, Crane A, Didt KD, Fallmann J, et al. Adipose tissue derived bacteria are associated with inflammation in obesity and type 2 diabetes. Gut 2020;69(10):1796–806. https://doi.org/10.1136/gutjnl-2019-320118.
[44] Anhê FF, Jensen BAH, Varin TV, Servant F, Van Blerk S, Richard D, et al. Type 2 diabetes influences bacterial tissue compartmentalisation in human obesity. Nat Metab 2020;2(3):233–42. https://doi.org/10.1038/s42255-020-0178-9.
[45] Stenkula KG, Erlanson-Albertsson C. Adipose cell size: importance in health and disease. Am J Physiol Integr Comp Physiol 2018;315:R284–95. https://doi.org/10.1152/ajpregu.00257.2017.
[46] Dickson RP, Erb-Downward JR, Martinez FJ, Huffnagle GB. The microbiome and the respiratory tract. Annu Rev Physiol 2016;78:481–504. https://doi.org/10.1146/annurev-physiol-021115-105238.

[47] Feng Q, Liang S, Jia H, Stadlmayr A, Tang L, Lan Z, et al. Gut microbiome development along the colorectal adenoma–carcinoma sequence. Nat Commun 2015;6:6528. https://doi.org/10.1038/ncomms7528.

[48] Moll JM, Myers PN, Zhang C, Eriksen C, Wolf J, Appelberg KS, et al. Gut microbiota perturbation in IgA deficiency is influenced by IgA-autoantibody status. Gastroenterology 2021. https://doi.org/10.1053/j.gastro.2021.02.053.

[49] Zhong H, Ren H, Lu Y, Fang C, Hou G, Yang Z, et al. Distinct gut metagenomics and metaproteomics signatures in prediabetics and treatment-naïve type 2 diabetics. EBioMedicine 2019. https://doi.org/10.1016/j.ebiom.2019.08.048.

[50] Xie H, Guo R, Zhong H, Feng Q, Lan Z, Qin B, et al. Shotgun metagenomics of 250 adult twins reveals genetic and environmental impacts on the gut microbiome. Cell Syst 2016;3:572–584.e3. https://doi.org/10.1016/j.cels.2016.10.004.

[51] David LA, CFC M, Carmody RN, Gootenberg DB, Button JE, Wolfe BE, et al. Diet rapidly and reproducibly alters the human gut microbiome. Nature 2013;505:559–63. https://doi.org/10.1038/nature12820.

[52] Li J, Jia H, Cai X, Zhong H, Feng Q, Sunagawa S, et al. An integrated catalog of reference genes in the human gut microbiome. Nat Biotechnol 2014;32:834–41. https://doi.org/10.1038/nbt.2942.

[53] Bouslimani A, Porto C, Rath CM, Wang M, Guo Y, Gonzalez A, et al. Molecular cartography of the human skin surface in 3D. Proc Natl Acad Sci U S A 2015;112:E2120–9. https://doi.org/10.1073/pnas.1424409112.

[54] Costea PI, Zeller G, Sunagawa S, Pelletier E, Alberti A, Levenez F, et al. Towards standards for human fecal sample processing in metagenomic studies. Nat Biotechnol 2017. https://doi.org/10.1038/nbt.3960.

[55] Tourlousse DM, Narita K, Miura T, Sakamoto M, Ohashi A, Shiina K, et al. Validation and standardization of DNA extraction and library construction methods for metagenomics-based human fecal microbiome measurements. Microbiome 2021;9:95. https://doi.org/10.1186/s40168-021-01048-3.

[56] Yang F, Sun J, Luo H, Ren H, Zhou H, Lin Y, et al. Assessment of fecal DNA extraction protocols for metagenomic studies. Gigascience 2020;9(7). https://doi.org/10.1093/gigascience/giaa071.

[57] Fang C, Zhong H, Lin Y, Chen B, Han M, Ren H, et al. Assessment of the cPAS-based BGISEQ-500 platform for metagenomic sequencing. Gigascience 2018;7:1–8. https://doi.org/10.1093/gigascience/gix133.

[58] Sender R, Fuchs S, Milo R. Revised estimates for the number of human and bacteria cells in the body. PLoS Biol 2016;14. https://doi.org/10.1371/journal.pbio.1002533, e1002533.

[59] Stephen AM, Cummings JH. The microbial contribution to human faecal mass. J Med Microbiol 1980;13:45–56. https://doi.org/10.1099/00222615-13-1-45.

[60] Lager S, de Goffau MC, Sovio U, Peacock SJ, Parkhill J, Charnock-Jones DS, et al. Detecting eukaryotic microbiota with single-cell sensitivity in human tissue. Microbiome 2018;6:151. https://doi.org/10.1186/s40168-018-0529-x.

[61] Patterson J, Carpenter EJ, Zhu Z, An D, Liang X, Geng C, et al. Impact of sequencing depth and technology on de novo RNA-Seq assembly. BMC Genomics 2019;20:604. https://doi.org/10.1186/s12864-019-5965-x.

[62] Browne PD, Nielsen TK, Kot W, Aggerholm A, Gilbert MTP, Puetz L, et al. GC bias affects genomic and metagenomic reconstructions, underrepresenting GC-poor organisms. Gigascience 2020;9. https://doi.org/10.1093/gigascience/giaa008.

[63] Fang C, Sun X, Fan F, Zhang X, Wang O, Zheng H, et al. High-resolution single-molecule long-fragment rRNA gene amplicon sequencing for uncultured bacterial and fungal communities. bioRxiv 2021. https://doi.org/10.1101/2021.03.29.437457.

[64] Kanehisa M, Goto S, Sato Y, Furumichi M, Tanabe M. KEGG for integration and interpretation of large-scale molecular data sets. Nucleic Acids Res 2012;40:D109–14. https://doi.org/10.1093/nar/gkr988.

[65] Valles-Colomer M, Falony G, Darzi Y, Tigchelaar EF, Wang J, Tito RY, et al. The neuroactive potential of the human gut microbiota in quality of life and depression. Nat Microbiol 2019. https://doi.org/10.1038/s41564-018-0337-x.

[66] Zhu F, Ju Y, Wang W, Wang Q, Guo R, Ma Q, et al. Metagenome-wide association of gut microbiome features for schizophrenia. Nat Commun 2020;11:1612. https://doi.org/10.1038/s41467-020-15457-9.

[67] Koh A, Molinaro A, Ståhlman M, Khan MT, Schmidt C, Mannerås-Holm L, et al. Microbially produced imidazole propionate impairs insulin signaling through mTORC1. Cell 2018;175:947–961.e17. https://doi.org/10.1016/j.cell.2018.09.055.

[68] Rao C, Coyte KZ, Bainter W, Geha RS, Martin CR, Rakoff-Nahoum S. Multi-kingdom ecological drivers of microbiota assembly in preterm infants. Nature 2021. https://doi.org/10.1038/s41586-021-03241-8.

[69] Moossavi S, Sepehri S, Robertson B, Bode L, Goruk S, Field CJ, et al. Composition and variation of the human milk microbiota are influenced by maternal and early-life factors. Cell Host Microbe 2019;25:324–335.e4. https://doi.org/10.1016/j.chom.2019.01.011.

[70] Kelly BJ, Gross R, Bittinger K, Sherrill-Mix S, Lewis JD, Collman RG, et al. Power and sample-size estimation for microbiome studies using pairwise distances and PERMANOVA. Bioinformatics 2015;31:2461–8. https://doi.org/10.1093/bioinformatics/btv183.

[71] Qin J, Li R, Raes J, Arumugam M, Burgdorf KSS, Manichanh C, et al. A human gut microbial gene catalogue established by metagenomic sequencing. Nature 2010;464:59–65. https://doi.org/10.1038/nature08821.

[72] Zou M, Jie Z, Cui B, Wang H, Feng Q, Zou Y, et al. Fecal microbiota transplantation results in bacterial strain displacement in patients with inflammatory bowel diseases. FEBS Open Bio 2019. https://doi.org/10.1002/2211-5463.12744.

[73] Fadlallah J, El Kafsi H, Sterlin D, Juste C, Parizot C, Dorgham K, et al. Microbial ecology perturbation in human IgA deficiency. Sci Transl Med 2018;10. https://doi.org/10.1126/scitranslmed.aan1217, eaan1217.

[74] Wang J, Jia H. Metagenome-wide association studies: fine-mining the microbiome. Nat Rev Microbiol 2016;14:508–22. https://doi.org/10.1038/nrmicro.2016.83.

[75] Liu R, Hong J, Xu X, Feng Q, Zhang D, Gu Y, et al. Gut microbiome and serum metabolome alterations in obesity and after weight-loss intervention. Nat Med 2017;23(7):859–68. https://doi.org/10.1038/nm.4358.

[76] Anderson MJ. A new method for non-parametric multivariate analysis of variance. Austral Ecol 2001. https://doi.org/10.1111/j.1442-9993.2001.01070.pp.x.

[77] Anderson MJ, Walsh Daniel CI. PERMANOVA, ANOSIM, and the Mantel test in the face of heterogeneous dispersions: What null hypothesis are you testing? Ecol Monogr 2013. https://doi.org/10.1890/12-2010.1.

[78] Liu X, Tong X, Zhu J, Liu T, Jie Z, Zou Y, et al. Metagenome-genome-wide association studies reveal human genetic impact on the oral microbiome. Biorxiv 2021. https://doi.org/10.1101/2021.05.06.443017.

[79] Falony G, Joossens M, Vieira-Silva S, Wang J, Darzi Y, Faust K, et al. Population-level analysis of gut microbiome variation. Science 2016;352:560–4. https://doi.org/10.1126/science.aad3503.

[80] Zhernakova A, Kurilshikov A, Bonder MJ, Tigchelaar EF, Schirmer M, Vatanen T, et al. Population-based metagenomics analysis reveals markers for gut microbiome composition and diversity. Science 2016;352:565–9. https://doi.org/10.1126/science.aad3369.

[81] Wang J, Thingholm LB, Skiecevičienė J, Rausch P, Kummen M, Hov JR, et al. Genome-wide association analysis identifies variation in vitamin D receptor and other host factors influencing the gut microbiota. Nat Genet 2016;48:1396–406. https://doi.org/10.1038/ng.3695.

[82] Gu Y, Wang X, Li J, Zhang Y, Zhong H, Liu R, et al. Analyses of gut microbiota and plasma bile acids enable stratification of patients for antidiabetic treatment. Nat Commun 2017;8:1785. https://doi.org/10.1038/s41467-017-01682-2.
[83] Liu X, Tang S, Zhong H, Tong X, Jie Z, Ding Q, et al. A genome-wide association study for gut metagenome in Chinese adults illuminates complex diseases. Cell Discov 2021;7:9. https://doi.org/10.1038/s41421-020-00239-w.
[84] Mattiello F, Verbist B, Faust K, Raes J, Shannon WD, Bijnens L, et al. A web application for sample size and power calculation in case-control microbiome studies. Bioinformatics 2016;32:2038–40. https://doi.org/10.1093/bioinformatics/btw099.
[85] He Y, Wu W, Zheng H-M, Li P, McDonald D, Sheng H-F, et al. Regional variation limits applications of healthy gut microbiome reference ranges and disease models. Nat Med 2018;24:1532–5. https://doi.org/10.1038/s41591-018-0164-x.
[86] Deschasaux M, Bouter KE, Prodan A, Levin E, Groen AK, Herrema H, et al. Depicting the composition of gut microbiota in a population with varied ethnic origins but shared geography. Nat Med 2018;24:1526–31. https://doi.org/10.1038/s41591-018-0160-1.
[87] AlmståhI A, Wikström M, Stenberg I, Jakobsson A, Fagerberg-Mohlin B. Oral microbiota associated with hyposalivation of different origins. Oral Microbiol Immunol 2003;18:1–8.
[88] Karlsson FH, Tremaroli V, Nookaew I, Bergström G, Behre CJ, Fagerberg B, et al. Gut metagenome in European women with normal, impaired and diabetic glucose control. Nature 2013;498:99–103. https://doi.org/10.1038/nature12198.
[89] Jie Z, Xia H, Zhong S-L, Feng Q, Li S, Liang S, et al. The gut microbiome in atherosclerotic cardiovascular disease. Nat Commun 2017;8:845. https://doi.org/10.1038/s41467-017-00900-1.
[90] Kachroo N, Lange D, Penniston KL, Stern J, Tasian G, Bajic P, et al. Standardization of microbiome studies for urolithiasis: an international consensus agreement. Nat Rev Urol 2021;18:303–11. https://doi.org/10.1038/s41585-021-00450-8.
[91] Goodrich JK, Waters JL, Poole AC, Sutter JL, Koren O, Blekhman R, et al. Human genetics shape the gut microbiome. Cell 2014;159(4):789–99. https://doi.org/10.1016/j.cell.2014.09.053.
[92] Goodrich JK, Davenport ER, Beaumont M, Jackson MA, Knight R, Ober C, et al. Genetic determinants of the gut microbiome in UK twins. Cell Host Microbe 2016;19(5):731–43. https://doi.org/10.1016/j.chom.2016.04.017.

4

Epidemiology in the human body

4.1 Analogy to COVID-19

The COVID-19 pandemic has facilitated public understanding of epidemiology and contact-tracing. Besides exchanging microbes with the environment on a daily basis, the human microbiome also involves movements of microbes within the body and sometimes into human cells. From an ecological point-of-view (Chapter 2), the microbiome at a given body site is a result of multiple source communities, and individual-specific factors help shape the steady-state abundance of microbes. Controlling the source and improving the local condition would lead to more effective treatment.

We discussed sample collection in Chapter 3, and the same principles would apply to all kinds of samples that may need to be collected to track down the disease-relevant microbes. Successful examples of comprehensive studies are as yet scarce, but we will hopefully have more to talk about in the years to come.

Worked sample 4.1

If you are traveling to a remote village, what samples from the environment and from the locals do you plan to study, before you look for these new microbes in your own microbiome? After how many days shall we expect to see a change in the skin, oral, or the fecal microbiome?

Worked sample 4.2

As we have mentioned in Chapter 3 and learned through experience with the COVID-19 pandemic, humans constantly contaminate the environment with the microbes we carry. This Worked Sample is also a warm-up for more on taxonomy in Chapter 5.

Investigating Human Diseases with the Microbiome: Metagenomics Bench to Bedside.
https://doi.org/10.1016/B978-0-323-91369-0.00003-0

The following study on the microbiome in New York subways (Fig. 4.1) had to rely on HMP data back then [1]. Can we better assign the metagenomic data to the most likely human body sites now? Shall we consider data from different ethnic groups in the analyses of some stations? What is now the closest match for the *Pseudoalteromonas* species identified in the previously flooded station?

4.2 Sources of potential pathogens in the infant gut

When a baby gets diarrhea, and we are lucky enough to catch some samples for metagenomic shotgun sequencing, is the diarrhea due to something that the baby ate or drank [2]? Was there a change in formula milk (contains bacterial spores from cows, in addition to nutrients)? Will direct breastfeeding be better than pumping milk [3]? Shall we pay more attention to the mother's gut microbiome (Fig. 4.2)? In addition to the transmission through the environment, some of the mother's gut microbes could somehow get to the mammary gland and become part of the breast milk (~ 10^6 cells/mL bacteria, ~ 3×10^5 cells/mL fungi according to PCR (polymerase chain reaction) [4,5]). Babies often spit up milk along with microbes back to the mother, so the baby's own oral microbiome is also part of the loop (Fig. 4.2). Oral-gut transmission is also more common in infants than in adults [6]. When only breast milk and areolar skin (nipple ring) microbiota from the mother are analyzed by amplicon sequencing, bacteria from breast milk contributed nearly 30% of the infant gut microbiota in the first month for infants who were more than 75% breast-fed, dropping below 10% contribution after the first month of age [7].

4.3 Ectopic presence of commensal microbes

Many of the microbes in the human body may be harmless if they stay where they belong. However, due to a rare occasion (to be captured for a more complete understanding, Chapter 8), or due to a routine process that is defended against under normal conditions, the microbes may stay and grow in the wrong place.

Colonization of salivary bacteria from Crohn's disease or Ulcerative Colitis patients (two major types of Inflammatory Bowel Diseases (IBDs)) into the gut of germfree mice has been shown to induce inflammation [8]. Some of the human disease biomarkers identified by MWAS are likely of oral origin, and their colonization in the gut could involve genetic factors [9,10], in addition to age and other physiological conditions.

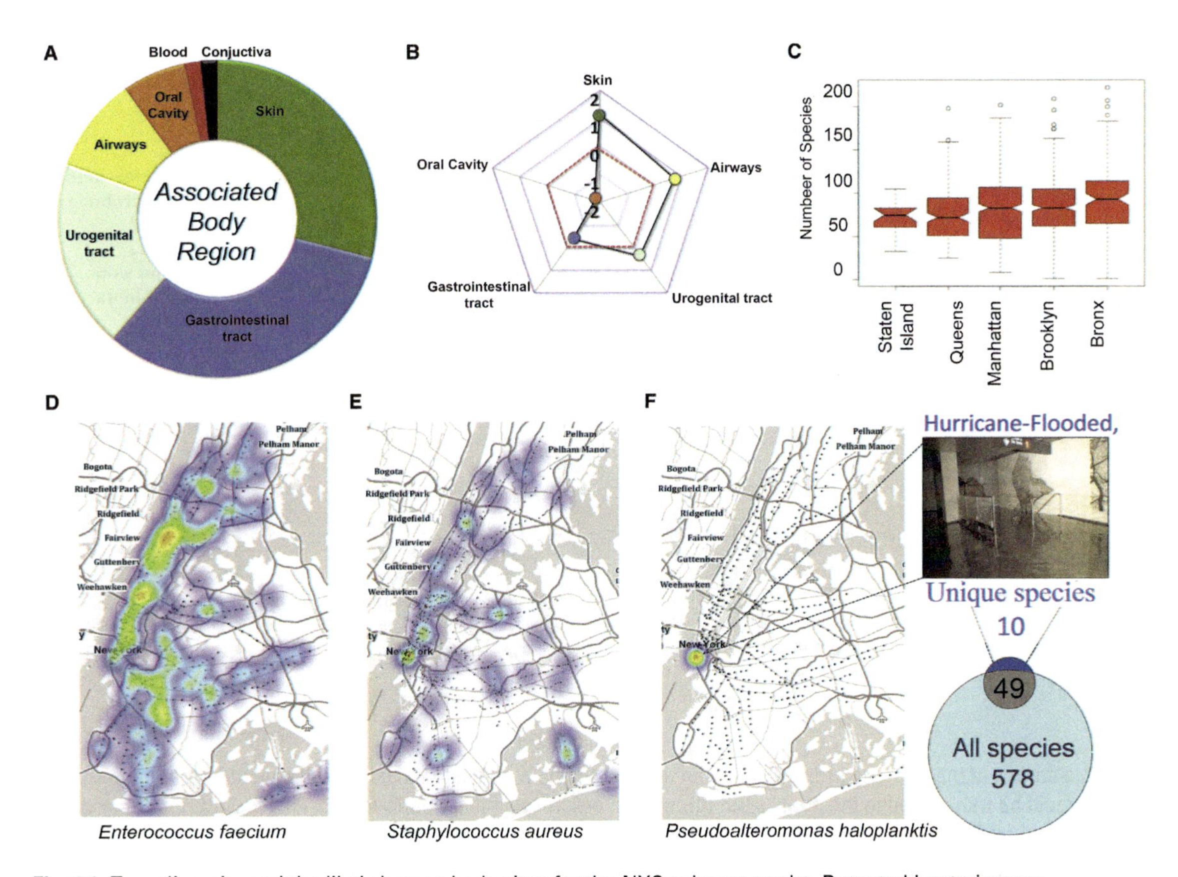

Fig. 4.1 Taxa diversity and the likely human body sites for the NYC subway swabs. Detected bacteria were annotated relative to the most commonly associated body part from the Human Microbiome Project (HMP) dataset. (A) Of the 67 PathoMap species that matched the HMP dataset, the proportions were greatest for the GI tract *(blue)*, skin *(green)*, and urogenital tract *(white)*. The entire circle represents 100% of the 67 species, and the sizes of each color represent the proportion of each type of bacteria. (B) To account for the database proportions from the HMP, we calculated the log2 of the observed versus expected numbers of species found for each category, which indicated that skin was the most predominant type of bacteria on the subway system. (C) Boxplot of the number of species found per borough. Middle line of each section shows the median, and the top and bottom of each box show the 75th and 25th percentiles, respectively. Notches show a significant difference between groups (95% confidence interval). (D and E) Heatmaps of NYC showing the density for *Enterococcus faecium* (D) and *Staphylococcus aureus* (E). Small red dots indicate the presence of a fully resequenced mecA gene. (F) Analysis of a subway station (picture on top shows the station) flooded during Hurricane Sandy. The Venn diagram compares the unique set of 10 species in the data from that station that did not appear in any other station or area of NYC, but 52 species overlapped with the set of 627 species present in the subway system. (A–F) The entire NYC MTA subway system, a total of 468 stations, was swabbed in triplicate over the course of the summer of 2013 and some additional samples taken for culturing and testing and in response to reviewers in 2014. Two surfaces were swabbed in each station, and one surface was swabbed within the train. Samples were collected from turnstiles and emergency exits, Metro Card kiosks, wooden and metal benches, stairwell handrails, and trashcans. The turnstiles and kiosks were prioritized at each station due to the level of human-surface interaction at these particular sites. In the train, the doors, poles, handrails, and seats were swabbed. Credit: Fig. 5 of Afshinnekoo E., Meydan C., Chowdhury S., Jaroudi D., Boyer C., Bernstein N., et al. Geospatial resolution of human and bacterial diversity with city-scale metagenomics. Cell Syst 2015;1:1–15. doi:10.1016/j.cels.2015.01.001.

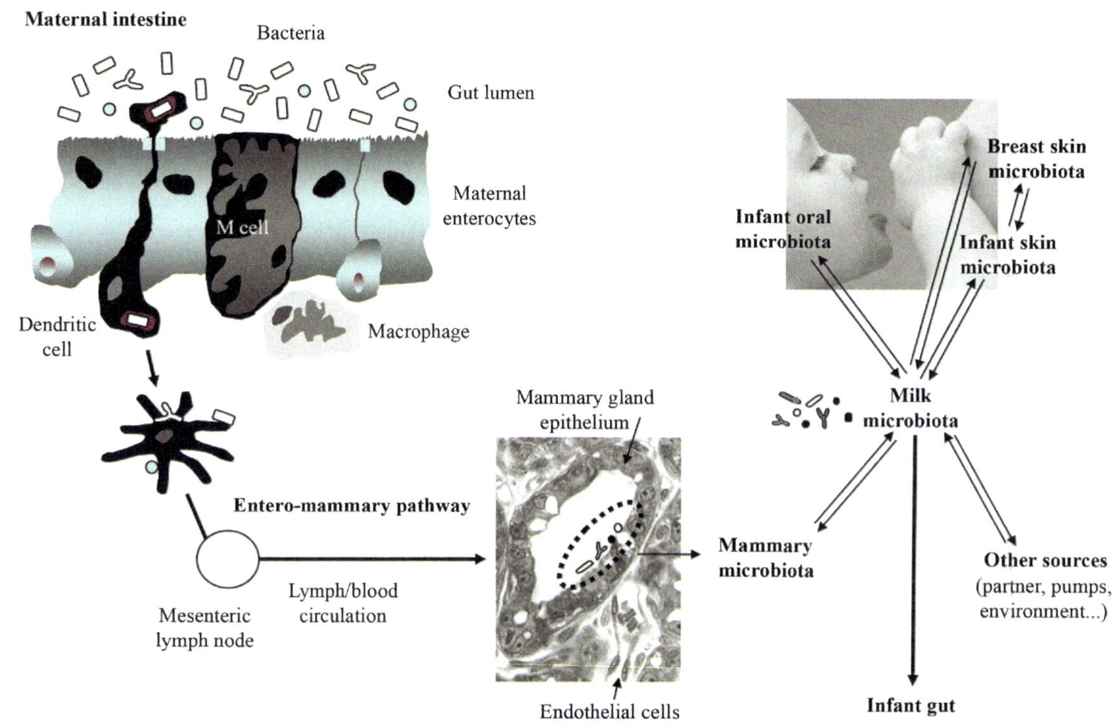

Fig. 4.2 Sources of the bacteria present in human milk, including a model to explain how some maternal bacterial strains could be transferred to the infant's gut through an entero-mammary pathway. Credit: Fig. 1 of Rodríguez JM. The origin of human milk bacteria: is there a bacterial entero-mammary pathway during late pregnancy and lactation? Adv Nutr 2014;5:779–84. https://doi.org/10.3945/an.114.007229.

Translocation of gut bacteria into secondary lymph nodes has been studied in mouse models of cancer, which could facilitate immune response and treatment (Chapter 7, Table 7.2).

Impaired gut epithelium barrier function (leaky gut) with insufficient butyrate production has been implicated in many diseases, while the inflammation is typically attributed to LPS (lipopolysaccharides). Bacteria DNA has been detected in the blood (e.g., Refs. [11–13]), but we do not know whether some of these bacteria may be alive in healthy adults (occult sepsis), e.g., hiding inside immune cells like some well-studied pathogens. For example, *Streptococcus pneumonia* is able to replicate within splenic macrophages [14]. *Streptococcus pyogenes* has been shown to remain extracellular, when transiting between multiple lymph nodes in lymphatic vessels to enter the bloodstream [15]. Vancomycin-resistant *Enterococcus* (VRE) is known to dominate the gut microbiome before bloodstream infection (sepsis) in patients undergoing allogeneic hematopoietic stem cell transplantation (allo-HCT) [16]. *Candida* spp. fungi have also been shown to expand in the gut of allo-HCT patients before they cause infection in the blood (Fig. 4.3; Remember the microbial cell number question in Chapter 1).

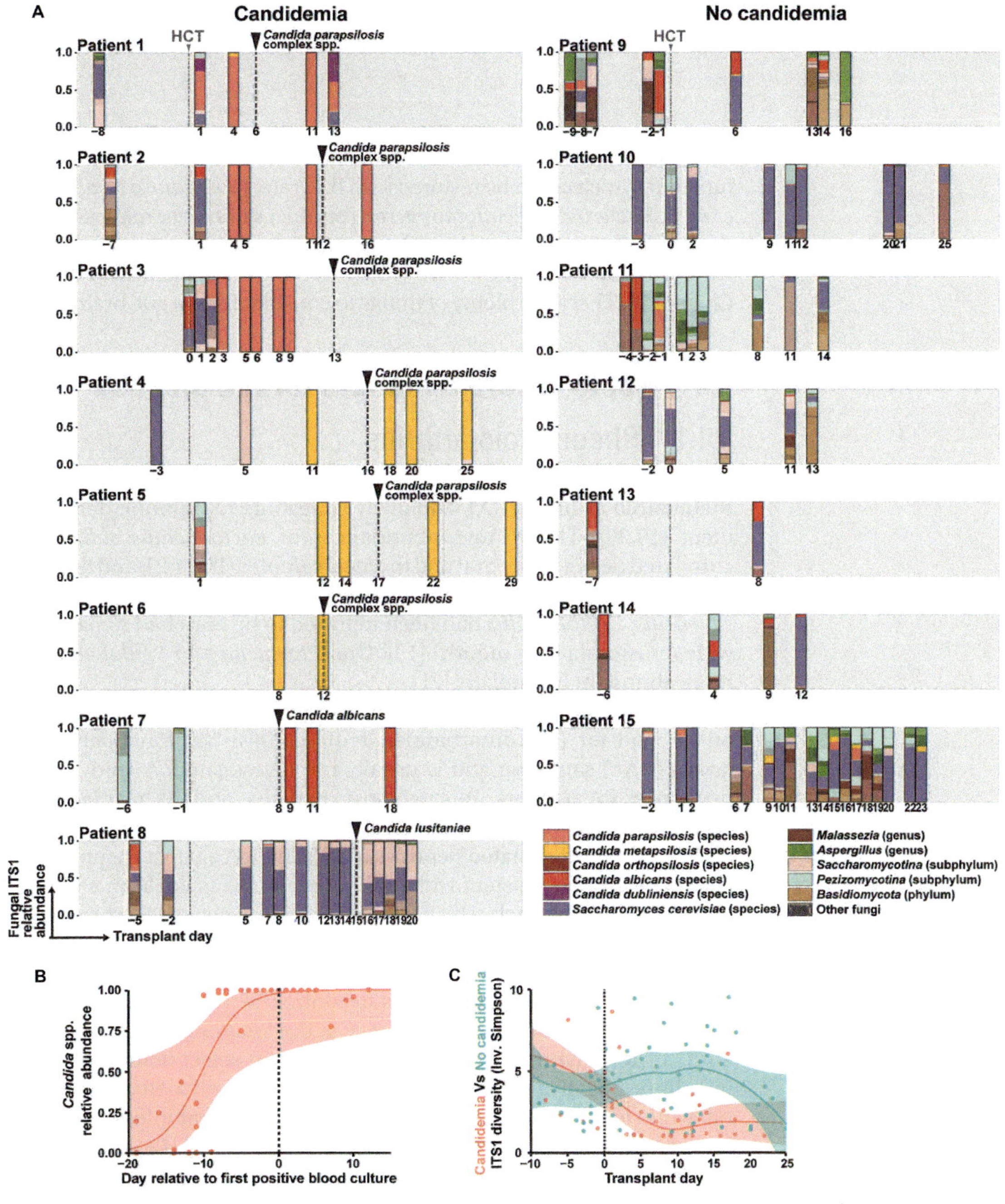

Fig. 4.3 Mycobiota dynamics in allo-HCT patients. (A) Species-level taxonomy of fecal mycobiota (average: 7 samples per patient, range: 2–18) from allo-HCT patients with (left column) and without (right column) candidemia, colored according to the legend. Frequent species, for example, *Candida* spp. and *Saccharomyces cerevisiae*, were individually color coded. The *gray box* indicates day −10 to day +30 of transplantation and the *gray dashed line* indicates the day of transplantation. The number below each bar graph indicates the day of sampling. The *black dashed line* and *arrow* indicate the day of the first fungal bloodstream infection in the candidemia group. (B) Quantification of total relative abundance of pathogenic *Candida* spp. of each fecal sample ($n=37$) from patient 1 to patient 7; a *solid line* represents the dynamic trend, with the *shaded area* indicating the 95% confidence interval. (C) α-Diversity of mycobiota in each sample, measured by the inverse Simpson index. *Red dots and line*: candidemia group ($n=51$); *turquoise dots and line*: noncandidemia group ($n=57$). Credit: Fig. 2 of Zhai B, Ola M, Rolling T, Tosini NL, Joshowitz S, Littmann ER, et al. High-resolution mycobiota analysis reveals dynamic intestinal translocation preceding invasive candidiasis. Nat Med 2020;26:59–64. https://doi.org/10.1038/s41591-019-0709-7.

Besides, germfree mice show impaired blood-brain barrier (BBB) function, which could be restored with butyrate, *Clostridium tyrobutyricum*, or *Bacteroides thetaiotaomicron* (produces the other major SCFAs, acetate and propionate) [17]. The hormonal cycle also impacts BBB, with estrogen being protective [18] (More on the menstrual cycle in Chapter 8). There are plenty of things to consider for the gut-brain axis.

4.4 Get to where it matters for the disease

4.4.1 Rheumatoid arthritis

The fecal, saliva, and dental microbiomes have been studied for rheumatoid arthritis (RA), and likely contribute to immune derangement [19,20]. The relative abundances of *Lactobacillus salivarius* correlated between the oral and the fecal samples (Fig. 4.2), and the RA-enriched strain likely differed from the known probiotic *Lactobacillus salivarius*; *Lactobacillus* had been reported to be enriched in patients with xerostomia (dry mouth) [19]. Oral *Prevotella* and *Veillonella* are more abundant in smokers [21].

The lung microbiome, although relevant for the effect of smoking on RA and for RA comorbidities, requires invasive bronchoalveolar lavage (BAL) sampling and is usually not investigated. A study of 20 new-onset RA patients, 10 sarcoidosis patients, and 28 healthy controls found BAL *Veillonella* to correlate with immunoglobulin A (IgA) against cyclic citrullinated peptides (CCP) and IgA against rheumatoid factor (RF) [22], consistent with oral microbiome results from a different cohort [19]. An unclassified Oxalobacteraceae negatively correlated with DAS28 (Disease Activity Score), reminiscent of a bacterium close to *Oxalobacter formigenes* that was relatively depleted in the fecal microbiome of RA patients compared to controls in a separate study [19].

Evidence for bacteria in the synovial fluid is available in the literature. PCR (polymerase chain reaction) against select dental bacteria in culture-negative synovial fluid samples frequently detected bacteria, while leukocytes were PCR negative [23]. For example, *Prevotella intermedia* was detected in 19/19 dental plaques, 14/19 serum, and 17/19 synovial fluid of patients with refractory RA (medication status unknown); *Porphyromonas gingivalis* was detected in 15/19 dental plaques, 8/19 serum, and 11/19 synovial fluid samples of refractory RA; *Aggregatibacter actinomycetemcomitans* was detected in 4/19 dental plaques, 0/19 serum, and 3/19 synovial fluid samples of refractory RA [23]. A 16S rRNA gene amplicon study on 110 synovial fluid samples from RA patients and 42 synovial fluid samples from osteoarthritis (OA) patients, identified *Veillonella dispar*, *Haemophilus parainfluenzae*, *Prevotella copri*, *Atopobium* sp. and *Treponema amylovorum* to be more abundant in RA, and *Bacteroides caccae* to be more abundant in OA [24]. These RA synovial fluid-enriched bacteria (or another species in the same genus) have all been previously reported as significant biomarkers in the gut or oral microbiome in RA patients or controls [19,20,25,26]. All these synovial fluid samples were culture-negative according to standard microbiology practices. The study also collected synovial tissue samples, and the results did not overlap with the synovial fluid bacterial biomarkers [24].

Aggregatibacter actinomycetemcomitans, but not other better-known periodontitis pathogens (*Porphyromonas gingivalis, Tannerella forsynthia, Treponema denticola, Fusobacterium nucleatum, Parvimonas micra, Prevotella intermedia*), triggered hypercitrullination (to proteins) in neutrophils, through a pore-forming toxin leukotoxin A (LtxA) expressed by *A. actinomycetemcomitans* [26]. Antibody against LtxA was enriched in RA patients and in periodontitis patients and showed some overlap with anticitrullinated protein antibodies (ACPA)-positive and with rheumatoid factor (RF)-positive RA [26]. LPS (lipopolysaccharide) from the periodontal bacteria *Aggregatibacter actinomycetemcomitans* (renamed from *Actinobacillus actinomycetemcomitans*), *P. intermedia*, and *Porphyromonas gingivalis* has also tested positive for their ability to induce osteoclast differentiation in vitro [27], potentially contributing to bone erosion, a debilitating aspect of RA progression.

In addition to Th17 (T helper 17) cell activation [28], recent single-cell studies indicated the recruitment of neutrophils and other immune cells by oral mucosal fibroblasts in periodontitis [29]. Neutrophils are the major population in White Blood Cell (WBC) counts reported by routine blood tests and showed positive associations with RA-enriched microbes such as *Lactobacillus salivarius*, *Veillonella* spp. and negative associations with control-enriched microbes such as *Lactococcus* sp. [19] (Fig. 4.4).

4.4.2 Cardiometabolic diseases

Patients with liver cirrhosis have been reported to show an overabundance of potentially oral bacteria in feces, which were relieved after withdraw of proton pump inhibitors (PPI, e.g., omeprazole) [30–33]. Yet we do not know whether or not these bacteria are in the liver. PPI use would allow more salivary bacteria to survive the stomach, and increased the relative abundance of fecal *Streptococcus*, in correlation with increased serum gastrin level [33]. Periodontal therapy showed favorable results in a group of cirrhosis patients [34].

Many cardiovascular events could be linked to a dental cleaning. A myriad of bacteria has been reported in atherosclerotic plaques [35] (Table 4.1), all of which can be part of the oral microbiome [77,78]. Atherosclerotic plaque species such as *Klebsiella pneumoniae* can

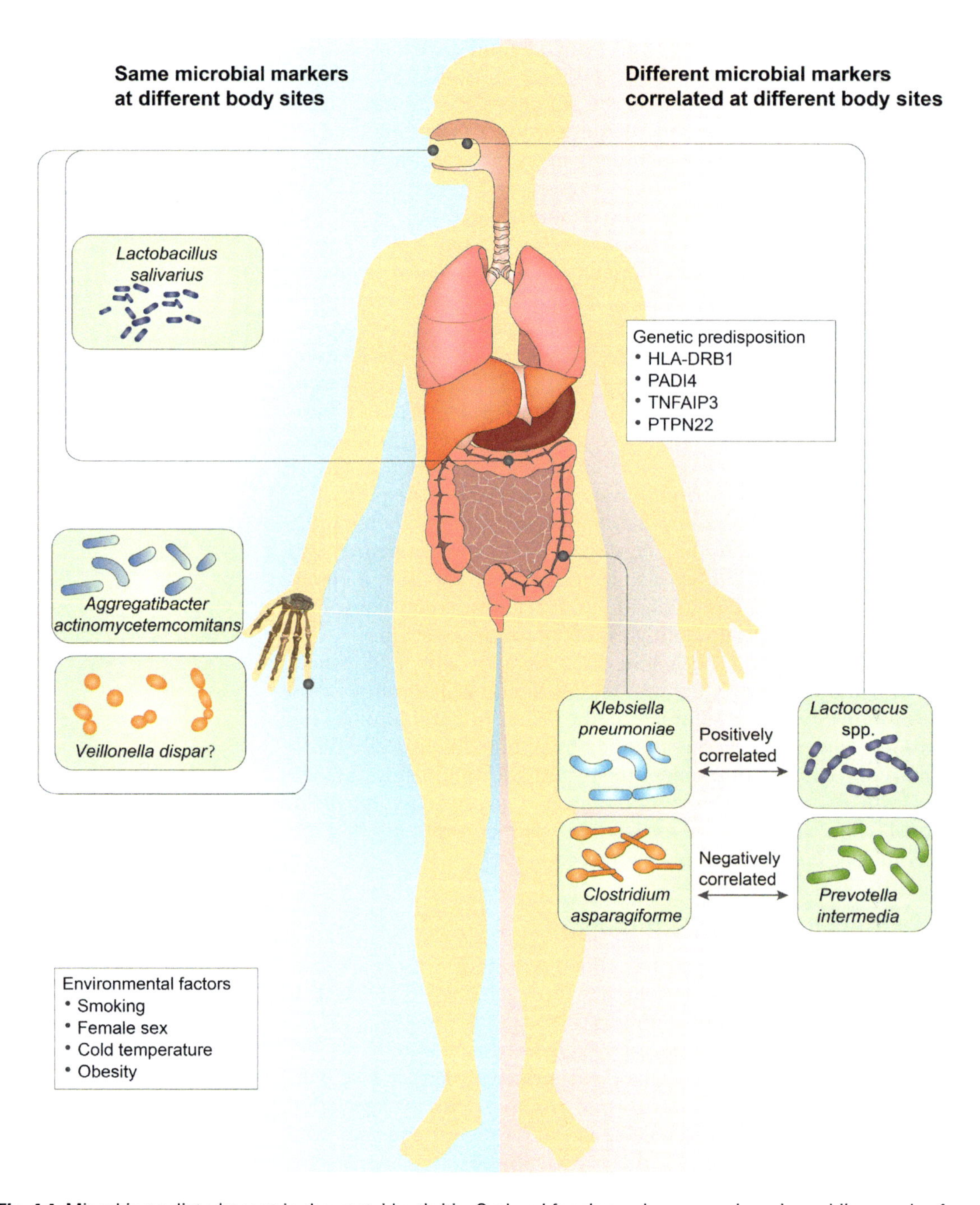

Fig. 4.4 Microbiome disturbances in rheumatoid arthritis. Oral and fecal samples are noninvasive, while samples from the joints or from the lungs are more difficult to obtain. This would also be a question for immune cell populations. The synovial fluid results are not yet shotgun metagenomics and are not matched with oral metagenomic data from the same patients. Credit: Added two bacteria onto Fig. 4 of Wang J, Jia H. Metagenome-wide association studies: fine-mining the microbiome. Nat Rev Microbiol 2016;14:508–22. https://doi.org/10.1038/nrmicro.2016.83, Huijue Jia, Chen Chen of BGI-Shenzhen.

be of either gut or oral origin [49], which contribute to LPS and TMA (trimethylamine, metabolized into TMAO (trimethylamine-*N*-oxide) in the liver implicated in atherosclerotic cardiovascular diseases [79,80]. More personalized monitoring of the microbiome would be recommended (Chapter 8).

Table 4.1 Atherosclerotic plaque-associated bacteria and methods of detection.

Atherosclerotic plaque-associated bacteria	Detection platform	Percentage of bacteria present in atherosclerotic plaque samples	Reference
Aggregatibacter actinomycetemcomitans [Phylum: Proteobacteria]	PCR	71.4% [5/7]	[36]
	16S rRNA	66.67% [28/42]	[37,38]
	mAb	17% [5/29]	[39,40]
	16S rRNA	21.87% [7/32]	[41]
	16S rRNA	18% [9/50]	[42]
	16S rRNA	25.9% [7/27]	[43]
	RT-PCR	46.2% [18/39]	[44,45]
	16S rRNA	29.4% [15/51]	[46,47]
Chlamydiae *pneumoniae* [Phylum: Chlamydiae]	mAb	20.6% [6/29]	[39,40]
	16S rRNA	35.4% [11/31]	[39,41]
	16S rDNA	18% [9/50]	[39,42]
	ICC/PCR	48% [11/23]	[48]
	16S rRNA	51.5% [17/33]	[49]
	MIF IgA	32.6% [63/193]	[50]
	MIF IgG	61.7% [119/193]	[50]
	16S rRNA	26% [12/46]	[51]
	PCR	42% [102/241 sections (10 samples)]	[52]
	PCR	69% [11/16]	[53]
	Immunofluorescence	79% [71/90]	[54]
	PCR	70% [42/60]	[55]
	IgG antibody	61.7% [50/81]	[56]
Campylobacter rectus [Phylum: Proteobacteria]	16S rRNA	9.52% [4/42]	[37,38,57]
	PCR	11.7% [6/51]	[44,46]
	16S rRNA	21.51% [11/51]	[44,46,58]
	16S rRNA	15.7% [8/51]	[59]
	16S rRNA	21.51% [11/51]	[43]
Enterobacter hormaechei [Phylum: Proteobacteria]	16S rRNA	50% [134/268]	[60]
	16S rRNA	40% [2/5]	[61]

Continued

Table 4.1 Atherosclerotic plaque-associated bacteria and methods of detection.—cont'd

Atherosclerotic plaque-associated bacteria	Detection platform	Percentage of bacteria present in atherosclerotic plaque samples	Reference
Eikenella corrodens [Phylum: Proteobacteria]	16S rRNA	54.76% [23/42]	[37,38]
	PCR	15.6% [8/51]	[57]
	16S rRNA	27.45% [14/51]	[59]
Fusobacterium nucleatum [Phylum: Fusobacteria]	16S rRNA	50% [21/42]	[37,38]
	Monoclonal antibody	34% [10/29]	[39,40]
	PCR	21% [4/19]	[62]
Fusobacterium necrophorum [Phylum: Fusobacteria]	–	–	[63–65]
Helicobacter pylori [Phylum: Proteobacteria]	IgA	55.4% [107/193]	[50]
	IgM	44.6% [86/193]	[50]
	16S rRNA	37% [17/46]	[51]
	IHC	57.8% [22/38]	[66]
	PCR	92.16% [47/51]	[67]
	IgG	67.9% [55/81]	[56]
Mycoplasma pneumoniae [Phylum: Tenericutes]	Seropositivity	14% [396]	[68]
	–	–	[69]
Porphyromonas endodontalis [Phylum: Bacteriodetes]	–	–	[70]
Porphyromonas gingivalis [Phylum: Bacteriodetes]	16S rRNA	78.57% [33/42]	[37,38]
	PCR	71.43% [5/7]	[36]
	16S rRNA	67% [134]	[60]
	mAb	52% [15/29]	[39,40]
	16S rRNA	22.27% [6/22]	[39,41]
	16S rRNA	26% [13/50]	[39,42]
	PCR	47.4% [9/19]	[62]
	PCR	51% [27/53]	[71,72]
	PCR	43.1% [22/51]	[57]
	16S rRNA	45.1% [23/51]	[44,46]
	16S rRNA	21.6% [11/51]	[44,46,58]
	RT-PCR	53.8% [21/39]	[44,45]
	16S rRNA	45.1% [23/51]	[59]
	16S rRNA	7.4% [2/27]	[43]
Prevotella intermedia [Phylum: Bacteroidetes]	mAb	41% [12/29]	[39,40]
	16S rRNA	9.37% [3/32]	[39,41]
	16S rRNA	14% [7/50]	[39,42]
	PCR	21% [4/19]	[62]
	PCR	15% [8/53]	[72,73]
	PCR	19.6% [10/51]	[57]
	RT-PCR	79.3% [23/29]	[44,45]
	PCR	71.43% [5/7]	[36]
	16S rRNA	3.7% [1/27]	[43]

Table 4.1 Atherosclerotic plaque-associated bacteria and methods of detection.—cont'd

Atherosclerotic plaque-associated bacteria	Detection platform	Percentage of bacteria present in atherosclerotic plaque samples	Reference
Prevotella nigrescens [Phylum: Bacteriodetes]	PCR	15.6% [8/51]	[57]
	RT-PCR	17.9% [7/39]	[44,45]
Pseudomonas aeruginosa [Phylum: Proteobacteria]	16S rRNA	40% [6/15]	[74][a]
Pseudomonas luteola [Phylum: Proteobacteria]	16S rRNA	100% [15/15]	[75]
Streptococcus gordonii	PCR	19.4% [–]	[43][b]
Streptococcus mitis	PCR	19.4% [–]	
Streptococcus mutans	PCR	74.1% [20/27]	
Streptococcus oralis	PCR	3.7% [1/27]	
Streptococcus sanguinis [Phylum: Firmicutes]	PCR	25.9% [7/27]	
Treponema denticola [Phylum: Spirochaetes]	PCR	43% [23/53]	[44,71]
	16S rRNA	44.4% [12/27]	[43]
	PCR	35.2% [18/51]	[57]
	16S rRNA	49.01% [25/51]	[44,46]
	16S rRNA	27.4% [14/51]	[44,46,47]
	16S rRNA	23.1% [6/26]	[43,47]
	16S rRNA	49.01% [25/51]	[59]
Tannerella forsythia [Phylum: Bacteriodetes]	16S rRNA	61.9% [26/42]	[37]
	PCR	100% [7/7]	[36]
	mAb	34% [10/29]	[39,40]
	16S rRNA	30% [15/50]	[39,42]
	PCR	10.5% [2/19]	[62]
	PCR	19.6% [10/51]	[57]
	16S rRNA	5.9% [3/51]	[38,44,46]
	RT-PCR	25.6% [10/39]	[44,45]
Veillonella [Phylum: Firmicutes]	16S rRNA	10% [2/20]	[76]
	16S rRNA	100% [13/13]	[75]

[a]The author did not find this result in the review and did not receive a response from [35].
[b]The original table from [35] indicated 16S rRNA, which was used for other bacteria in [43]; Streptococcus species were distinguished from one another using PCR against the glucosyltransferase gene [43].
Credit: From Table 1 of Chhibber-Goel J, Singhal V, Bhowmik D, Vivek R, Parakh N, Bhargava B, et al. Linkages between oral commensal bacteria and atherosclerotic plaques in coronary artery disease patients. NPJ Biofilms Microbiomes 2016;2:7. https://doi.org/10.1038/s41522-016-0009-7.

Members of the human saliva microbiome and subgingival microbiome have been found in mouse placenta following injection through the tail vein [81]. In addition to human cases of preterm birth and term stillbirth, *Fusobacterium* has also been implicated in preeclampsia. Mice transplanted with feces from preeclampsia cases showed bacteria including *Fusobacterium* in their placenta, along with elevated expression of proinflammatory cytokines and chemokines such as IL-6 (interleukin-6), IL-1b, Ccl3 (CC-type chemokine 3), Ccl4 [82].

Lactobacillus crispatus is a dominant bacterium in the healthy vagina correlated with hormones including testosterone [83,84], and its abundance in the fecal microbiome correlated with that in the vagina [85]. However, *Lactobacillus crispatus* together with other Lactobacilli appeared more abundant in the fecal microbiome of atherosclerotic cardiovascular disease (ACVD) patients (Chapter 2, Fig. 2.14); the *Lactobacillus crispatus*-dominant vaginal microbiome was more prevalent in women who used statin, compared to those with high cholesterol who did not take statin and to those with normal cholesterol [86]. Further study would be needed for the potential modulation of *Lactobacillus crispatus* by statin in cardiovascular diseases, and the likely sex difference in *Lactobacillus crispatus* distribution in the human body.

Worked sample 4.3

With your practical and theoretical knowledge of the microbiome from the previous Chapters, what question would you like to investigate for the lungs (Fig. 4.5), and how would you design the collection of samples and other information?

4.5 Interkingdom interactions in the microbiome in diseases

Commensal bacteria typically prefer a neutral pH and body temperature, while fungi can tolerate lower pH, dryness, and other not so encouraging conditions. Unfortunately, amplicon sequencing for bacteria would not detect fungi, and amplicon sequencing for fungi would not detect bacteria (Chapter 3, Fig. 3.4). Judging from shotgun metagenomic data of a typical sequencing amount (if the samples were properly extracted), fungi are usually of low abundance in the gut, vagina, and mouth of healthy individuals [85,87,88], while being better known on the skin [89]. Mycobiota dysbiosis is seen in infectious diseases and other conditions involving various body sites (Fig. 4.6). Perhaps with more data, we would be able to predict a fungal boom when it is

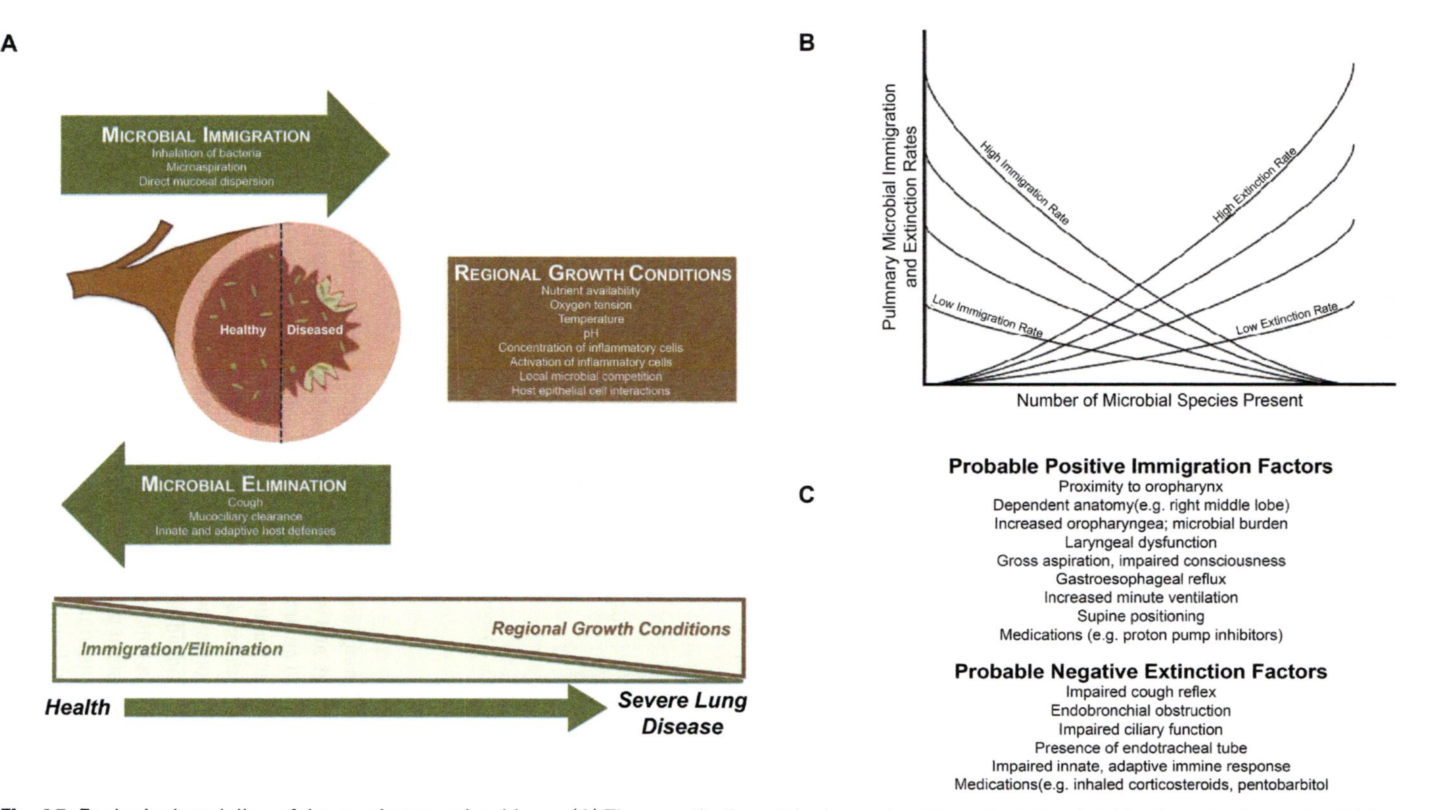

Fig. 4.5 Ecological modeling of the respiratory microbiome. (A) The constitution of the lung microbiome is determined by three factors: microbial immigration from the mouth and the upper respiratory system, microbial elimination locally, and the relative reproduction rates of its members. In a healthy lung, the microbiome is determined primarily by immigration and elimination; in advanced lung disease which impaired both immigration and elimination, the microbiome is determined primarily by regional growth conditions, and the same species could develop into different lineages. (B) The adapted island model of lung biogeography. Community richness in health for a given site in the respiratory tract is a function of immigration and elimination factors. Speculated positive immigration and negative extinction factors for the lung microbiota are shown. Credit: Similar to Fig. 2 of Dickson RP, Erb-Downward JR, Martinez FJ, Huffnagle GB. The microbiome and the respiratory tract. Annu Rev Physiol 2016;78:481–504. https://doi.org/10.1146/annurev-physiol-021115-105238. Panel A was from Fig. 2 of Stefka AT, Feehley T, Tripathi P, Qiu J, McCoy K, Mazmanian SK, et al. Commensal bacteria protect against food allergen sensitization. Proc Natl Acad Sci U S A 2014;111:13145–150. https://doi.org/10.1073/pnas.1412008111. Panel B was from Fig. 1C,D of Dickson RP, Erb-Downward JR, Huffnagle GB. Towards an ecology of the lung: new conceptual models of pulmonary microbiology and pneumonia pathogenesis. Lancet Respir Med 2014;2:238–46. https://doi.org/10.1016/S2213-2600(14)70028-1.

present in low abundance along with bacteria that may or may not be able to maintain their foothold at the site [90].

In addition to SCFAs and secondary bile acids produced by commensal bacteria [91,92], bacteriophages and anelloviruses might also contribute to the treatment of *Clostridium difficile* (now *Peptoclostridium difficile*) using fecal microbiome transplant (FMT) [93].

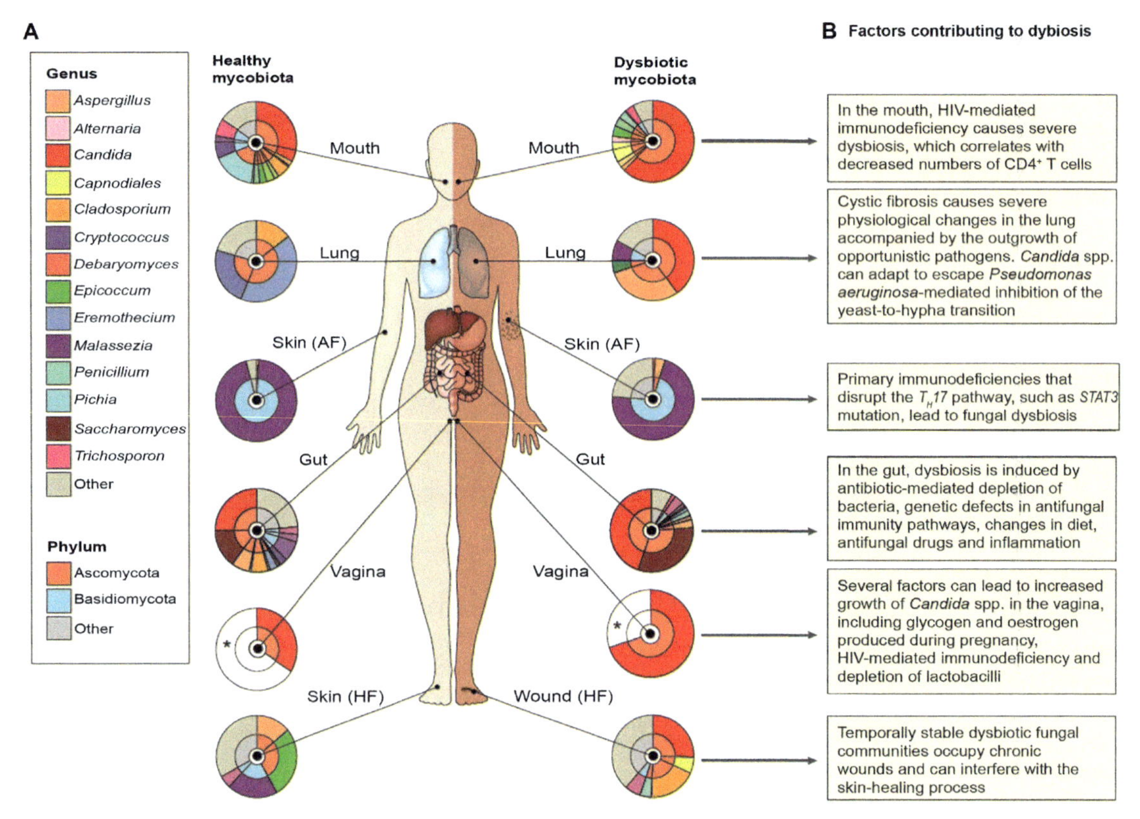

Fig. 4.6 Mycobiota in health and dysbiosis. (A) During homeostasis, diverse fungal communities reside on all human barrier surfaces, such as the mouth, lung, skin, gut, and vagina (left side). The pie charts represent the relative abundance of the observed taxa at the phylum and genus levels (*inner and outer circles*, respectively). Of note, the data for the vagina are estimates that are based on culture-dependent studies, due to a lack of sequencing-based studies related to disease conditions (indicated with an *asterisk*). "Other" refers to sequences with $< 5\%$ relative abundance. During disease states, these fungal communities are perturbed (right side). Dysbiotic fungal communities are observed in the oral cavity and the vagina in individuals with HIV; in the lungs of individuals with cystic fibrosis; on the skin of individuals with primary immunodeficiency and chronic wounds; and in the gut of patients with Crohn's disease. (B) Factors contributing to fungal dysbiosis at different barrier surfaces. AF, antecubital fossa; HF, hind foot; STAT3, signal transducer and activator of transcription 3. Credit: Fig. 1 of Iliev ID, Leonardi I. Fungal dysbiosis: immunity and interactions at mucosal barriers. Nat Rev Immunol 2017. https://doi.org/10.1038/nri.2017.55.

4.6 Other omics data that hint at a difference in microbiome

Besides sequencing the microbes themselves, other omics data could also provide very useful information regarding what might have gone abnormal. In animal models such as the pig, metabolites entering and leaving each organ have been systematically studied (Tables 4.2 and 4.3, [94]). Metabolomics technology is also being developed for single-cell measurements [95]. Microbes within each organ, together with the host enzymes, might have contributed to the level of specific metabolites, e.g., amino acids, SCFAs (short-chain fatty acids). These metabolites could further contribute to differential growth or inhibition of microbes [96]. An overarching analogy is that the arteries, veins, and lymphatic circulation are like the sewage system in the 1854 Chlora outbreak in London. It will be important to track down the source of the microbiome culprit.

Table 4.2 Organ-specific metabolite production and consumption in the pig.

Organ	Exemplary discovery	Key evidence
Liver	Clears unsaturated fatty acids	Compared to the most abundant saturated fatty acids, oleate (C18:1) and linoleate (C18:2) show greater uptake and TCA contribution
	Produces amino acids	Significant release of amino acids
(Small) Intestine	Consumes glucose and amino acids	Greatest absolute uptake flux, of any organ, of both glucose and amino acids
Pancreas	Produces TCA intermediates	Significant release of citrate, ketoglutarate, succinate, fumarate, and malate
Spleen	Produces nucleosides	Significant release of cytidine, deoxycytidine, deoxyuridine, guanosine, inosine, thymidine, uridine, and xanthosine
	Produces unsaturated very long chain fatty acids	Significant release of C22:1, C22:2, C22:3, C22:4, C22:5, C22:6, C24:1, C24:2, C24:3, C24:4, and C24:5
Brain	Produces acetate	$> 2 \times$ increase in acetate in jugular vein blood

Continued

Table 4.2 Organ-specific metabolite production and consumption in the pig—cont'd

Organ	Exemplary discovery	Key evidence
Leg muscle	Consumes short-chain acylcarnitines	Significant uptake of C2:0, C3:0, C4:0, C5:0, and C5:1 carnitines
	Produces long chain acylcarnitines	Significant release of C8:0, C10:0, C12:0, C12:1, C14:1, C14:2, C16:0, C16:1, C18:1, C18:2, and C20:4 carnitines
Heart	Consumes long chain fatty acids	Significant uptake of C16:0, C16:1, C18:0, C18:1, C18:2, C20:1, C20:2, C22:4, C24:0, and C24:1
Lung	Produces saturated very long chain fatty acids	Significant release of C22:0 and C24:0
Kidney	Consumes citrate	Only organ with significant citrate uptake; TCA contribution from citrate $>10\times$ higher than any other organ
	Maintains circulating pyruvate/ lactate ratio	Significant increase in pyruvate relative to lactate in renal vein blood
	Produces amino acids	Significant release of amino acids
	Consumes medium and long chain acylcarnitines	significant uptake of C5:0, C6:0, C8:0, C10:0, C10:1, C12:0, C12:1, C14:0, C14:1, C14:2, and C16:1 (without release into urine)

Credit: Table 1 of Jang C, Hui S, Zeng X, Cowan AJ, Wang L, Chen L, et al. Metabolite exchange between mammalian organs quantified in pigs. Cell Metab 2019:1–13. https://doi.org/10.1016/j.cmet.2019.06.002.

Table 4.3 Top three metabolites produced and consumed by each organ in pigs.

	Production		Consumption	
Organ	Metabolite	Log2 (V/A)	Metabolite	Log2 (V/A)
Liver	Glutamate	0.64 ± 0.11	Bile acids (5)	-2.89 ± 0.19
	Triethanolamine	0.49 ± 0.17	Phenylpropionic acid (2)	-2.29 ± 0.12
	Acetoacetate	0.38 ± 0.09	Short-chain fatty acids (3)	-2.02 ± 0.83
Portal (intestine)	Bile acids (6)	3.28 ± 0.21	2-Methylhippuric acid	-0.69 ± 0.15
	Phenylpropionic acid (2)	2.84 ± 0.32	Glucose	-0.31 ± 0.05
	Short-chain fatty acids (3)	2.82 ± 1.15	Glutamine	-0.28 ± 0.02
Colon	Short-chain fatty acids (3)	4.65 ± 1.21	2-Methylhippuric acid	-0.60 ± 0.21
	Lithocholic acid	4.04 ± 1.10	5-Hydroxylysine	-0.41 ± 0.05
	Phenylpropionic acid (2)	3.42 ± 0.48	Glucose	-0.39 ± 0.05
Pancreas	Xanthine	1.05 ± 0.26	5-Hydroxylysine	-0.79 ± 0.19
	Capryloyl glycine	0.51 ± 0.09	*N*-carbamoylsarcosine	-0.39 ± 0.09
	TCA intermediates (5)	0.36 ± 0.17	Amino acids (8)	-0.36 ± 0.01

Table 4.3 Top three metabolites produced and consumed by each organ in pigs—cont'd

	Production		Consumption	
Organ	**Metabolite**	**Log2 (V/A)**	**Metabolite**	**Log2 (V/A)**
Spleen	*O*-phosphorylethanolamine	1.11±0.22	Adenosine	– 0.61±0.14
	Nucleosides (9)	0.52±0.03	Dihydroxymandelic acid	– 0.33±0.07
	C22 and C24 very long-chain fatty acids (11)	0.35±0.008	C5 acylcarnitine	– 0.26±0.02
Head (brain)	Synephrine	1.89±0.68	Dihydroxymandelic acid	– 0.38±0.12
	Gluconolactone and gluconate	1.66±0.03	2-Methylhippuric acid	– 0.34±0.11
	Acetate	1.46±0.39	Glutamate	– 0.30±0.09
Leg (muscle)	Hypotaurine	0.69±0.12	Glutamate	– 1.41±0.33
	Branched chain hydroxyl acids (2)	0.65±0.12	Ketone bodies (2)	– 0.58±0.14
	Medium and long-chain acylcarnitines (11)	0.57±0.02	Short-chain acylcarnitines (5)	– 0.36±0.05
Lung	2-Phenylpropionic acid	0.48±0.14	5-Keto-D-gluconic acid	– 0.31±0.07
	Aconitate	0.26±0.03	Kynurenate	– 0.22±0.03
	C22:0 and C24:0 fatty acids	0.24±0.02	3-Hydroxyanthranilic acid	– 0.17±0.02
Kidney	Glycocyamine	1.87±0.12	*N*-formyl-L-methionine	– 2.66±0.32
	Serine	0.73±0.12	Medium-chain acylcarnitines (4)	– 2.61±0.19
	Allantoate	0.53±0.12	*N*-acetyl amino acids (9)	– 1.27±0.87
Heart	Hypotaurine	0.34±0.11	3-Phenylpropionic acid	– 0.71±0.24
	Glutamate	0.26±0.04	Unsaturated long-chain fatty acids (11)	– 0.53±0.06
	Biotin	0.25±0.06	Hydroxyindoleacetic acid	– 0.47±0.12
Ear (skin)	Guanine	0.82±0.19	Hydroxyhippuric acid	– 0.35±0.08
	Taurine	0.53±0.08	Indole metabolites (2)	– 0.23±0.02
	Long-chain acylcarnitines (3)	0.20±0.04	Serine	– 0.15±0.01

Ranking is based on multiplying the log2 (t value) and log2 (Vein/Artery) to reflect both statistical significance and fold change. Numbers in parentheses refer to the number of metabolites in that category showing statistically significant arterio-venous differences across the indicated organ. All arterio-venous differences included in the table are statistically significant (FDR < 0.05).
Credit: From Table 2 of Jang C, Hui S, Zeng X, Cowan AJ, Wang L, Chen L, et al. Metabolite exchange between mammalian organs quantified in pigs. Cell Metab 2019:1–13. https://doi.org/10.1016/j.cmet.2019.06.002.

Studies on the immune cell populations in each organ are still limited. Hopefully, in the near future, they will be mapped throughout the body, together with the microbiome. Epitopes from engulfed bacteria have been shown to be presented by MHC I in melanoma cells [97] (Fig. 4.7).

For cancer and aging-related diseases, DNA mutation patterns in tissue samples might also suggest the presence of certain microbes [98].

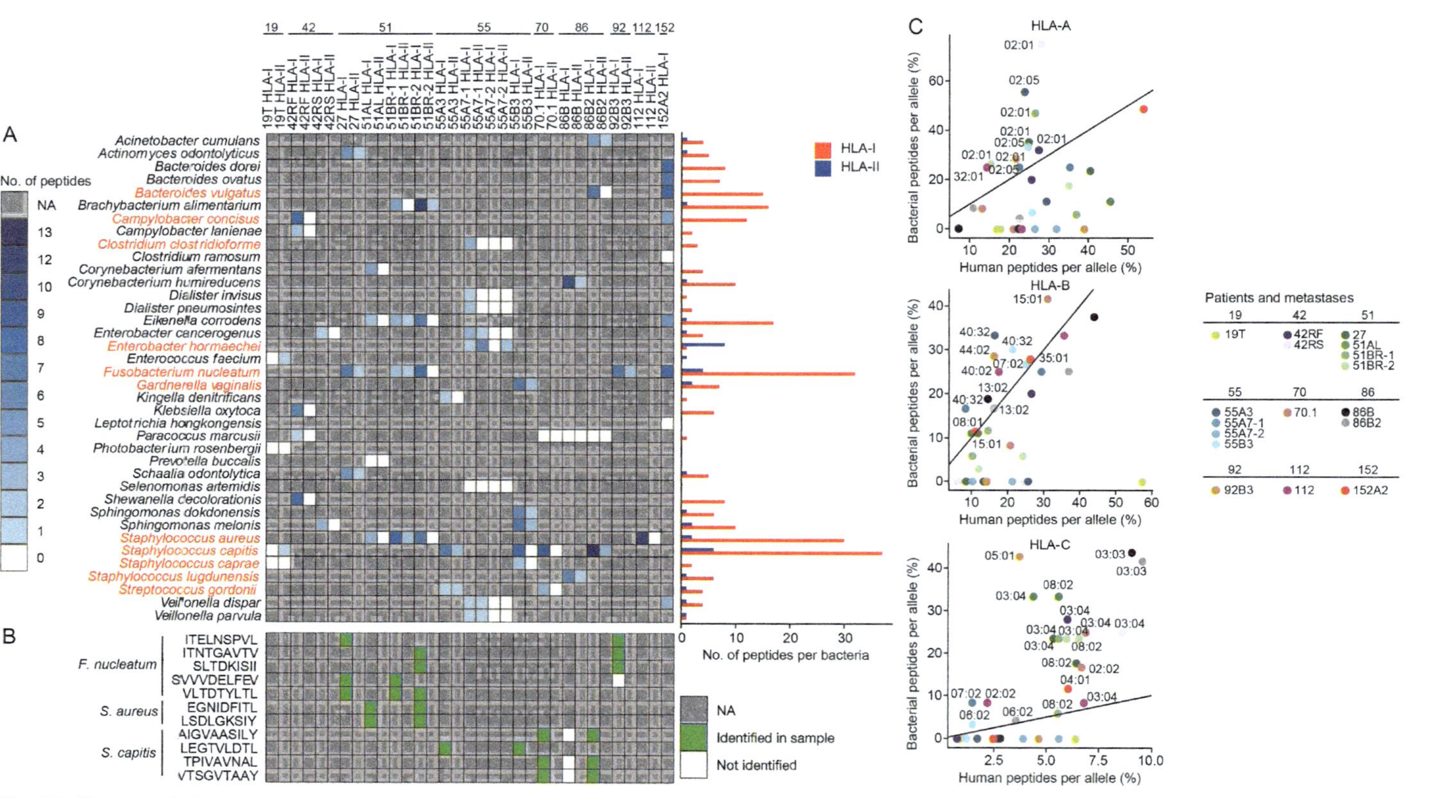

Fig. 4.7 Characteristics of bacterial peptides from melanoma samples. Peptidome was analyzed by mass spectrometry and matched with the proteome of bacteria that were identified by 16S rRNA gene amplicon sequencing, with filtering and validation steps [97]. (A) The number of bacterial peptides presented on HLA-I and HLA-II in each patient sample (patient number indicated at the top) is indicated in a *blue* color scale (left). *White* indicates that no peptides were identified in the sample, and *gray* indicates that the bacterium was not identified in this metastasis (NA, not applicable). The total number of bacterial HLA-I and HLA-II peptides from each bacterium is noted in the bar plot on the right. Species names marked in *red* are known to be intracellular bacteria (Supplementary Table 6) of [97]. (B) Bacterial peptides that were identified in a few metastases from the same patient or in different patients are indicated. Peptides identified in the sample are marked *green*, and *white* denotes peptides that were not identified in the sample (although the metastasis has the required HLA allele for this peptide presentation and the species of bacteria). *Gray* indicates samples that lack the HLA allele and bacteria to produce the peptide. (C) For each metastasis, the percentages of bacterial and human peptides that match each HLA-A (left), HLA-B (middle), or HLA-C (right) allele of the patient is indicated. The allele with the best percent rank binding prediction (by NetMHCpan) was assigned to each peptide; the full allele list is indicated in Extended Data Fig. 6 of [97]. Credit: Fig. 2 of Kalaora S, Nagler A, Nejman D, Alon M, Barbolin C, Barnea E, et al. Identification of bacteria-derived HLA-bound peptides in melanoma. Nature 2021;592:138–43. https://doi.org/10.1038/s41586-021-03368-8.

Worked sample 4.4

After removing a kidney stone, can the patient's fecal, urinary, or oral microbiomes be matched with the type of stone (calcium oxalate (dihydrate, monohydrate, and more complexities [99]), calcium phosphate, struvite, or uric acid stones)? What lifestyle factors shall we inquire about, and can we give the patient some useful advice? (Get ready for Chapters 7 and 8).

4.7 Summary

This chapter focuses on identifying sources of a given microbiome, whether it is from the environment, from members of the family, or from elsewhere in the same person. Knowledge from Chapter 3 will be put into use in all kinds of samples. Many bacteria can enter the lymph nodes or enter circulation. Fungi can lurk in the gut or other mucosal sites before serious symptoms elsewhere. For rheumatoid arthritis, the synovial fluid contained bacterial DNA identified in the oral or fecal microbiome. Many oral microbes have been found in atherosclerotic plaques. The flow of metabolites, and the tissue-resident immune cells, may also provide clues for where to look for the microbial culprits, in order to better understand and treat the diseases (Chapters 6 and 7).

References

[1] Afshinnekoo E, Meydan C, Chowdhury S, Jaroudi D, Boyer C, Bernstein N, et al. Geospatial resolution of human and bacterial diversity with city-scale metagenomics. Cell Syst 2015;1:1–15. https://doi.org/10.1016/j.cels.2015.01.001.

[2] Ugboko HU, Nwinyi OC, Oranusi SU, Oyewale JO. Childhood diarrhoeal diseases in developing countries. Heliyon 2020;6. https://doi.org/10.1016/j.heliyon.2020.e03690, e03690.

[3] Moossavi S, Sepehri S, Robertson B, Bode L, Goruk S, Field CJ, et al. Composition and variation of the human milk microbiota are influenced by maternal and early-life factors. Cell Host Microbe 2019;25:324–335.e4. https://doi.org/10.1016/j.chom.2019.01.011.

[4] Boix-Amorós A, Collado MC, Mira A. Relationship between milk microbiota, bacterial load, macronutrients, and human cells during lactation. Front Microbiol 2016;7. https://doi.org/10.3389/fmicb.2016.00492.

[5] Boix-Amorós A, Martinez-Costa C, Querol A, Collado MC, Mira A. Multiple approaches detect the presence of fungi in human breastmilk samples from healthy mothers. Sci Rep 2017;7:13016. https://doi.org/10.1038/s41598-017-13270-x.

[6] Ferretti P, Pasolli E, Tett A, Asnicar F, Gorfer V, Fedi S, et al. Mother-to-infant microbial transmission from different body sites shapes the developing infant gut microbiome. Cell Host Microbe 2018;24:133–145.e5. https://doi.org/10.1016/j.chom.2018.06.005.

[7] Pannaraj PS, Li F, Cerini C, Bender JM, Yang S, Rollie A, et al. Association between breast milk bacterial communities and establishment and development of the infant gut microbiome. JAMA Pediatr 2017;171:647. https://doi.org/10.1001/jamapediatrics.2017.0378.

[8] Atarashi K, Suda W, Luo C, Kawaguchi T, Motoo I, Narushima S, et al. Ectopic colonization of oral bacteria in the intestine drives T H 1 cell induction and inflammation. Science 2017;358:359–65. https://doi.org/10.1126/science.aan4526.
[9] Liu X, Tang S, Zhong H, Tong X, Jie Z, Ding Q, et al. A genome-wide association study for gut metagenome in Chinese adults illuminates complex diseases. Cell Discov 2021;7:9. https://doi.org/10.1038/s41421-020-00239-w.
[10] Rühlemann MC, Hermes BM, Bang C, Doms S, Moitinho-Silva L, Thingholm LB, et al. Genome-wide association study in 8,956 German individuals identifies influence of ABO histo-blood groups on gut microbiome. Nat Genet 2021. https://doi.org/10.1038/s41588-020-00747-1.
[11] Poore GD, Kopylova E, Zhu Q, Carpenter C, Fraraccio S, Wandro S, et al. Microbiome analyses of blood and tissues suggest cancer diagnostic approach. Nature 2020;579:567–74. https://doi.org/10.1038/s41586-020-2095-1.
[12] Anhê FF, Jensen BAH, Varin TV, Servant F, Van Blerk S, Richard D, et al. Type 2 diabetes influences bacterial tissue compartmentalisation in human obesity. Nat Metab 2020;2:233–42. https://doi.org/10.1038/s42255-020-0178-9.
[13] Li B, He Y, Ma J, Huang P, Du J, Cao L, et al. Mild cognitive impairment has similar alterations as Alzheimer's disease in gut microbiota. Alzheimers Dement 2019;1–10. https://doi.org/10.1016/j.jalz.2019.07.002.
[14] Ercoli G, Fernandes VE, Chung WY, Wanford JJ, Thomson S, Bayliss CD, et al. Intracellular replication of *Streptococcus pneumoniae* inside splenic macrophages serves as a reservoir for septicaemia. Nat Microbiol 2018;1. https://doi.org/10.1038/s41564-018-0147-1.
[15] Siggins MK, Lynskey NN, Lamb LE, Johnson LA, Huse KK, Pearson M, et al. Extracellular bacterial lymphatic metastasis drives *Streptococcus pyogenes* systemic infection. Nat Commun 2020;11:1–12. https://doi.org/10.1038/s41467-020-18454-0.
[16] Ubeda C, Taur Y, Jenq RR, Equinda MJ, Son T, Samstein M, Viale A, Socci ND, van den Brink MRM, Kamboj M, Pamer EG. Vancomycin-resistant Enterococcus domination of intestinal microbiota is enabled by antibiotic treatment in mice and precedes bloodstream invasion in humans. J Clin Invest 2010. https://doi.org/10.1172/JCI43918.
[17] Braniste V, Al-Asmakh M, Kowal C, Anuar F, Abbaspour A, Tóth M, et al. The gut microbiota influences blood-brain barrier permeability in mice. Sci Transl Med 2014;6. https://doi.org/10.1126/scitranslmed.3009759, 263ra158.
[18] Sohrabji F. Guarding the blood-brain barrier: a role for estrogen in the etiology of neurodegenerative disease. Gene Expr 2007;13:311–9. https://doi.org/10.3727/000000006781510723.
[19] Zhang X, Zhang D, Jia H, Feng Q, Wang D, Di Liang D, et al. The oral and gut microbiomes are perturbed in rheumatoid arthritis and partly normalized after treatment. Nat Med 2015;21:895–905. https://doi.org/10.1038/nm.3914.
[20] Wang J, Jia H. Metagenome-wide association studies: fine-mining the microbiome. Nat Rev Microbiol 2016;14:508–22. https://doi.org/10.1038/nrmicro.2016.83.
[21] Bostanci N, Krog MC, Hugerth LW, Bashir Z, Fransson E, Boulund F, et al. Dysbiosis of the human oral microbiome during the menstrual cycle and vulnerability to the external exposures of smoking and dietary sugar. Front Cell Infect Microbiol 2021;11. https://doi.org/10.3389/fcimb.2021.625229.
[22] Scher JU, Joshua V, Artacho A, Abdollahi-Roodsaz S, Öckinger J, Kullberg S, et al. The lung microbiota in early rheumatoid arthritis and autoimmunity. Microbiome 2016;4:60. https://doi.org/10.1186/s40168-016-0206-x.
[23] Martinez-Martinez RE, Abud-Mendoza C, Patiño-Marin N, Rizo-Rodríguez JC, Little JW, Loyola-Rodríguez JP. Detection of periodontal bacterial DNA in serum and synovial fluid in refractory rheumatoid arthritis patients. J Clin Periodontol 2009;36:1004–10. https://doi.org/10.1111/j.1600-051X.2009.01496.x.

[24] Zhao Y, Chen B, Li S, Yang L, Zhu D, Wang Y, et al. Detection and characterization of bacterial nucleic acids in culture-negative synovial tissue and fluid samples from rheumatoid arthritis or osteoarthritis patients. Sci Rep 2018;8:14305. https://doi.org/10.1038/s41598-018-32675-w.
[25] Scher JU, Sczesnak A, Longman RS, Segata N, Ubeda C, Bielski C, et al. Expansion of intestinal Prevotella copri correlates with enhanced susceptibility to arthritis. Elife 2013;2:e01202. https://doi.org/10.7554/eLife.01202.
[26] Konig MF, Abusleme L, Reinholdt J, Palmer RJ, Teles RP, Sampson K, et al. *Aggregatibacter actinomycetemcomitans*-induced hypercitrullination links periodontal infection to autoimmunity in rheumatoid arthritis. Sci Transl Med 2016;8:369ra176. https://doi.org/10.1126/scitranslmed.aaj1921.
[27] Ito HO, Shuto T, Takada H, Koga T, Aida Y, Hirata M, et al. Lipopolysaccharides from *Porphyromonas gingivalis, Prevotella intermedia* and *Actinobacillus actinomycetemcomitans* promote osteoclastic differentiation in vitro. Arch Oral Biol 1996;41:439–44. https://doi.org/10.1016/0003-9969(96)00002-7.
[28] Xiao E, Mattos M, Vieira GHA, Chen S, Corrêa JD, Wu Y, et al. Diabetes enhances IL-17 expression and alters the oral microbiome to increase its pathogenicity. Cell Host Microbe 2017;22:120–128.e4. https://doi.org/10.1016/j.chom.2017.06.014.
[29] Williams DW, Greenwell-Wild T, Brenchley L, Dutzan N, Overmiller A, Sawaya AP, et al. Human oral mucosa cell atlas reveals a stromal-neutrophil axis regulating tissue immunity. Cell 2021. https://doi.org/10.1016/j.cell.2021.05.013.
[30] Qin N, Yang F, Li A, Prifti E, Chen Y, Shao L, et al. Alterations of the human gut microbiome in liver cirrhosis. Nature 2014;513:59–64. https://doi.org/10.1038/nature13568.
[31] Bajaj JS, Betrapally NS, Hylemon PB, Heuman DM, Daita K, White MB, et al. Salivary microbiota reflects changes in gut microbiota in cirrhosis with hepatic encephalopathy. Hepatology 2015;62:1260–71. https://doi.org/10.1002/hep.27819.
[32] Bajaj JS, Acharya C, Fagan A, White MB, Gavis E, Heuman DM, et al. Proton pump inhibitor initiation and withdrawal affects gut microbiota and readmission risk in cirrhosis. Am J Gastroenterol 2018;113:1177–86. https://doi.org/10.1038/s41395-018-0085-9.
[33] Bajaj JS, Cox IJ, Betrapally NS, Heuman DM, Schubert ML, Ratneswaran M, et al. Systems biology analysis of omeprazole therapy in cirrhosis demonstrates significant shifts in gut microbiota composition and function. Am J Physiol Gastrointest Liver Physiol 2014;307:G951–7. https://doi.org/10.1152/ajpgi.00268.2014.
[34] Bajaj JS, Matin P, White MB, Fagan A, Golob Deeb J, Acharya C, et al. Periodontal therapy favorably modulates the oral-gut-hepatic axis in cirrhosis. Am J Physiol Liver Physiol 2018. https://doi.org/10.1152/ajpgi.00230.2018.ajpgi.00230.2018.
[35] Chhibber-Goel J, Singhal V, Bhowmik D, Vivek R, Parakh N, Bhargava B, et al. Linkages between oral commensal bacteria and atherosclerotic plaques in coronary artery disease patients. NPJ Biofilms Microbiomes 2016;2:7. https://doi.org/10.1038/s41522-016-0009-7.
[36] Rath SK, Mukherjee M, Kaushik R, Sen S, Kumar M. Periodontal pathogens in atheromatous plaque. Indian J Pathol Microbiol 2014;57:259–64. https://doi.org/10.4103/0377-4929.134704.
[37] Figuero E, Sánchez-Beltrán M, Cuesta-Frechoso S, Tejerina JM, del Castro JA, Gutiérrez JM, et al. Detection of periodontal bacteria in atheromatous plaque by nested polymerase chain reaction. J Periodontol 2011;82:1469–77. https://doi.org/10.1902/jop.2011.100719.
[38] Ao M, Miyauchi M, Inubushi T, Kitagawa M, Furusho H, Ando T, et al. Infection with *Porphyromonas gingivalis* exacerbates endothelial injury in obese mice. PLoS One 2014;9. https://doi.org/10.1371/journal.pone.0110519, e110519.

[39] Bartova J, Sommerova P, Lyuya-Mi Y, Mysak J, Prochazkova J, Duskova J, et al. Periodontitis as a risk factor of atherosclerosis. J Immunol Res 2014;2014:1–9. https://doi.org/10.1155/2014/636893.

[40] Ford PJ, Gemmell E, Chan A, Carter CL, Walker PJ, Bird PS, et al. Inflammation, heat shock proteins and periodontal pathogens in atherosclerosis: an immunohistologic study. Oral Microbiol Immunol 2006;21:206–11. https://doi.org/10.1111/j.1399-302X.2006.00276.x.

[41] Taylor-Robinson D, Aduse-Opoku J, Sayed P, Slaney JM, Thomas BJ, Curtis MA. Oro-dental bacteria in various atherosclerotic arteries. Eur J Clin Microbiol Infect Dis 2002;21:755–7. https://doi.org/10.1007/s10096-002-0810-5.

[42] Haraszthy VI, Zambon JJ, Trevisan M, Zeid M, Genco RJ. Identification of periodontal pathogens in atheromatous plaques. J Periodontol 2000;71:1554–60. https://doi.org/10.1902/jop.2000.71.10.1554.

[43] Nakano K, Inaba H, Nomura R, Nemoto H, Takeda M, Yoshioka H, et al. Detection of cariogenic Streptococcus mutans in extirpated heart valve and atheromatous plaque specimens. J Clin Microbiol 2006;44:3313–7. https://doi.org/10.1128/JCM.00377-06.

[44] Teles R, Wang C-Y. Mechanisms involved in the association between periodontal diseases and cardiovascular disease. Oral Dis 2011;17:450–61. https://doi.org/10.1111/j.1601-0825.2010.01784.x.

[45] Gaetti-Jardim E, Marcelino SL, Feitosa ACR, Romito GA, Avila-Campos MJ. Quantitative detection of periodontopathic bacteria in atherosclerotic plaques from coronary arteries. J Med Microbiol 2009;58:1568–75. https://doi.org/10.1099/jmm.0.013383-0.

[46] Mahendra J, Mahendra L, Kurian V, Jaishankar K, Mythilli R. 16S rRNA-based detection of oral pathogens in coronary atherosclerotic plaque. Indian J Dent Res 2010;21:248. https://doi.org/10.4103/0970-9290.66649.

[47] Ishihara K, Nabuchi A, Ito R, Miyachi K, Kuramitsu HK, Okuda K. Correlation between detection rates of periodontopathic bacterial DNA in coronary stenotic artery plaque [corrected] and in dental plaque samples. J Clin Microbiol 2004;42:1313–5. https://doi.org/10.1128/JCM.42.3.1313-1315.2004.

[48] Kuo C, Campbell LA. Is infection with *Chlamydia pneumoniae* a causative agent in atherosclerosis? Mol Med Today 1998;4:426–30. https://doi.org/10.1016/s1357-4310(98)01351-3.

[49] Ott SJ, El Mokhtari NE, Musfeldt M, Hellmig S, Freitag S, Rehman A, et al. Detection of diverse bacterial signatures in atherosclerotic lesions of patients with coronary heart disease. Circulation 2006;113:929–37. https://doi.org/10.1161/CIRCULATIONAHA.105.579979.

[50] Schumacher A, Seljeflot I, Lerkerød AB, Sommervoll L, Otterstad JE, Arnesen H. Does infection with *Chlamydia pneumoniae* and/or *Helicobacter pylori* increase the expression of endothelial cell adhesion molecules in humans? Clin Microbiol Infect 2002;8:654–61. https://doi.org/10.1046/j.1469-0691.2002.00439.x.

[51] Farsak B, Yildirir A, Akyön Y, Pinar A, Oç M, Böke E, et al. Detection of *Chlamydia pneumoniae* and *Helicobacter pylori* DNA in human atherosclerotic plaques by PCR. J Clin Microbiol 2000;38:4408–11. https://doi.org/10.1128/JCM.38.12.4408-4411.2000.

[52] Cochrane M, Pospischil A, Walker P, Gibbs H, Timms P. Distribution of *Chlamydia pneumoniae* DNA in atherosclerotic carotid arteries: significance for sampling procedures. J Clin Microbiol 2003;41:1454–7. https://doi.org/10.1128/JCM.41.4.1454-1457.2003.

[53] Jackson LA, Campbell LA, Kuo CC, Rodriguez DI, Lee A, Grayston JT. Isolation of *Chlamydia pneumoniae* from a carotid endarterectomy specimen. J Infect Dis 1997;176:292–5. https://doi.org/10.1086/517270.

[54] Muhlestein JB, Hammond EH, Carlquist JF, Radicke E, Thomson MJ, Karagounis LA, et al. Increased incidence of Chlamydia species within the coronary arteries of patients with symptomatic atherosclerotic versus other forms of cardiovascular disease. J Am Coll Cardiol 1996;27:1555–61. https://doi.org/10.1016/0735-1097(96)00055-1.
[55] Dobrilovic N, Vadlamani L, Meyer M, Wright CB. Chlamydia pneumoniae in atherosclerotic carotid artery plaques: high prevalence among heavy smokers. Am Surg 2001;67:589–93.
[56] Oshima T, Ozono R, Yano Y, Oishi Y, Teragawa H, Higashi Y, et al. Association of *Helicobacter pylori* infection with systemic inflammation and endothelial dysfunction in healthy male subjects. J Am Coll Cardiol 2005;45:1219–22. https://doi.org/10.1016/j.jacc.2005.01.019.
[57] Mahendra J, Mahendra L, Nagarajan A, Mathew K. Prevalence of eight putative periodontal pathogens in atherosclerotic plaque of coronary artery disease patients and comparing them with noncardiac subjects: a case-control study. Indian J Dent Res 2015;26:189. https://doi.org/10.4103/0970-9290.159164.
[58] Okuda K, Kato T, Ishihara K. Involvement of periodontopathic biofilm in vascular diseases. Oral Dis 2004;10:5–12. https://doi.org/10.1046/j.1354-523x.2003.00979.x.
[59] Mahendra J, Mahendra L, Kurian VM, Jaishankar K, Mythilli R. Prevalence of periodontal pathogens in coronary atherosclerotic plaque of patients undergoing coronary artery bypass graft surgery. J Oral Maxillofac Surg 2009;8:108–13. https://doi.org/10.1007/s12663-009-0028-5.
[60] Serra e Silva Filho W, Casarin RCV, Nicolela EL, Passos HM, Sallum AW, Gonçalves RB. Microbial diversity similarities in periodontal pockets and atheromatous plaques of cardiovascular disease patients. PLoS One 2014;9. https://doi.org/10.1371/journal.pone.0109761, e109761.
[61] Rafferty B, Dolgilevich S, Kalachikov S, Morozova I, Ju J, Whittier S, et al. Cultivation of *Enterobacter hormaechei* from human atherosclerotic tissue. J Atheroscler Thromb 2011;18:72–81. https://doi.org/10.5551/jat.5207.
[62] Latronico M, Segantini A, Cavallini F, Mascolo A, Garbarino F, Bondanza S, et al. Periodontal disease and coronary heart disease: an epidemiological and microbiological study. New Microbiol 2007;30:221–8.
[63] Moore C, Addison D, Wilson JM, Zeluff B. First case of *Fusobacterium necrophorum* endocarditis to have presented after the 2nd decade of life. Tex Heart Inst J 2013;40:449–52.
[64] Samant JS, Peacock JE. Fusobacterium necrophorum endocarditis case report and review of the literature. Diagn Microbiol Infect Dis 2011;69:192–5. https://doi.org/10.1016/j.diagmicrobio.2010.09.014.
[65] Stuart G, Wren C. Endocarditis with acute mitral regurgitation caused by *Fusobacterium necrophorum*. Pediatr Cardiol 1992;13:230–2. https://doi.org/10.1007/BF00838782.
[66] Ameriso SF, Fridman EA, Leiguarda RC, Sevlever GE. Detection of *Helicobacter pylori* in human carotid atherosclerotic plaques. Stroke 2001;32:385–91. https://doi.org/10.1161/01.str.32.2.385.
[67] Martínez Torres A, Martínez GM. *Helicobacter pylori*: ¿un nuevo factor de riesgo cardiovascular? Rev Española Cardiol 2002;55:652–6. https://doi.org/10.1016/S0300-8932(02)76673-6.
[68] Momiyama Y, Ohmori R, Taniguchi H, Nakamura H, Ohsuzu F. Association of *Mycoplasma pneumoniae* infection with coronary artery disease and its interaction with chlamydial infection. Atherosclerosis 2004;176:139–44. https://doi.org/10.1016/j.atherosclerosis.2004.04.019.
[69] Higuchi-dos-Santos MH, Pierri H, de Higuchi ML, Nussbacher A, Palomino S, Sambiase NV, et al. *Chlamydia pneumoniae* e *Mycoplasma pneumoniae* nos nódulos de calcificação da estenose da valva aórtica. Arq Bras Cardiol 2005;84. https://doi.org/10.1590/S0066-782X2005000600002.

[70] Kong H-J, Choi K-K, Park S-H, Lee J-Y, Choi G-W. Gene expression of human coronary artery endothelial cells in response to Porphyromonas endodontalis invasion. J Korean Acad Conserv Dent 2009;34:537. https://doi.org/10.5395/JKACD.2009.34.6.537.
[71] Toyofuku T, Inoue Y, Kurihara N, Kudo T, Jibiki M, Sugano N, et al. Differential detection rate of periodontopathic bacteria in atherosclerosis. Surg Today 2011;41:1395–400. https://doi.org/10.1007/s00595-010-4496-5.
[72] Curran SA, Hollan I, Erridge C, Lappin DF, Murray CA, Sturfelt G, et al. Bacteria in the adventitia of cardiovascular disease patients with and without rheumatoid arthritis. PLoS One 2014;9. https://doi.org/10.1371/journal.pone.0098627, e98627.
[73] Igari K, Kudo T, Toyofuku T, Inoue Y, Iwai T. Association between periodontitis and the development of systemic diseases. Oral Biol Dent 2014;2:4. https://doi.org/10.7243/2053-5775-2-4.
[74] Hans M, Madaan HV. Epithelial antimicrobial peptides: guardian of the oral cavity. Int J Pept 2014;2014:370297. https://doi.org/10.1155/2014/370297.
[75] Koren O, Spor A, Felin J, Fak F, Stombaugh J, Tremaroli V, et al. Human oral, gut, and plaque microbiota in patients with atherosclerosis. Proc Natl Acad Sci U S A 2011;108:4592–8. https://doi.org/10.1073/pnas.1011383107.
[76] Ismail F, Baetzner C, Heuer W, Stumpp N, Eberhard J, Winkel A, et al. 16S rDNA-based metagenomic analysis of human oral plaque microbiota in patients with atherosclerosis and healthy controls. Indian J Med Microbiol 2012;30:462–6. https://doi.org/10.4103/0255-0857.103771.
[77] Mark Welch JL, Ramírez-Puebla ST, Borisy GG. Oral microbiome geography: micron-scale habitat and niche. Cell Host Microbe 2020;28:160–8. https://doi.org/10.1016/j.chom.2020.07.009.
[78] Zhu J. Over 50,000 metagenomically assembled draft genomes for the human oral microbiome reveal new taxa and a male-specific bacterium; 2021. p. 2790.
[79] Jie Z, Xia H, Zhong S-L, Feng Q, Li S, Liang S, et al. The gut microbiome in atherosclerotic cardiovascular disease. Nat Commun 2017;8:845. https://doi.org/10.1038/s41467-017-00900-1.
[80] Zhu W, Gregory JC, Org E, Buffa JA, Gupta N, Wang Z, et al. Gut microbial metabolite TMAO enhances platelet hyperreactivity and thrombosis risk. Cell 2016;165:111–24. https://doi.org/10.1016/j.cell.2016.02.011.
[81] Fardini Y, Chung P, Dumm R, Joshi N, Han YW. Transmission of diverse oral bacteria to murine placenta: evidence for the oral microbiome as a potential source of intrauterine infection. Infect Immun 2010;78:1789–96. https://doi.org/10.1128/IAI.01395-09.
[82] Chen X, Li P, Liu M, Zheng H, He Y, Chen M-XX, et al. Gut dysbiosis induces the development of pre-eclampsia through bacterial translocation. Gut 2020;69:513–22. https://doi.org/10.1136/gutjnl-2019-319101.
[83] Ravel J, Gajer P, Abdo Z, Schneider GM, Koenig SSK, Mcculle SL, et al. Vaginal microbiome of reproductive-age women. Proc Natl Acad Sci U S A 2010;108:4680–7. http://www.pnas.org/cgi/doi/10.1073/pnas.1002611107.
[84] Fredricks DN, Fiedler TL, Marrazzo JM. Molecular identification of bacteria associated with bacterial vaginosis. N Engl J Med 2005;353:1899–911. https://doi.org/10.1056/NEJMoa043802.
[85] Jie Z, Chen C, Hao L, Li F, Song L, Zhang X, et al. Life history recorded in the vagino-cervical microbiome along with multi-omics. Genomics Proteomics Bioinformatics 2021. https://doi.org/10.1016/j.gpb.2021.01.005.
[86] Abdelmaksoud AA, Girerd PH, Garcia EM, Brooks JP, Leftwich LM, Sheth NU, et al. Association between statin use, the vaginal microbiome, and *Gardnerella vaginalis* vaginolysin-mediated cytotoxicity. PLoS One 2017;12. https://doi.org/10.1371/journal.pone.0183765, e0183765.

[87] Qin J, Li R, Raes J, Arumugam M, Burgdorf KSS, Manichanh C, et al. A human gut microbial gene catalogue established by metagenomic sequencing. Nature 2010;464:59–65. https://doi.org/10.1038/nature08821.
[88] The Human Microbiome Project Consortium. Structure, function and diversity of the healthy human microbiome. Nature 2012;486:207–14. https://doi.org/10.1038/nature11234.
[89] Byrd AL, Belkaid Y, Segre JA. The human skin microbiome. Nat Rev Microbiol 2018;16:143–55. https://doi.org/10.1038/nrmicro.2017.157.
[90] Rao C, Coyte KZ, Bainter W, Geha RS, Martin CR, Rakoff-Nahoum S. Multi-kingdom ecological drivers of microbiota assembly in preterm infants. Nature 2021. https://doi.org/10.1038/s41586-021-03241-8.
[91] Buffie CG, Bucci V, Stein RR, McKenney PT, Ling L, Gobourne A, et al. Precision microbiome reconstitution restores bile acid mediated resistance to Clostridium difficile. Nature 2014;517:205–8. https://doi.org/10.1038/nature13828.
[92] Hryckowian AJ, Van Treuren W, Smits SA, Davis NM, Gardner JO, Bouley DM, et al. Microbiota-accessible carbohydrates suppress *Clostridium difficile* infection in a murine model. Nat Microbiol 2018;3:662–9. https://doi.org/10.1038/s41564-018-0150-6.
[93] Zuo T, Wong SH, Lam LYK, Lui R, Cheung K, Tang W, et al. Bacteriophage transfer during fecal microbiota transplantation is associated with treatment response in *Clostridium difficile* infection. Gut 2017. https://doi.org/10.1136/gutjnl-2017-313952.
[94] Jang C, Hui S, Zeng X, Cowan AJ, Wang L, Chen L, et al. Metabolite exchange between mammalian organs quantified in pigs. Cell Metab 2019;1–13. https://doi.org/10.1016/j.cmet.2019.06.002.
[95] Rappez L, Stadler M, Triana S, Phapale P, Heikenwalder M, Alexandrov T. Spatial single-cell profiling of intracellular metabolomes in situ. BioRxiv 2019. https://doi.org/10.1101/510222.
[96] Liu X, Tong X, Zou Y, Lin X, Zhao H, Tian L, et al. Inter-determination of blood metabolite levels and gut microbiome supported by Mendelian randomization. BioRxiv 2020. https://doi.org/10.1101/2020.06.30.181438. 2020.06.30.
[97] Kalaora S, Nagler A, Nejman D, Alon M, Barbolin C, Barnea E, et al. Identification of bacteria-derived HLA-bound peptides in melanoma. Nature 2021;592:138–43. https://doi.org/10.1038/s41586-021-03368-8.
[98] Barrett M, Hand CK, Shanahan F, Murphy T, O'Toole PW. Mutagenesis by microbe: the role of the microbiota in shaping the cancer genome. Trends Cancer 2020;6:277–87. https://doi.org/10.1016/j.trecan.2020.01.019.
[99] Sivaguru M, Saw JJ, Wilson EM, Lieske JC, Krambeck AE, Williams JC, et al. Human kidney stones: a natural record of universal biomineralization. Nat Rev Urol 2021;2021:1–29. https://doi.org/10.1038/s41585-021-00469-x.

5

The evolving microbial taxonomy

5.1 Approaching a closed reference set for routine applications

Direct assignment of metagenomic sequencing reads or assembled genes to an existing set of reference sequences would be a quick way of analyzing the data that could serve clinical needs. 16S rRNA gene amplicon sequencing is currently not a closed reference approach (Fig. 5.1) [2,3]; sequences are clustered into operational taxonomic units (OTUs), and an inferred unique "seed sequence" was then used to map to families, genera or species that are already in the database, leaving a varying portion of unannotable OTUs. Metagenomic shotgun sequencing, which can be completely free of PCR amplification biases [4], further includes eukaryotes and viruses and could map to any part of the microbial genomes. According to Dr. Junjie Qin, when he asked Mr. Shenghui Li in the early 2010s to show taxonomic information for each gene that clustered together according to covariations in abundance among hundreds of samples, it turned out that the genes in the same cluster belonged to the same bacterial species [5]. The underlying physical linkage in the microbial genome that was captured by the covariations should be at the strain level, but half of the clusters were unknown species already, and there was much to be improved both computationally and experimentally.

Nowadays, contigs assembled from a single metagenomic sample could be binned according to sequence composition such as tetranucleotide frequency, but covariations in multiple samples could refine the contigs' coverage information and add to the number of assembled genomes in medium to high quality. The metagenomic assembly algorithms are confused by similar sequences from related microbes [6,7], both in the assembly stage (e.g., sequences of two strains in the same sample) and in the binning of contigs into genomes. Reference genome dataset for the human fecal microbiome was constructed from both cultured isolates and metagenome-assembled genomes (MAGs) [8], then the genomes will be dereplicated on species (or strain) level for genome-resolved analyses (Fig. 5.1).

Investigating Human Diseases with the Microbiome: Metagenomics Bench to Bedside.
https://doi.org/10.1016/B978-0-323-91369-0.00004-2

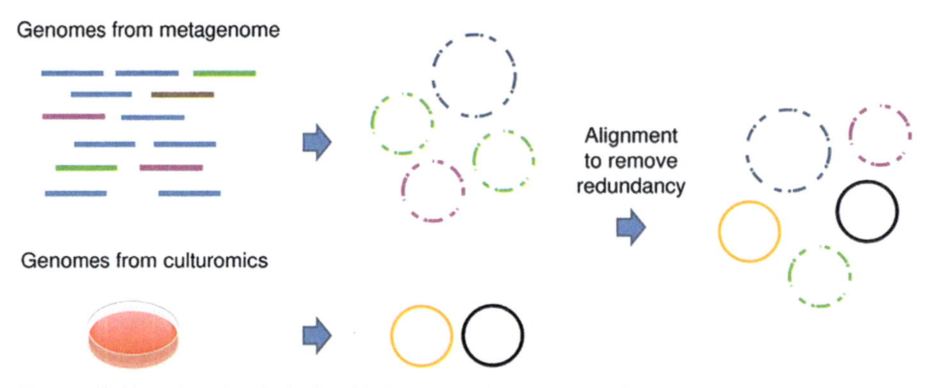

Fig. 5.1 Unbiased retrieval of microbial genomes in metagenomic samples to approach the complete representation of the community. Genomes assembled from high-throughput metagenomic shotgun sequencing data are not yet perfect [1], and are shown in *dashed lines*. Credit: Huijue Jia, Jie Zhu.

The human gut microbiome at the phylum level mostly consists of Firmicutes, Bacteroidetes, Proteobacteria, Actinobacteria, and Fusobacteria, plus Fibrobacteres, Spirochaetes, Lentisphaera, and more mysterious phyla such as Deinococcus-Thermus, Cyanobacteria, Chloroflexi (Fig. 5.2, e.g., db.cngb.org/microbiome/). For the oral microbiome, we see more phyla from the Candidas Phyla Radiation (CPR) (Figs. 5.2 and 5.3), the most famous being the TM7x genus (Saccharibacteria phylum, formerly TM7) which is at least an episymbiont for *Actinomyces odontolyticus* [11–13]. A recent study supports the idea that CPRs evolved through genome reduction from a free-living form, instead of being at the basal place shown in Fig. 5.2 [14], consistent with their obligate symbiont lifestyle and the size constraints [15,16]. Besides the methanogens (Chapter 2, Box 2.4), the human skin contains archaea called *Thaumarchaeota* that can oxidize ammonia to produce nitrite [17].

Due to the traditional view that much of the microbiome is "unculturable" and the difficulty in obtaining the optimal condition, some of the abundant genera and species in the human microbiome got more than 1 draft genomes only recently [18–20]. Comprehensive culturomics for the lungs, for example, would need to take into account the temperature, gas, and pH gradients at the different positions (Chapter 3, Fig. 3.6). The match between culturomics and high-throughput sequencing may have an even longer way to go for fungi (Fig. 5.4), not to mention viruses. While metagenomic sequencing should be able to detect everything in a sample (Chapter 1, Fig. 1.2), it is possible that some low-abundant taxa can be better picked up by specific culturing, which could be facilitated by genomic and other omics information [10,18,21,22].

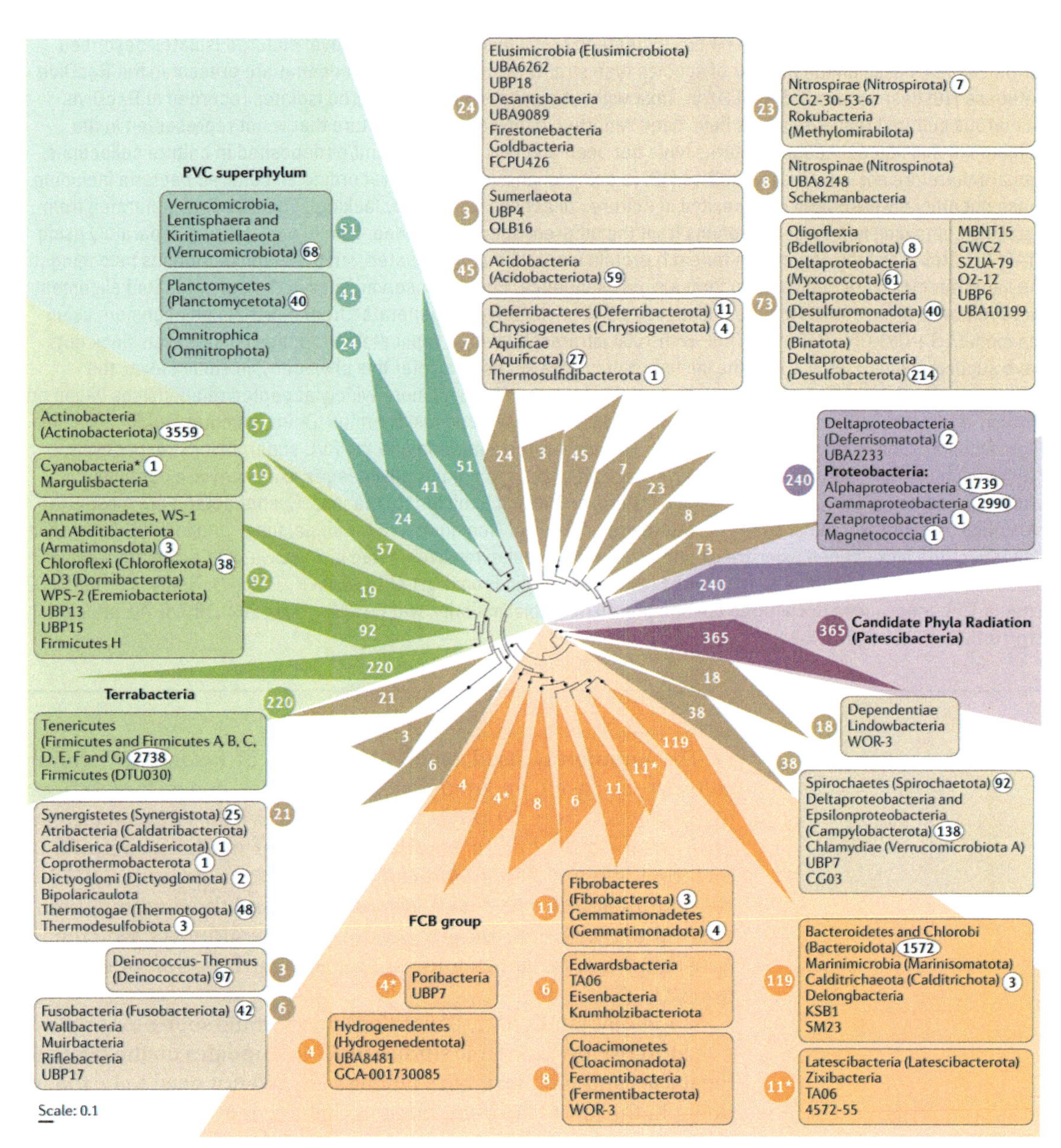

Fig. 5.2 Number of isolated genomes for each bacterial phylum. Without distinguishing between environments, hosts, and body sites, cultured bacteria are currently biased toward Bacteroidetes, Proteobacteria, Firmicutes, and Actinobacteria. A phylogenetic species tree for bacteria, inferred from concatenated alignments of a minimum of 5 out of a total 15 ribosomal proteins per species, encoded by 1541 bacterial genomes that were obtained from the Genome Taxonomy Database [9]. Numbers in *white* font in colored circles are the number of individual taxa in each collapsed clade and are also used to connect corresponding taxa names to clades. Numbers in *black* font in *white*

(Continued)

Fig. 5.2, cont'd ellipses next to taxa names indicate the total number of species level cultured isolates described for those taxa, based on the number of species type strains assigned to each clade that are present in the BacDive database [10] (last accessed 6 April 2020). Taxa without numbers have no cultured isolates recorded in BacDive. Numerous cultured representatives have been reported in the scientific literature that is not represented in the numbers in this figure, because cultures have not been officially described and/or deposited in culture collections, and are therefore not included in BacDive [10] (a comprehensive database recording all cultured bacteria including those not officially described or deposited in culture collections is currently lacking). The tree was generated from datasets containing homologous proteins from the different species included, which were aligned separately using MAFFT (L-INS-i) and the alignments for each protein were then concatenated, such that those proteins belonging to the same species were combined to form a single sequence. Poorly conserved sites in the concatenated alignment were removed using trimAl with the option—gt 0.5. A phylogeny was generated from this trimmed alignment using the model LG + C60 + F + R10 in IQ-TREE with 1000 ultrafast bootstrap replicates. Branches labeled with black dots have support values $\geq$ 95%. Given the limited protein data set used to infer this phylogeny, in some cases, the deeper relationships between some species or groups may not reflect more widely accepted relationships based on more in-depth and better-supported analyses. Particularly, Deinococcus-Thermus (Deinococcota) and Chlamydiae (Verrucomicrobiota A) do not group with other lineages of Terrabacteria and the PVC superphylum, respectively. *Although numerous cultured representatives for numerous cyanobacterial lineages exist, they are particularly underrepresented in BacDive. Unlike most bacteria, and owing to historical reasons, Cyanobacteria are mostly classified using the Botanical code (i.e., International Code of Nomenclature for algae, fungi, and plants). As a result, Cyanobacteria lack defined type strains and are therefore not extensively listed in BacDive, and a comprehensive database of existing Cyanobacteria cultures is lacking. Credit: From Fig.1 of Lewis WH, Tahon G, Geesink P, Sousa DZ, Ettema TJG. Innovations to culturing the uncultured microbial majority. Nat Rev Microbiol 2021;19:225–40. https://doi.org/10.1038/s41579-020-00458-8.

The commonly used MetaPhlAn series of taxonomic profiling software is based on predetermined marker genes for each taxonomic level, from 50 phyla all the way down to 7677 species and more strains [23–25]. To establish the set of markers, MetaPhlAn2 included 300 archaea genomes, 12,926 bacteria genomes, 3565 virus genomes, and 112 eukaryote genomes, and the total number increased to 99.2k high-quality genomes in MetaPhlAn3, which has a new set of marker genes and estimates the proportion of unknown taxa [23,26].

Hopefully, family and genus information would soon be more accurate for most metagenomic studies [6,7], and updates on the reference databases and bioinformatics pipelines would not need to be too frequent in the future. As mentioned in Chapter 1, even the Neanderthal oral and fecal microbiome showed many of the same genera we have, and we would love to try to assemble the microbial genomes for such rare samples [27–30], however fragmented. For specific applications (Chapters 7 and 8), a smaller reference database could mean faster and less confusing results.

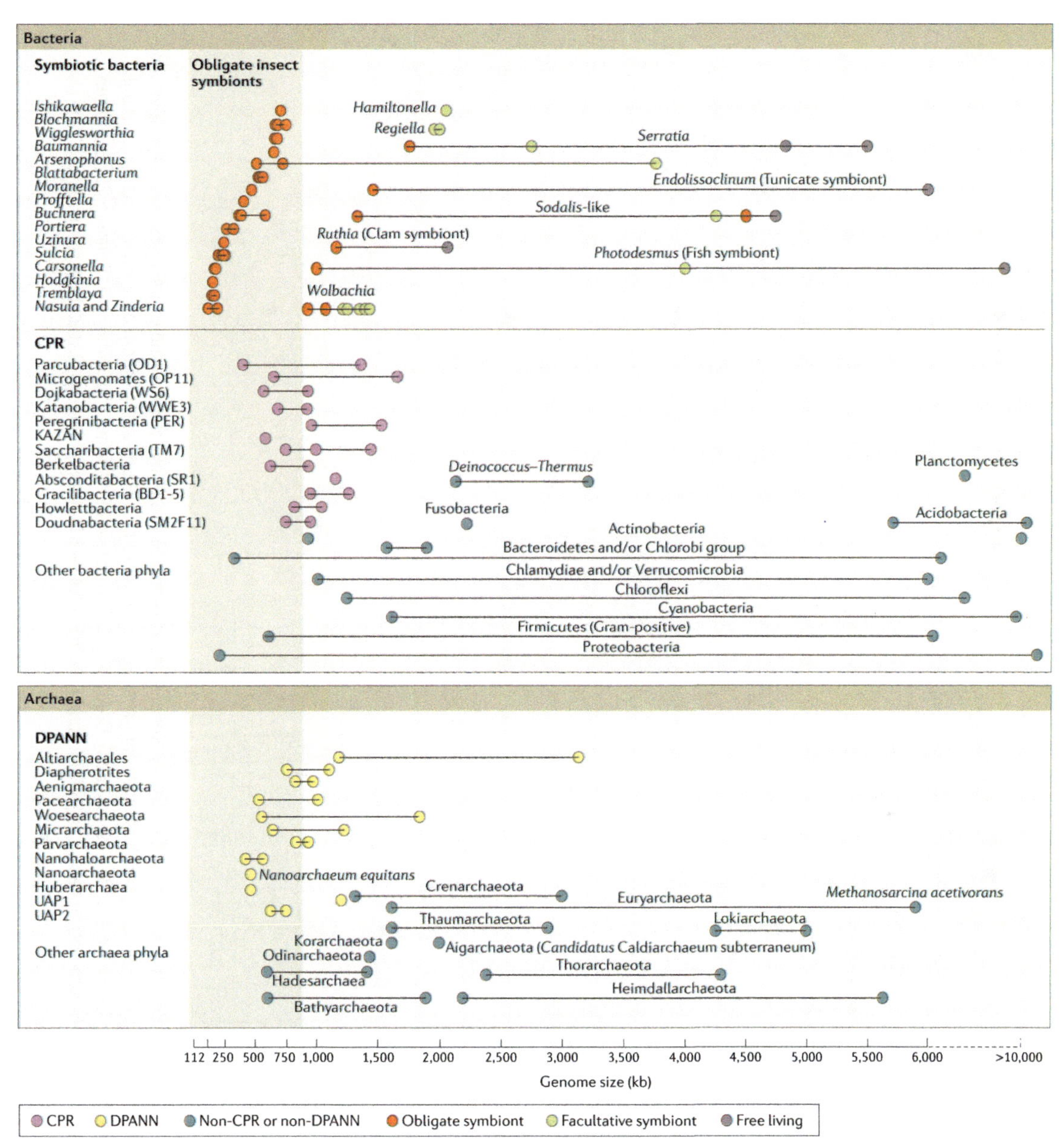

Fig. 5.3 The size ranges for CPR (candidate phyla radiation) bacteria and DPANN ("Candidatus Diapherotrites," "Candidatus Parvarchaeota," "Candidatus Aenigmarchaeota," Nanoarchaeota, "Candidatus Nanohaloarchaeota," and other lineages) archaea genomes compared with size ranges for the genomes of known bacterial symbionts as well as other bacteria and archaea. The top panel shows data for well-studied bacteria that are obligate symbionts *(orange dots)*, facultative symbionts *(green dots)*, and free-living *(gray dots)*. The middle panel provides information for CPR *(purple dots)*, and the bottom panel provides genome size information for DPANN *(yellow dots)*. The middle and bottom panels also show the size ranges for other bacteria and archaea *(blue dots)*. CPR and DPANN genome sizes overlap with those of obligate symbionts. Credit: Fig. 2 of Castelle CJ, Brown CT, Anantharaman K, Probst AJ, Huang RH, Banfield JF. Biosynthetic capacity, metabolic variety and unusual biology in the CPR and DPANN radiations. Nat Rev Microbiol 2018;16:629–45. https://doi.org/10.1038/s41579-018-0076-2.

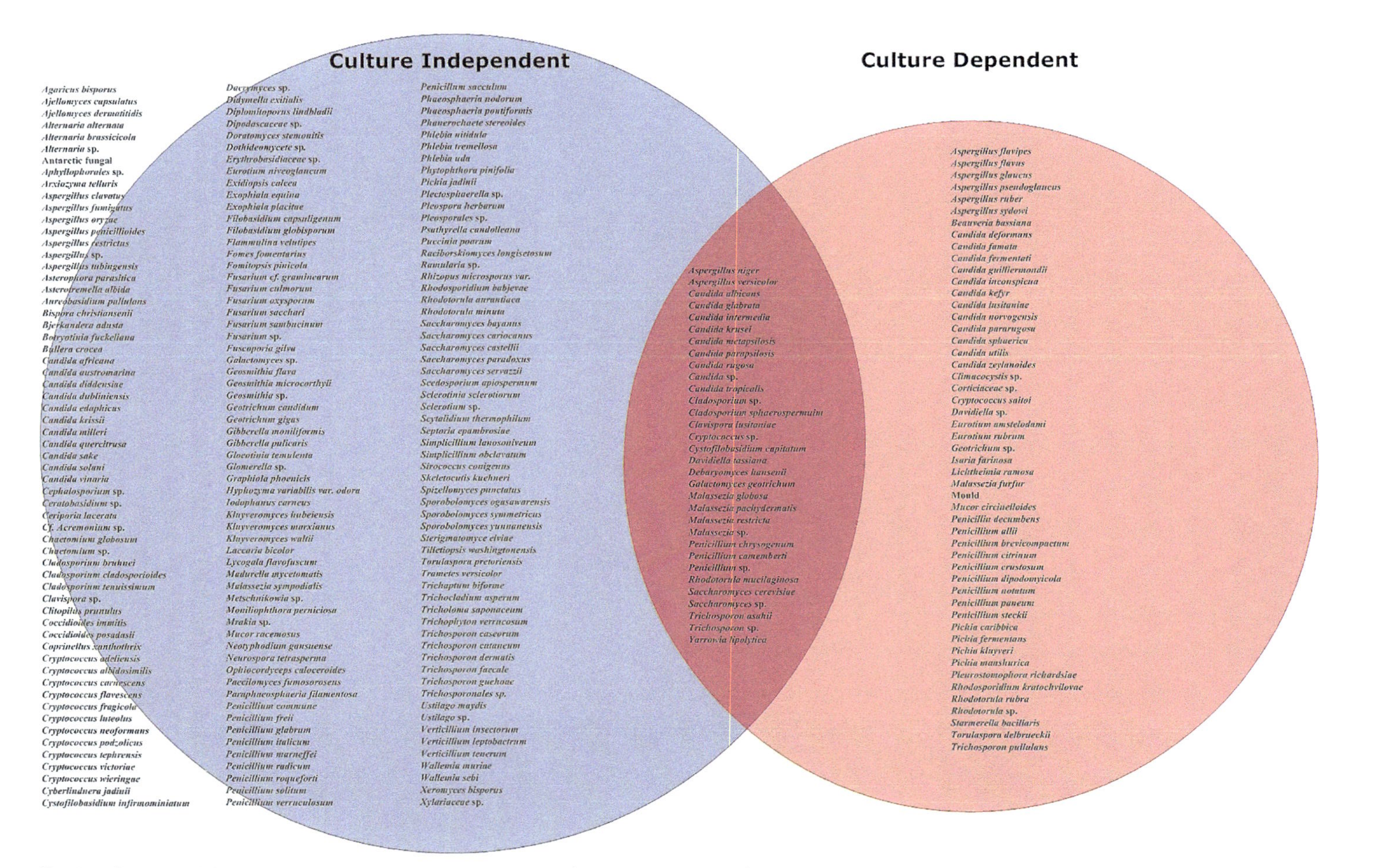

Fig. 5.4 Commonalities and differences between fungal data (at the species level) reported in gut mycobiome studies using culture-dependent and culture-independent methodologies. This Venn diagram highlights fungal species detected by culture-independent only, culture-dependent only, and species that have been detected by both methods (intersection). Credit: Fig. 2 of Huseyin CE, O'Toole PW, Cotter PD, Scanlan PD. Forgotten fungi—the gut mycobiome in human health and disease. FEMS Microbiol Rev 2017;41:479–511. https://doi.org/10.1093/femsre/fuw047.

Worked sample 5.1

Below are the relative abundances of microbial species in one nasal sample (Ju et al. unpublished) using existing softwares, including the k-mer (strings of k nucleotides in the genome)-based Kraken2+Bracken [31], and the marker gene-based mOTU2 [24], MetaPhlAn2 and MetaPhlAn3. Please get a sense of the different methods and databases, their sensitivity, and accuracy [6,7]. Would you prefer genus-level results instead?

For each method, try to rank the most abundant species 1, 2, 3, ... , 30, ... Do the different methods look more consistent now?

With more samples, do you see why the most often used correlation coefficient for metagenomic data (between taxa, which would be compositional, i.e., constrained by a total of 1 (Chapter 3), and between a taxon and another phenotype or a taxon at a different body site) is rank-based, such as Spearman's rho (e.g., Chapter 2, Fig. 2.14, the *Bacteroides* vs *Prevotella* is preserved after adjustment for compositional effect [32]), instead of the faster and not so appropriate Pearson's correlation (including SparCC which takes into account the compositional nature of relative abundance data [33])?

What would be the taxa you care more about for your study of the human microbiome, and do you think specific improvements would be needed? Can you guess which method would more easily accommodate newly assembled genomes?

Broad category	Species	Kraken2+ Bracken	mOTU2	MetaPhlAn2	MetaPhlAn3
Bacteria	*Acinetobacter baumannii*	0.157951	0	0	0
Bacteria	*Anaerococcus* species incertae sedis (uncertain placement of species) [meta mOTU v25 12712]	0	0.454865	0	0
Bacteria	Bacilli sp. [ref mOTU v25 00344]	0	0.239723	0	0
Bacteria	*Bacillus cereus*	0.130481	0	0	0
Bacteria	Bacteria sp. [ref mOTU v25 00259]	0	0.155647	0	0
Bacteria	Bacteria sp. [ref mOTU v25 00964]	0	0.131472	0	0
Bacteria	*Brachybacterium paraconglomeratum*	0	0.116516	0	0
Bacteria	*Corynebacterium accolens*	0	24.27697	9.31315	32.04989
Bacteria	*Corynebacterium ammoniagenes*	0.1528	0	0	0
Bacteria	*Corynebacterium aurimucosum*	0.448099	0.490033	0	0
Bacteria	*Corynebacterium camporealensis*	0.458401	0	0	0
Bacteria	*Corynebacterium casei*	0.357106	0	0	0
Bacteria	*Corynebacterium diphtheriae*	0.559695	0	0	0
Bacteria	*Corynebacterium flavescens*	0.375992	0	0	0
Bacteria	*Corynebacterium glutamicum*	0.923669	0	0	0
Bacteria	*Corynebacterium jeikeium*	0.243794	0	0	0

Continued

Broad category	Species	Kraken2+ Bracken	mOTU2	MetaPhlAn2	MetaPhlAn3
Bacteria	*Corynebacterium kroppenstedtii*	1.222402	1.712234	2.09802	1.23582
Bacteria	*Corynebacterium minutissimum*	0.293582	0	0	0
Bacteria	*Corynebacterium phocae*	0.1528	0	0	0
Bacteria	*Corynebacterium propinquum*	0	0.421024	2.12244	0
Bacteria	*Corynebacterium pseudogenitalium*	0	0	0.71378	0
Bacteria	*Corynebacterium resistens*	0.104728	0	0	0
Bacteria	*Corynebacterium simulans*	0.882464	0	0	0
Bacteria	*Corynebacterium singulare*	0.336504	0	0	0
Bacteria	*Corynebacterium* sp. [ref mOTU v25 03067]	0	1.811982	0	0
Bacteria	*Corynebacterium* sp. [ref mOTU v25 00802]	0	0.109947	0	0
Bacteria	*Corynebacterium stationis*	0.14765	0	0	0
Bacteria	*Corynebacterium striatum*	1.857638	0	0	0
Bacteria	*Corynebacterium ureicelerivorans*	0.108162	0	0	0
Bacteria	*Cutibacterium* (formerly *Propionibacterium*) *acnes*	12.71332	9.133923	20.45757	11.96932
Viruses	*Propionibacterium* phage BruceLethal	0.157951	0	0	0
Viruses	*Propionibacterium* phage Moyashi	0.243794	0	0	0
Viruses	*Propionibacterium* phage P101A	0	0	5.99864	0
Viruses	*Propionibacterium* phage PA1-14	0.140782	0	0	0
Viruses	*Propionibacterium* phage PHL009	0.255812	0	0	0
Viruses	*Propionibacterium* phage PHL010M04	0.456684	0	0	0
Viruses	*Propionibacterium* phage PHL030	0.441232	0	0	0
Viruses	*Propionibacterium* phage PHL055	0.104728	0	0	0
Viruses	*Propionibacterium* phage PHL070	0.489304	0	0	0
Viruses	*Propionibacterium* phage PHL082	0.118463	0	0	0
Viruses	*Propionibacterium* phage PHL085	0.396594	0	0	0
Viruses	*Propionibacterium* phage PHL116	0.400028	0	0	0
Viruses	*Propionibacterium* phage PHL132	0.209456	0	0	0
Viruses	*Propionibacterium* phage PHL141	0.157951	0	0	0
Viruses	*Propionibacterium* phage PHL152	0.679875	0	0	0
Viruses	*Propionibacterium* phage PHL171	0.127047	0	0	0
Viruses	*Propionibacterium* phage QueenBey	0.427497	0	0	0
Viruses	*Propionibacterium* virus Attacne	0.260962	0	0	0
Viruses	*Propionibacterium* virus Lauchelly	0.496171	0	0	0
Viruses	*Propionibacterium* virus Ouroboros	0.454967	0	0	0
Viruses	*Propionibacterium* virus P100A	0.108162	0	0	0
Viruses	*Propionibacterium* virus PHL071N05	0.199155	0	0	0
Viruses	*Propionibacterium* virus PHL114L00	0.204306	0	0	0
Viruses	*Propionibacterium* virus Pirate	0.338221	0	0	0
Viruses	*Propionibacterium* virus Solid	0.116746	0	0	0

Continued

Broad category	Species	Kraken2 + Bracken	mOTU2	MetaPhlAn2	MetaPhlAn3
Viruses	*Propionibacterium* virus Stormborn	0.257528	0	0	0
Bacteria	*Cutibacterium granulosum*	0.990626	0.557325	0	1.05931
Bacteria	*Erythrobacteraceae* bacterium CCH12-C2	0	0.11621	0	0
Bacteria	*Haemophilus parainfluenzae*	0.111596	0	0.16822	0
Bacteria	*Klebsiella michiganensis/oxytoca*	0	0.124238	0	0
Bacteria	*Klebsiella oxytoca*	0.121897	0	0	0
Bacteria	*Lautropia mirabilis*	0	0.118127	0.16232	0
Bacteria	*Lawsonella clevelandensis*	0.175119	6.701121	0	0
Eukaryota-Fungi	*Malassezia restricta*	0	0	0	15.47571
Eukaryota-Fungi	*Malassezia* species incertae sedis [meta mOTU v25 12989]	0	15.03185	0	0
Bacteria	*Moraxellaceae* sp. [ref mOTU v25 06002]	0	0.109049	0	0
Bacteria	*Morococcus cerebrosus*	0	0.105722	0	0
Bacteria	*Neisseria elongata*	0.582014	0.283177	0.38064	0.42907
Bacteria	*Neisseria macacae*	0	0	0.1394	0.11948
Bacteria	*Neisseria meningitidis*	0.103011	0.231828	0	0
Bacteria	*Neisseria mucosa*	0.255812	0	0	0
Bacteria	*Neisseria sicca*	0.281564	0	0.43785	0.89061
Bacteria	*Neisseria sicca/macacae*	0	0.216081	0	0
Bacteria	*Neisseria* sp. [ref mOTU v25 04798]	0	0.169086	0	0
Bacteria	*Neisseria* sp. HMSC064E01	0	0.295144	0	0
Bacteria	*Neisseria* sp. oral taxon 014	0	0.102625	0	0
Bacteria	*Neisseria* unclassified	0	0	0.60812	0
Bacteria	*Prevotella melaninogenica*	0	0.196181	0	0
Bacteria	*Pseudomonas stutzeri*	0.18027	0.133901	0	0
Bacteria	Sphingomonadales bacterium RIFCSPHIGHO2 01 FULL 65 20	0	0.101038	0	0
Bacteria	*Staphylococcus aureus*	0.479003	0	0	0
Bacteria	*Staphylococcus capitis*	0.20774	0	0	0
Bacteria	*Staphylococcus epidermidis*	39.33489	33.71282	55.81221	35.78187
Bacteria	*Staphylococcus hominis*	0.307317	0	0.22698	0.16856
Viruses	*Staphylococcus* phage StB27	0.259245	0	0	0
Viruses	*Staphylococcus* virus IPLAC1C	0.382859	0	0	0
Viruses	*Staphylococcus* virus SEP9	7.926725	0	0	0
Viruses	*Staphylococcus* virus Sextaec	12.46781	0	0	0
Bacteria	*Streptococcus mitis*	0.173403	0	0	0.15237
Bacteria	*Streptococcus mitis/oralis/pneumoniae*	0	0	0.25495	0
Bacteria	*Streptococcus sanguinis/cristatus*	0	0.106522	0	0
Bacteria	*Streptococcus* sp. [ref mOTU v25 00283]	0	0.412096	0	0

mOTU2 was based on 5232 ref mOTUs which have reference genomes, and 2494 meta mOTUs which were supported by metagenomic data from the human or ocean microbiome.

Worked sample 5.2

Take a sample of the tongue dorsum and a sample of the saliva from the same person for metagenomic shotgun sequencing, or find some published data.

How many species and higher taxa do you get in each sample?

Could you assemble some high-quality genomes, and how different are they between the tongue and the saliva samples?

Is the lactic acid-metabolizing *Veillonella* found in the gut of elite runners [34] different from the *Veillonella* in autoimmune diseases such as rheumatoid arthritis (Section 4.4.1)?

5.2 Sparser data with increasing taxonomic resolution

We talked about statistical practices with metagenomic studies in Chapter 3. While knowing the species (Box 5.1) or strains are important for functional characterizations, finer grains of taxonomy means that more samples would have zero abundance for a taxon. This sparsity is a problem in statistics. There are efforts to distinguish sampling zeros (e.g., not detected due to low sequencing amount) from real absence (structural zeros) [35]. Before better statistical methods are developed for metagenomic data, we are going to see some larger *P*-values with a higher-taxonomic resolution, for which the effective sample size is smaller. For example, the phylum Proteobacteria showed up multiple times in an MR (Mendelian Randomization) analysis of fecal microbiome and plasma metabolites, often more "significant" than the genera or species [36]. Below the genus level, the issue of the same sequencing read mapping to multiple genomes can strongly affect the relative abundance values [6]. So we are not sure whether multiples species or strains are similarly associated with a human gene, or there is finer work to do in the abundance profile, before we try it.

The finer taxonomic resolution would also affect correlations (between taxa, or between a taxon and another omics feature) calculated from multiple samples. A taxon that was ranked as highly abundant (e.g., Worked sample 5.1) would be broken into smaller pieces with more fluctuating relative abundances and may no longer show up in Spearman's correlation or other statistical measures.

To select biomarkers in metagenome-wide association studies (MWAS), machine learning algorithms such as random forest and LASSO (Least absolute shrinkage and selection operator) can handle the sparse (zero-inflated) data [37], and the microbial markers selected by the models are more like to be validated in a different cohort, instead of being overfit to the training set. Moreover, 10- or 5-fold cross-validation is often used with random forest models (RFcv) and run multiple times, i.e., 1/10 of the samples were randomly left out in the training, and used as a test set, the second time another 1/10 of the samples were left out for validation, the third time... (e.g., Refs. [38–40]).

Recent algorithm developments in neural networks (Artificial Intelligence) could potentially better handle the different layers of relationships in metagenomic data, along with other omics, facilitating association analyses, biomarker discovery for diagnosis and prognosis, as well as MAG assembly and functional annotations.

Box 5.1 The species concept for bacteria

Defining species was not only a problem for bacteria or viruses. Charles Darwin wrote in the Origin of Species published in 1859 [43,44]:

"To sum up, I believe that species come to be tolerably well-defined objects, and do not at any one period present an inextricable chaos of varying and intermediate links ...

... if my theory be true, numberless intermediate varieties, linking most closely all the species of the same group together, must assuredly have existed; but the very process of natural selection constantly tends...to exterminate the parent-forms and the intermediate links.

... it will be seen that I look upon the term species, as one arbitrarily given for the sake of convenience to a set of individuals closely resembling each other, and that it does not essentially differ from the term variety, which is given to less distinct and more fluctuating forms.

In short, we shall have to treat species in the same manner as those naturalists treat genera, who admit that genera are merely artificial combinations made for convenience. This may not be a cheering prospect; but we shall at least be freed from the vain search for the undiscovered and undiscoverable essence of the term species."

Alfred Russel Wallace, natural selection's co-discoverer, (he made other comments earlier) later reached a species definition that was more like an ecotype, but is maintained across generations:

"A species ... is a group of living organisms, separated from all other such groups by a set of distinctive character(istic)s, having relations to the environment not identical with those of any other group of organisms, and having the power of continuously reproducing its like"—Wallace [45].

A major difficulty for defining species in prokaryotic organisms is that sexual reproduction cannot be used as a criterion to mark boundaries between species. The microbiome has interestingly be shown to facilitate reproductive isolation, mate discrimination, and hybrid infertility/lethality in insects, thereby contributing to speciation [46].

Also of concern is that the microbial genome may be more plastic. Horizontal gene transfer occurs through mobile elements that can cross taxonomic boundaries, yet does not appear to happen too frequently [16,47]. Sequences of core genes do show a clear boundary at the species level (Fig. 5.5). So, bacteria species have distinct genomic features that would not easily shift into a different species through the accumulation of mutations or through horizontal gene transfer. The core genes also define the metabolic and cell wall traits that are traditionally assayed for in microbiology (e.g., Chapter 1, Fig. 1.10).

Ecotypes, a concept from the niche theory, are not necessarily genetic, and also include heterogeneous gene expression at the single-cell level in a community [48–50]. When not aligned with heritable genomic features, ecotypes are more for functional studies under a variety of conditions and are rather auxiliary for defining microbial species. For species in the human microbiome, functional studies would probably need to include interactions with the host immune system, e.g., antigenic properties can be predicted from the microbial genome if we have first accumulated experimental evidence.

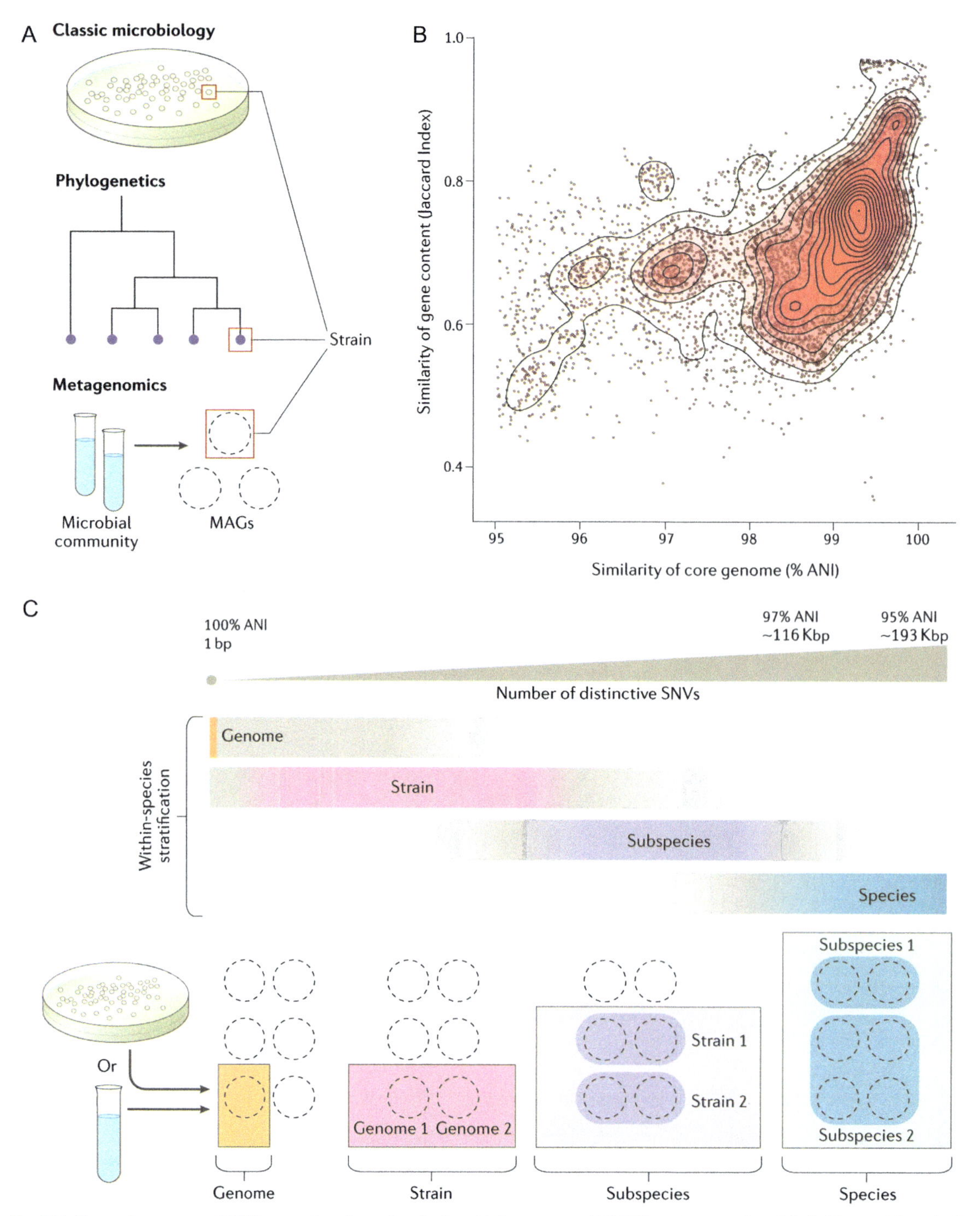

Fig. 5.5 Never the same—Within species diversity of microbial genomes. (A) Different operational definitions of "strain," based on the field of investigation: a cultured isolate in classic microbiology, a leaf node in a phylogenetic tree, and a metagenome-assembled genome (MAG) in metagenomics. (B) Each point is a pairwise comparison of one isolate genome versus all other conspecific isolate genomes. The data are from 155 bacterial species, each with at least 10 sequenced isolate genomes. The opacity of the *red*-colored topographical overlay indicates the density of points. The plot shows the relationship between the similarity of the core genome, measured by average nucleotide identity (ANI), versus the similarity of gene content, measured by the Jaccard Index. Genomes with higher similarity between their core gene sequences tend to have more genes in common (Spearman correlation $R=0.57$, $P<2.2\times10^{-16}$).

(Continued)

Fig. 5.5, cont'd However, a high ANI does not necessarily imply a highly similar gene content, with many genomes with an over 99% core genome ANI having less than 70% of genes in common. Most within-species ANI values are greater than 97%; the few data points below 95% ANI are not shown (83% and 4% of data points, respectively). (C) Spatial distribution of key terminology used to stratify variation within bacterial species, ranging from a single nucleotide variant (SNV) in the whole genome to the species-level threshold (97% ANI). The colored portions of the bars reflect the recommended scope of use for each term, and the *gray* portions indicate the common, often unspecific, scope of use. Broadly speaking, conspecific genomes have identical nucleotides at homologous positions across 97% of their genome (97% ANI), which corresponds to differences in the order of 116,000 SNVs based on an average bacterial genome size of 3.87 Mb. The bottom panel illustrates the hierarchy of these terms, with a species potentially containing multiple subspecies, a subspecies containing multiple strains and a strain containing multiple (nonidentical) genomes. These genomes can be sequenced from cultured isolates or through assembly

Worked sample 5.3

Based on 1267 fecal samples profiled according to a reference gene catalog of 9,879,896 genes (the shorter version of "redundant genes" removed at merging, according to 95% identity), we previously estimated that every two individuals share ~ 1/3 of their gut microbial genes. Each sample contained an average of 762,665 genes and any two samples had in common an average of 250,382 genes (32.8% of 762,665 genes) [41].

At the taxonomic level, and focusing on a particular group of people, what is your current thinking for the number of gut microbial phyla, ..., families, genera, and species shared between two individuals? Could you see different patterns in spore-forming bacteria [42], and bacteria that rely more on vertical transmission (e.g., between mother and infant)? (Dispersal limitation, Chapter 2, Fig. 2.3).

What about the microbiome in other body sites?

5.3 Evolutionary history below the species level

Although reproductive isolation does not work for bacteria (Box 5.1), the boundary for species is clear at the genome level based on core genes (Fig. 5.5). Some genera are eventually split from a older genus name and renamed according to genomic distance and functional differences. *Prevotella copri* (Chapter 2), which for years only had a single draft genome, contains at least 4 clades and is now referred to as the *Prevotella copri* complex (Fig. 5.6), to indicate that it is not a homogenous species. For a DNA polymerase error rate of 10^{-8} during genome replication and a repair rate lower than that of eukaryotes, *Escherichia coli* accumulates about 1 mutation in every 1.85×10^9 nucleotides for a genome of 4.6 Mb [51]. *Bacteroides fragilis* have a genome size of 5.2 Mb, and repeated isolation of the species from the same individuals showed

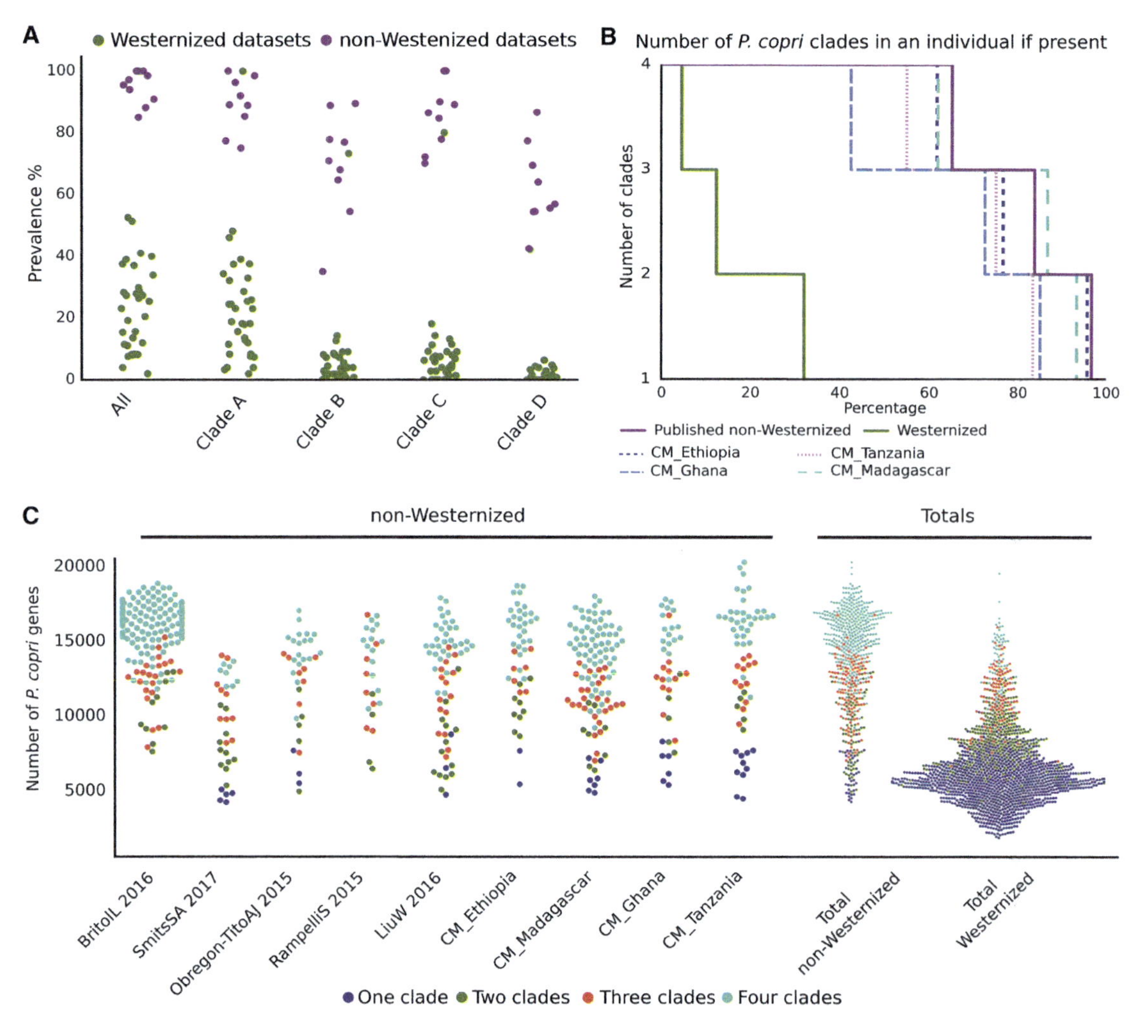

Fig. 5.6 Prevalence of the *Prevotella copri* complex and its association with non-Westernized populations. (A) *Prevotella copri* prevalence in non-Westernized and Westernized datasets. "All" refers to the prevalence of any of the four clades being present. (B) Percentage of individuals harboring multiple *Prevotella copri* clades. (C) *Prevotella copri* complex pangenome sizes for non-Westernized individuals by dataset compared to Westernized individuals. Protein coding genes specific to each clade of the *Prevotella copri* complex were defined as present in > 95% of the *Prevotella copri* genomes of a given clade but absent in all others. This gave for Clade A $n=430$ markers, for Clade B $n=954$, for Clade C $n=479$, and for Clade D $n=585$. Credit: Fig. 3 of Tett A, Huang KD, Asnicar F, Fehlner-Peach H, Pasolli E, Karcher N, et al. The Prevotella copri complex comprises four distinct clades underrepresented in westernized populations. Cell Host Microbe 2019. https://doi.org/10.1016/j.chom.2019.08.018. of a metagenomic sample, creating a MAG that represents the consensus genome of a population of cells. Credit: Fig. 2 of Van Rossum T, Ferretti P, Maistrenko OM, Bork P. Diversity within species: interpreting strains in microbiomes. Nat Rev Microbiol 2020;18:491–506. https://doi.org/10.1038/s41579-020-0368-1. Panel B was adapted by Van Rossum et al. from Maistrenko OM, Mende DR, Luetge M, Hildebrand F, Schmidt TSB, Li SS, et al. Disentangling the impact of environmental and phylogenetic constraints on prokaryotic within-species diversity. ISME J 2020;14:1247–59. https://doi.org/10.1038/s41396-020-0600-z.

an accumulation of 1 SNP per year [52] (Figs. 5.7 and 5.8), suggesting that this gut mucosal-resident bacterium may only replicate about once every day $(1.85\times10^9/(5.2\times10^6)/365=0.97\approx1)$, if starting from a homogenous population. Greater changes are probably not well tolerated at this niche, e.g., colonization-deficient *Bacteroides fragilis* Δ*ccf* (Chapter 1 Fig. 1.1C), while horizontal gene transfer could occasionally be observed [52]. Those that replicate more often or have a larger population could accumulate SNPs faster. For *Pseudomonas aeruginosa* in the lungs of cystic fibrous patients, distinct lineages could form in different regions of the lungs [53].

Horizontal gene transfers through plasmids, prophages, or other mobile elements are more frequent for functions that provide a clear advantage under stress, e.g., antibiotic resistance [54,55]. Such functions can become a burden when the stress is no longer present, and the abundant strains do not have to be the most resistant ones when the stress is present [56].

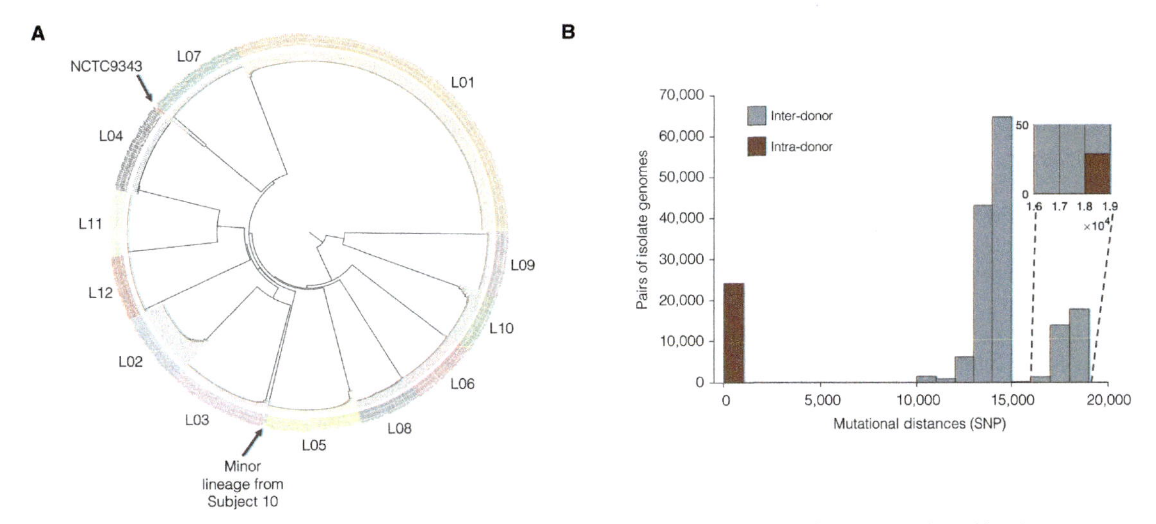

Fig. 5.7 Example of stable lineage and individual specificity of a gut bacterium. Each of the 12 healthy subject's *Bacteroides fragilis* population is dominated by a single lineage. Samples from 7 of the subjects span 2 years. (A) Phylogenetic reconstruction shows that isolates cluster by subject ($n=602$). Isolates are colored according to the subject, which grouped in lineages (L01 to L12). The arrow on top-left marks the NCBI reference genome assembly for *B. fragilis* CCUG4856T (NCTC9343). The arrow next to L05 indicates a single sample from Subject 10 that did not cluster in L10. (B) Isolates from the same subjects generally differ by <100 single nucleotide differences (SNPs), while isolates from different subjects differ by $>10{,}000$ SNPs. Inset: intra-subject pairs separated by $>18{,}000$ SNPs all involve the outlier isolate from subject 10. Credit: Fig. 1 of Zhao S, Lieberman TD, Poyet M, Kauffman KM, Gibbons SM, Groussin M, et al. Adaptive evolution within gut microbiomes of healthy people. Cell Host Microbe 2019. https://doi.org/10.1016/j.chom.2019.03.007.

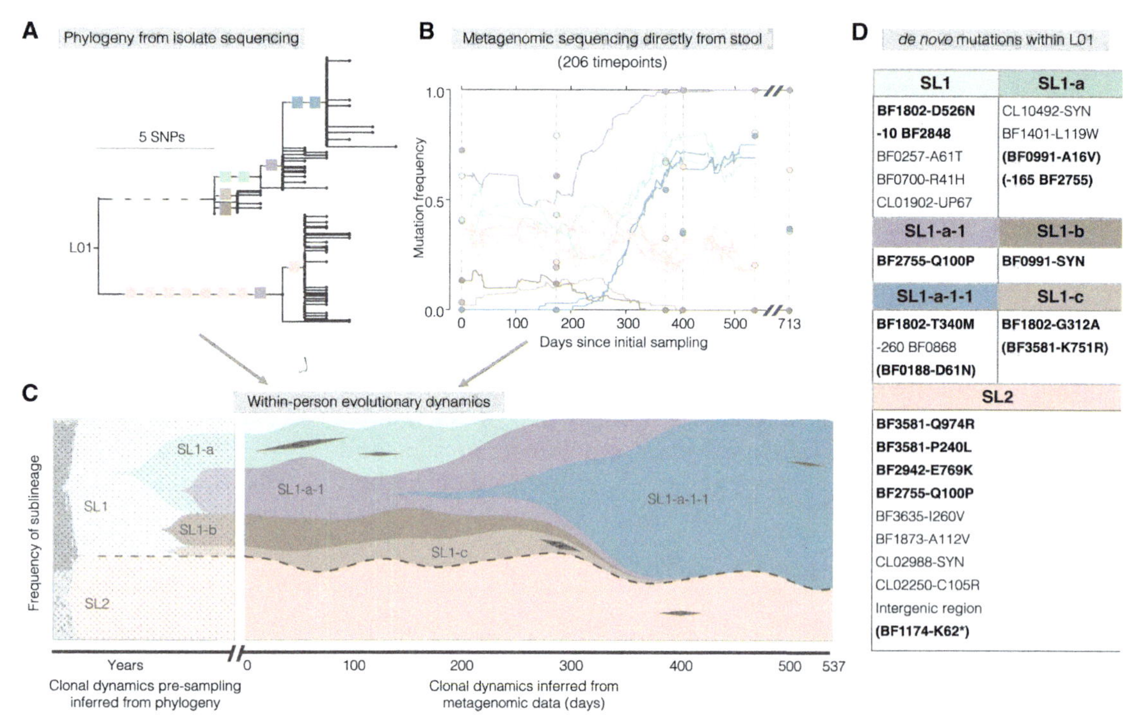

Fig. 5.8 Evolutionary dynamics over a 1.5-year sampling period for one volunteer reveals a steady increase in mutational frequencies and a stable coexistence of two sublineages of *Bacteroides fragilis*. Mobile genetic elements are not shown. (A–C) We combined 206 stool metagenomes and 187 isolate whole genomes to infer evolutionary dynamics within L01 (Fig. 5.7). (A) Branches with at least 4 isolates are labeled with colored *squares* that represent individual SNPs. One SNP was inferred to have happened twice and is indicated in both locations *(purple)*. (B) Frequencies of labeled SNPs were inferred from metagenomes. *Circles* represent SNP frequencies inferred from isolated genomes. (C) We combined these data types to infer the trajectory of sublineages prior to and during sampling. Sublineages are labeled with names and colored as in (A). The two major sublineages, SL1 and SL2, are separated by a *dashed line*. *Black* diamonds represent transient SNPs from polysaccharide utilization (PULs) and cell-envelope biosynthesis genes. (D) The identity of SNPs is shown in (A–C). SNPs in the 16 genes under positive selection are bolded and transient mutations in these genes are indicated with parentheses. Negative numbers indicate mutations upstream of the start of the gene. Credit: Cropped from Fig. 5 of Zhao S, Lieberman TD, Poyet M, Kauffman KM, Gibbons SM, Groussin M, et al. Adaptive evolution within gut microbiomes of healthy people. Cell Host Microbe 2019. https://doi.org/10.1016/j.chom.2019.03.007.

5.4 Whole-cell modeling to predict functional differences from genomic differences?

With a small genome of 525 genes, and over 900 publications, *Mycoplasma genitalium* became the first organism to be computationally modeled for every major process in the cell (Fig. 5.9) [57], with 27.5% of the parameter values found for actual experiments with *M. genitalium* instead of other bacteria. The

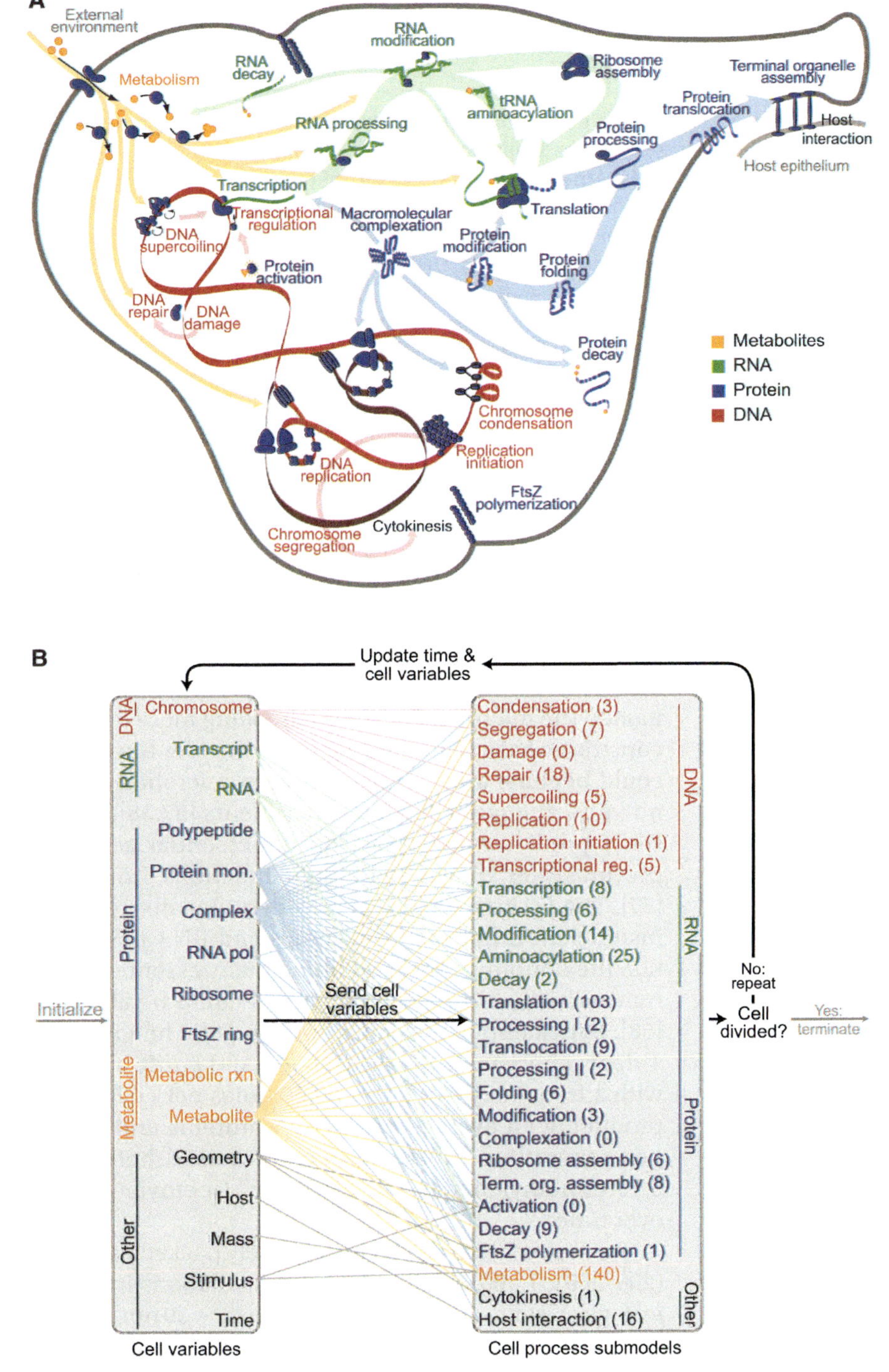

Fig. 5.9 A whole-cell model for *Mycoplasma genitalium*. (A) *M. genitalium* whole-cell model integrates 28 submodels of diverse cellular processes. Diagram schematically depicts the 28 submodels as colored words—grouped by category as metabolic *(orange)*, RNA *(green)*, protein *(blue)*, and DNA *(red)*—in the context of a single *M. genitalium* cell with its characteristic flask-like shape. Submodels are connected through common metabolites, RNA, protein, and the chromosome, which are depicted as *orange, green, blue, and red arrows*, respectively.

(Continued)

Fig. 5.9, cont'd (B) The model integrates cellular function submodels through 16 cell variables. First, simulations are randomly initialized to the beginning of the cell cycle (left *gray arrow*). Next, for each 1 s time step *(dark black arrows)*, the submodels retrieve the current values of the cellular variables, calculate their contributions to the temporal evolution of the cell variables, and update the values of the cellular variables. This is repeated thousands of times during the course of each simulation. For clarity, cell functions and variables are grouped into five physiologic categories: DNA *(red)*, RNA *(green)*, protein *(blue)*, metabolite *(orange)*, and other *(black)*. Colored lines between the variables and submodels indicate the cell variables predicted by each submodel. The number of genes associated with each submodel is indicated in parentheses. Finally, simulations are terminated upon cell division when the septum diameter equals zero (right *gray arrow*). Credit: From Fig.1 of Karr JR, Sanghvi JC, Macklin DN, Gutschow MV., Jacobs JM, Bolival B, et al. A whole-cell computational model predicts phenotype from genotype. Cell 2012;150:389–401. https://doi.org/10.1016/j.cell.2012.05.044.

recent model for *Escherichia coli* considered 1214 genes (43% of the well-annotated *Escherichia coli* genes) and found > 19,000 measured parameter values from decades of literature on *Escherichia coli* itself [58]. This is still a luxury for most members of the human microbiome, which might have only one publication that named the microbe. Such models, being more defined compared to constraint-based models that only consider the metabolic flux [59], could become a key bridge between microbial genomes and phenotypes, and guide further experiments [57,58]. The *M. genitalium* study predicted essential and nonessential genes and identified previously unknown redundant functions from the discrepancy [57]. The *Escherichia coli* model identified discrepancies such as an insufficient number of ribosomes and RNA polymerases to maintain the doubling time, and found many essential proteins to be not transcribed within a cell cycle, and could be absent in some cells [58]. For example, the genes encoding the heterodimeric 4-amino-4-deoxychorismate synthase, *pabA* and *pabB*, are each transcribed with a frequency of 0.94 and 0.66 times per cell cycle, respectively, producing an average of 34 PabA proteins and 101 PabB proteins per generation; in the absence of PabAB heterodimers, cellular 5,10-dimethylene tetrahydrofolate (methylene-THF) decreases over time.

An *Escherichia coli* cell is so densely packed with macromolecules (200–300 g/L [60], higher than the simulations in Fig. 1.3C) that it is in a glass state—molecules larger than ~ 30 nm, such as ribosomes (21 nm each), large enzyme complexes, plasmids, and phage particles, would stay in place with no diffusion, in contrast to freely diffusing small molecules [61] (Fig. 5.10). Metabolic activity fluidizes the glass state and increases movement [61]. The nucleus in mammalian cells

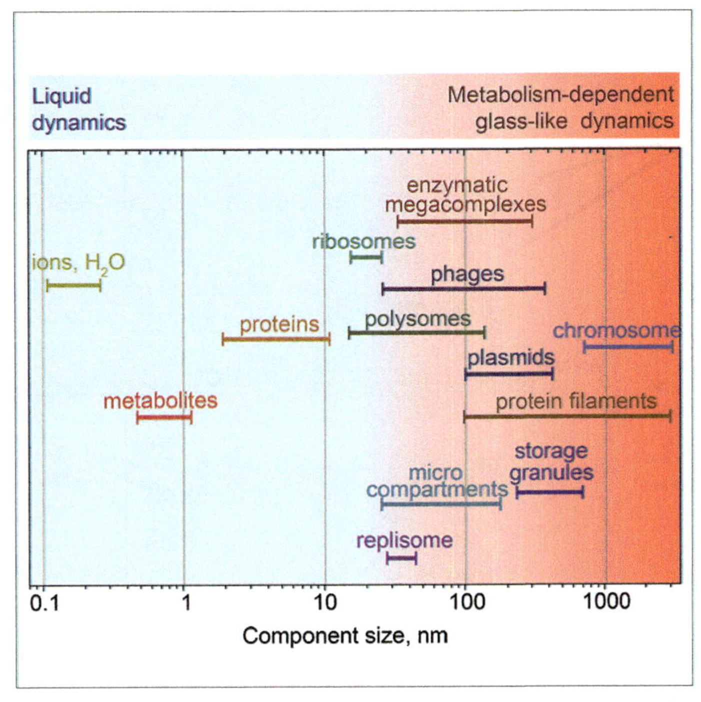

Fig. 5.10 Size of molecules, and mobility in bacterial cells dependent on active metabolism. Credit: Graphical Abstract of Parry BR, Surovtsev IV., Cabeen MT, O'Hern CS, Dufresne ER, Jacobs-Wagner C. The bacterial cytoplasm has glass-like properties and is fluidized by metabolic activity. Cell 2014;156:183–94. https://doi.org/10.1016/j.cell.2013.11.028.

is also much like that. If whole-cell models have enough experimental data to predict physical properties, they would eventually be able to predict dynamics in colonies, and in complex interactions with the host epithelium.

Metabolic models of multiple species can be validated by in vitro experiments and cohort data. *Bacteroides thetaiotaomicron* by itself produces acetate and propionate. In the presence of *Faecalibacterium prausnitzii* or *Eubacterium rectale*, the 4-species community (with *Bifidobacterium adolescentis* and *Ruminococcus bromii*) produces butyrate, acetate, and much less propionate than *Bacteroides thetaiotaomicron* alone [62] (Fig. 5.11, more on the short-chain fatty acids (SCFAs) in Chapter 6 Fig. 6.5). A model of butyrate production considered as many as 25 species and identified inhibition by hydrogen sulfate (H_2S), effect of pH, etc. [63].

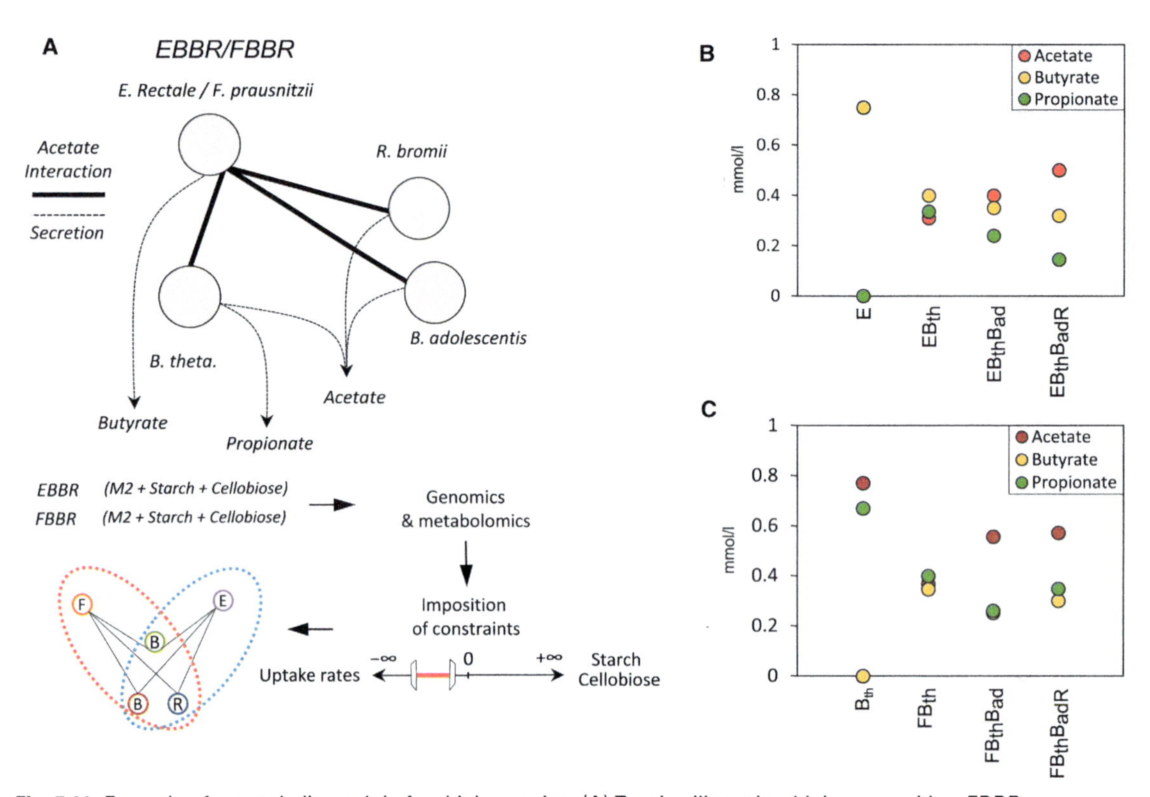

Fig. 5.11 Example of a metabolic model of multiple species. (A) Two in silico microbial communities, EBBR (*Eubacterium rectale* + *Bacteroides thetaiotaomicron* + *Bifidobacterium adolescentis* + *Ruminococcus bromii*) and FBBR (*Faecalibacteirum prausnitzii* + *B. thetaiotaomicron* + *B. adolescentis* + *R. bromii*), were designed and simulated using the CASINO (Community And Systems-level INteractive Optimization) Toolbox. The results were compared with data from triplicate in vitro experiments for EBBR and FBBR communities, grown in M2 medium supplemented with 0.2% (weight/volume) starch and 0.2% (weight/volume) cellobiose. In CASINO, the interactions of the bacteria as well as the phenotype of the community were identified using an optimization algorithm. Growth of each bacterium had local optimum, whereas the community had global optimum. The community optimum was detected by the intersection point of the fixed constraints for the community and the calculated dynamic constraints, which were obtained by summation of the local and community forces. (B and C) Network structure influenced SCFA production. The sensitivity of CASINO optimization was tested by evaluating the changes in the SCFA profile upon adding different species to the community. First, the most important receptor (receiving metabolites from the other microbes) and effector (producing metabolites consumed by receptors) in the communities were identified according to power centrality and degree centrality. 1 mmol/L of glucose was used for all the simulations, and the SCFA profiles were predicted. Following identification of the dominant receptor and effector, the key species, the other species were added to the community one by one until the EBBR (B) and FBBR (C) communities were reconstructed. Comparison between the simulations showed that the SCFA profile is very sensitive to the absence and presence of species with respect to their abundance and interactions. Credit: Fig. 2A, Fig. 3B of Shoaie S, Ghaffari P, Kovatcheva-Datchary P, Mardinoglu A, Sen P, Pujos-Guillot E, et al. Quantifying diet-induced metabolic changes of the human gut microbiome. Cell Metab 2015;22:320–31. https://doi.org/10.1016/j.cmet.2015.07.001.

5.5 Summary

The human microbiome involves the largest taxonomic division, from all domains of life all the way down to microbial strains, and single-nucleotide polymorphisms (SNPs). With reference genomes accumulating from both cultured isolates and metagenomic assembly for each body site, taxonomic profilers would have more unbiased marker sequences for each taxonomic level. Such developments would lead to more accurate abundance of information for genera and species, which underly the establishment of transmission routes (Chapter 4) and causal roles in diseases (Chapter 6). Potential functions, including growth on defined culture media, could be inferred from metagenomic assembled genomes. Variations below the species level, commonly referred to as strains, can also be tracked in the same person overtime, using either metagenomic sequencing or culturomics. With more in vitro characterization and multiomic studies on the commensal microbes, taxonomy according to bioinformatics and metabolic modeling would eventually assimilate traditional taxonomy according to functional measurements.

References

[1] Bowers RM, Kyrpides NC, Stepanauskas R, Harmon-Smith M, Doud D, Reddy TBK, et al. Minimum information about a single amplified genome (MISAG) and a metagenome-assembled genome (MIMAG) of bacteria and archaea. Nat Biotechnol 2017;35:725–31. https://doi.org/10.1038/nbt.3893.

[2] Větrovský T, Baldrian P, Morais D. SEED 2: a user-friendly platform for amplicon high-throughput sequencing data analyses. Bioinformatics 2018;34:2292–4. https://doi.org/10.1093/bioinformatics/bty071.

[3] Prodan A, Tremaroli V, Brolin H, Zwinderman AH, Nieuwdorp M, Levin E. Comparing bioinformatic pipelines for microbial 16S rRNA amplicon sequencing. PLoS One 2020;15. https://doi.org/10.1371/journal.pone.0227434, e0227434.

[4] Sun X, Hu Y-H, Wang J, Fang C, Li J, Han M, et al. Efficient and stable metabarcoding sequencing data using a DNBSEQ-G400 sequencer validated by comprehensive community analyses. Gigabyte 2021;2021:1–15. https://doi.org/10.46471/gigabyte.16.

[5] Qin J, Li Y, Cai Z, Li S, Zhu J, Zhang F, et al. A metagenome-wide association study of gut microbiota in type 2 diabetes. Nature 2012;490:55–60. https://doi.org/10.1038/nature11450.

[6] Sczyrba A, Hofmann P, Belmann P, Koslicki D, Janssen S, Dröge J, et al. Critical assessment of metagenome interpretation—a benchmark of metagenomics software. Nat Methods 2017;14:1063–71. https://doi.org/10.1038/nmeth.4458.

[7] Meyer F, Fritz A, Deng Z-L, Koslicki D, Gurevich A, Robertso G, et al. Critical Assessment of Metagenome Interpretation - the second round of challenges. bioRxiv 2021. https://doi.org/10.1101/2021.07.12.451567.

[8] Almeida A, Nayfach S, Boland M, Strozzi F, Beracochea M, Shi ZJ, et al. A unified catalog of 204,938 reference genomes from the human gut microbiome. Nat Biotechnol 2020. https://doi.org/10.1038/s41587-020-0603-3.

[9] Parks DH, Chuvochina M, Waite DW, Rinke C, Skarshewski A, Chaumeil P-A, et al. A standardized bacterial taxonomy based on genome phylogeny substantially revises the tree of life. Nat Biotechnol 2018;36:996–1004. https://doi.org/10.1038/nbt.4229.

[10] Reimer LC, Vetcininova A, Carbasse JS, Söhngen C, Gleim D, Ebeling C, et al. BacDive in 2019: bacterial phenotypic data for high-throughput biodiversity analysis. Nucleic Acids Res 2019;47:D631–6. https://doi.org/10.1093/nar/gky879.
[11] Duran-Pinedo AE, Chen T, Teles R, Starr JR, Wang X, Krishnan K, et al. Community-wide transcriptome of the oral microbiome in subjects with and without periodontitis. ISME J 2014;8:1659–72. https://doi.org/10.1038/ismej.2014.23.
[12] Utter DR, He X, Cavanaugh CM, McLean JS, Bor B. The saccharibacterium TM7x elicits differential responses across its host range. ISME J 2020. https://doi.org/10.1038/s41396-020-00736-6.
[13] Zhu J. Over 50,000 metagenomically assembled draft genomes for the human oral microbiome reveal new taxa and a male-specific bacterium; 2021. p. 2790.
[14] Coleman GA, Davín AA, Mahendrarajah TA, Szánthó LL, Spang A, Hugenholtz P, et al. A rooted phylogeny resolves early bacterial evolution. Science 2021;372. https://doi.org/10.1126/science.abe0511, eabe0511.
[15] Kempes CP, Wang L, Amend JP, Doyle J, Hoehler T. Evolutionary tradeoffs in cellular composition across diverse bacteria. ISME J 2016;10:2145–57. https://doi.org/10.1038/ismej.2016.21.
[16] Mira A, Ochman H, Moran NA. Deletional bias and the evolution of bacterial genomes. Trends Genet 2001;17:589–96. https://doi.org/10.1016/s0168-9525(01)02447-7.
[17] Probst AJ, Auerbach AK, Moissl-Eichinger C. Archaea on human skin. PLoS One 2013;8. https://doi.org/10.1371/journal.pone.0065388, e65388.
[18] Zou Y, Xue W, Luo G, Deng Z, Qin P, Guo R, et al. 1,520 reference genomes from cultivated human gut bacteria enable functional microbiome analyses. Nat Biotechnol 2019;37:179–85. https://doi.org/10.1038/s41587-018-0008-8.
[19] Forster SC, Kumar N, Anonye BO, Almeida A, Viciani E, Stares MD, et al. A human gut bacterial genome and culture collection for precise and efficient metagenomic analysis. Nat Biotechnol 2019;37. https://doi.org/10.1038/s41587-018-0009-7.
[20] Groussin M, Poyet M, Sistiaga A, Kearney SM, Moniz K, Noel M, et al. Elevated rates of horizontal gene transfer in the industrialized human microbiome. Cell 2021;184:2053–2067.e18. https://doi.org/10.1016/j.cell.2021.02.052.
[21] Cross KL, Campbell JH, Balachandran M, Campbell AG, Cooper SJ, Griffen A, et al. Targeted isolation and cultivation of uncultivated bacteria by reverse genomics. Nat Biotechnol 2019. https://doi.org/10.1038/s41587-019-0260-6.
[22] Lagier J-C, Dubourg G, Million M, Cadoret F, Bilen M, Fenollar F, et al. Culturing the human microbiota and culturomics. Nat Rev Microbiol 2018;16:540–50. https://doi.org/10.1038/s41579-018-0041-0.
[23] Truong DT, Franzosa EA, Tickle TL, Scholz M, Weingart G, Pasolli E, et al. MetaPhlAn2 for enhanced metagenomic taxonomic profiling. Nat Methods 2015;12:902–3. https://doi.org/10.1038/nmeth.3589.
[24] Milanese A, Mende DR, Paoli L, Salazar G, Ruscheweyh H-J, Cuenca M, et al. Microbial abundance, activity and population genomic profiling with mOTUs2. Nat Commun 2019;10:1014. https://doi.org/10.1038/s41467-019-08844-4.
[25] Ye SH, Siddle KJ, Park DJ, Sabeti PC. Benchmarking metagenomics tools for taxonomic classification. Cell 2019;178:779–94. https://doi.org/10.1016/j.cell.2019.07.010.
[26] Segata N. MetaPhlAn3; 2021. https://doi.org/10.1101/2020.11.19.388223.
[27] Weyrich LS, Duchene S, Soubrier J, Arriola L, Llamas B, Breen J, et al. Neanderthal behaviour, diet, and disease inferred from ancient DNA in dental calculus. Nature 2017;544:357–61. https://doi.org/10.1038/nature21674.
[28] Wibowo MC, Yang Z, Borry M, Hübner A, Huang KD, Tierney BT, et al. Reconstruction of ancient microbial genomes from the human gut. Nature 2021. https://doi.org/10.1038/s41586-021-03532-0.

[29] Rampelli S, Turroni S, Mallol C, Hernandez C, Galván B, Sistiaga A, et al. Components of a Neanderthal gut microbiome recovered from fecal sediments from El salt. Commun Biol 2021;4:169. https://doi.org/10.1038/s42003-021-01689-y.
[30] Fellows Yates JA, Velsko IM, Aron F, Posth C, Hofman CA, Austin RM, et al. The evolution and changing ecology of the African hominid oral microbiome. Proc Natl Acad Sci 2021;118. https://doi.org/10.1073/pnas.2021655118, e2021655118.
[31] Wood DE, Lu J, Langmead B. Improved metagenomic analysis with kraken 2. Genome Biol 2019;20:257. https://doi.org/10.1186/s13059-019-1891-0.
[32] Cao Y, Lin W, Li H. Large covariance estimation for compositional data via composition-adjusted thresholding. J Am Stat Assoc 2018;1–45. https://doi.org/10.1080/01621459.2018.1442340.
[33] Friedman J, Alm EJ. Inferring correlation networks from genomic survey data. PLoS Comput Biol 2012;8. https://doi.org/10.1371/journal.pcbi.1002687, e1002687.
[34] Scheiman J, Luber JM, Chavkin TA, MacDonald T, Tung A, Pham L-D, et al. Meta-omics analysis of elite athletes identifies a performance-enhancing microbe that functions via lactate metabolism. Nat Med 2019;25:1104–9. https://doi.org/10.1038/s41591-019-0485-4.
[35] Deek RA, Li H. A zero-inflated latent Dirichlet allocation model for microbiome studies. Front Genet 2021;11. https://doi.org/10.3389/fgene.2020.602594.
[36] Liu X, Tong X, Zou Y, Lin X, Zhao H, Tian L, et al. Inter-determination of blood metabolite levels and gut microbiome supported by Mendelian randomization. BioRxiv 2020. https://doi.org/10.1101/2020.06.30.181438. 2020.06.30.
[37] Wang J, Jia H. Metagenome-wide association studies: fine-mining the microbiome. Nat Rev Microbiol 2016;14:508–22. https://doi.org/10.1038/nrmicro.2016.83.
[38] Zhang X, Zhang D, Jia H, Feng Q, Wang D, Di Liang D, et al. The oral and gut microbiomes are perturbed in rheumatoid arthritis and partly normalized after treatment. Nat Med 2015;21:895–905. https://doi.org/10.1038/nm.3914.
[39] Jie Z, Xia H, Zhong S-L, Feng Q, Li S, Liang S, et al. The gut microbiome in atherosclerotic cardiovascular disease. Nat Commun 2017;8:845. https://doi.org/10.1038/s41467-017-00900-1.
[40] Jie Z, Liang S, Ding Q, Li F, Tang S, Wang D, et al. A transomic cohort as a reference point for promoting a healthy gut microbiome. Med Microecol 2021. https://doi.org/10.1016/j.medmic.2021.100039.
[41] Li J, Jia H, Cai X, Zhong H, Feng Q, Sunagawa S, et al. An integrated catalog of reference genes in the human gut microbiome. Nat Biotechnol 2014;32:834–41. https://doi.org/10.1038/nbt.2942.
[42] Browne HP, Forster SC, Anonye BO, Kumar N, Neville BA, Stares MD, et al. Culturing of 'unculturable' human microbiota reveals novel taxa and extensive sporulation. Nature 2016;533:543–6. https://doi.org/10.1038/nature17645.
[43] Darwin C. On the origin of species by means of natural selection, or the preservation of Favoured races in the struggle for life. London: John Murray; 1859.
[44] Mallet J. Darwin and species. In: Ruse M, editor. Cambridge Encycl. Darwin Evol. Thought. Cambridge: Cambridge University Press; 2020. p. 109–15. https://doi.org/10.1017/CBO9781139026895.013.
[45] Wallace AR. The method of organic evolution. Fortn Rev 1895;435–45. NS.57.
[46] Perlmutter JI, Bordenstein SR. Microorganisms in the reproductive tissues of arthropods. Nat Rev Microbiol 2020;18:97–111. https://doi.org/10.1038/s41579-019-0309-z.
[47] Brito IL, Yilmaz S, Huang K, Xu L, Jupiter SD, Jenkins AP, et al. Mobile genes in the human microbiome are structured from global to individual scales. Nature 2016;535:435–9. https://doi.org/10.1038/nature18927.

[48] Fraser C, Alm EJ, Polz MF, Spratt BG, Hanage WP. The bacterial species challenge : ecological diversity. Science 2009;323:741–6.
[49] Sheridan PO, Martin JC, Lawley TD, Browne HP, Harris HMB, Bernalier-Donadille A, et al. Polysaccharide utilization loci and nutritional specialization in a dominant group of butyrate-producing human colonic Firmicutes. Microb Genom 2016;2. https://doi.org/10.1099/mgen.0.000043, e000043.
[50] Rosenthal AZ, Qi Y, Hormoz S, Park J, Li SH-J, Elowitz MB. Metabolic interactions between dynamic bacterial subpopulations. Elife 2018;7. https://doi.org/10.7554/eLife.33099.
[51] Drake JW, Charlesworth B, Charlesworth D, Crow JF. Rates of spontaneous mutation. Genetics 1998;148:1667–86.
[52] Zhao S, Lieberman TD, Poyet M, Kauffman KM, Gibbons SM, Groussin M, et al. Adaptive evolution within gut microbiomes of healthy people. Cell Host Microbe 2019. https://doi.org/10.1016/j.chom.2019.03.007.
[53] Jorth P, Staudinger BJ, Wu X, Hisert KB, Hayden H, Garudathri J, et al. Regional isolation drives bacterial diversification within cystic fibrosis lungs. Cell Host Microbe 2015;18:307–19. https://doi.org/10.1016/j.chom.2015.07.006.
[54] Kent AG, Vill AC, Shi Q, Satlin MJ, Brito IL. Widespread transfer of mobile antibiotic resistance genes within individual gut microbiomes revealed through bacterial hi-C. Nat Commun 2020;11:1–9. https://doi.org/10.1038/s41467-020-18164-7.
[55] Brito IL. Examining horizontal gene transfer in microbial communities. Nat Rev Microbiol 2021;1–12. https://doi.org/10.1038/s41579-021-00534-7.
[56] Baym M, Lieberman TD, Kelsic ED, Chait R, Gross R, Yelin I, et al. Spatiotemporal microbial evolution on antibiotic landscapes. Science 2016;353:1147–51. https://doi.org/10.1126/science.aag0822.
[57] Karr JR, Sanghvi JC, Macklin DN, Gutschow MV, Jacobs JM, Bolival B, et al. A whole-cell computational model predicts phenotype from genotype. Cell 2012;150:389–401. https://doi.org/10.1016/j.cell.2012.05.044.
[58] Covert M. Simultaneous cross-evaluation of heterogeneous e coli datasets via mechanistic simulation. Science 2020. https://doi.org/10.1126/science.eaav3751.
[59] Bordbar A, Monk JM, King ZA, Palsson BO. Constraint-based models predict metabolic and associated cellular functions. Nat Rev Genet 2014;15:107–20. https://doi.org/10.1038/nrg3643.
[60] Mika JT, Poolman B. Macromolecule diffusion and confinement in prokaryotic cells. Curr Opin Biotechnol 2011;22(1):117–26. https://doi.org/10.1016/j.copbio.2010.09.009.
[61] Parry BR, Surovtsev IV, Cabeen MT, O'Hern CS, Dufresne ER, Jacobs-Wagner C. The bacterial cytoplasm has glass-like properties and is fluidized by metabolic activity. Cell 2014;156:183–94. https://doi.org/10.1016/j.cell.2013.11.028.
[62] Shoaie S, Ghaffari P, Kovatcheva-Datchary P, Mardinoglu A, Sen P, Pujos-Guillot E, et al. Quantifying diet-induced metabolic changes of the human gut microbiome. Cell Metab 2015;22:320–31. https://doi.org/10.1016/j.cmet.2015.07.001.
[63] Clark RL, Connors BM, Stevenson DM, Hromada SE, Hamilton JJ, Amador-Noguez D, et al. Design of synthetic human gut microbiome assembly and butyrate production. Nat Commun 2021;12:3254. https://doi.org/10.1038/s41467-021-22938-y.

6

Blurring the line between opportunistic pathogens and commensals

6.1 Causal reasoning 101

When Dr. Robert Koch formulated what is now known as Koch's postulates in 1884, he had a lot to fight against (Figs. 6.1 and 6.2) [1,2]. Louis Pasteur demonstrated with a swan-neck flask in 1859 that bacteria (germs) from the air (10^{4-6} bacterial cells/m^3 [3], compare with the lung microbiome in Chapter 3, Fig. 3.6) spoiled gravy, instead of something that spontaneously generated in the gravy [4]. Antoni van Leeuwenhoek's early study on oral bacteria (Chapter 1, Box 1.1) also makes them sound quite benign and constitute an everyday presence. In Koch's time, there was no metagenomics, so microscopes would be the gold standard for telling this bacterium from that bacterium (Fig. 6.1), whether in a host animal or in culture. If bacteria are so omnipresent, and people did not wash their hands nearly as often as we do now, to link one bacterium to such a deadly disease as anthrax would have to pass the hardest scrutiny. And Pasteur did develop a vaccine against anthrax. The pioneering experiments by Dr. Robert Koch involved causing disease with the isolated live bacterium (or spores), and further showing that it was still that bacterium, under the microscope. Now we know that the association with the disease was unnecessarily stringent, and the effect of the intervention was too successful (Figs. 6.1 and 6.2), so well-known pathogens that have cost many lives did not fulfill Koch's postulates [2].

The biomedical field (and the finance field) has been heavily plagued by statistics from Sir. R.A. Fisher [5,6]. As physicists know very well, what really matters is not the sample size or the *P*-value, but the likely scenarios, and the probability for each scenario with the existing evidence. The probabilities are updated with new evidence (e.g., Pasteur's definitive victory over the "spontaneous generation" theory, making other explanations unlikely [4]). If a new explanation (scenario) emerges, the probabilities may need to be re-allocated among the scenarios, and some evidence may fit the new explanation better

Investigating Human Diseases with the Microbiome: Metagenomics Bench to Bedside.
https://doi.org/10.1016/B978-0-323-91369-0.00007-8

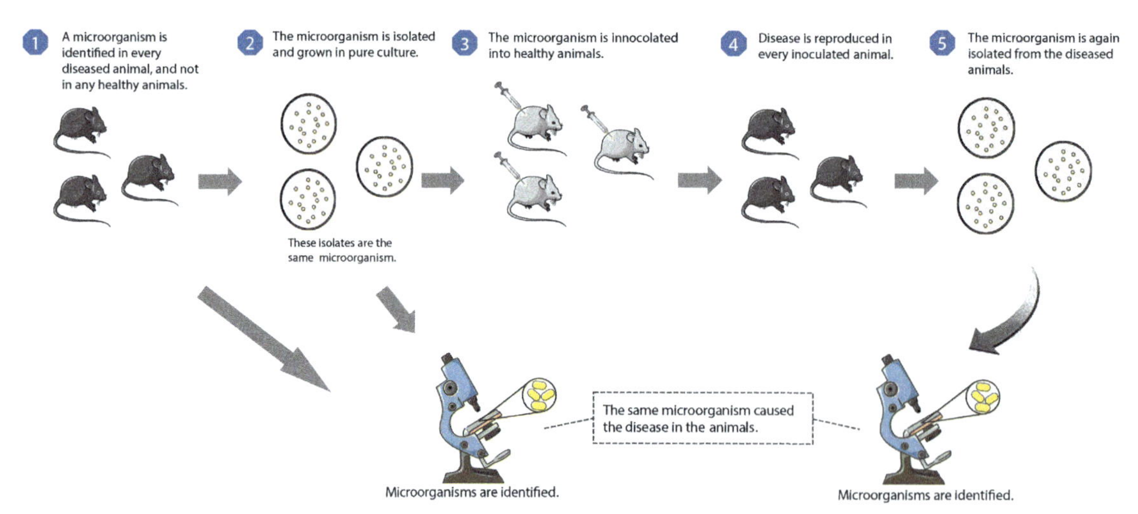

Fig. 6.1 Koch's experiment with *Bacillus anthracis*. The steps are shown according to the requirements of Koch's postulates (Fig. 6.2), and the technologies available—microscopy and isolated culture. Healthy animals are not shown, which according to Koch's postulate should not have the microorganism. Credit: Huijue Jia, Yanzheng Meng.

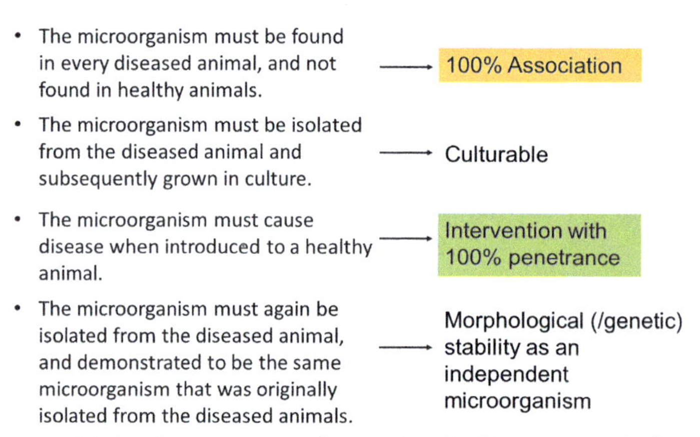

Fig. 6.2 Koch's postulates and their relevance to causal reasoning. Koch's postulates [1,2] are broken down into Association—Level 1 causal evidence, and Intervention—Level 2 causal evidence. The requirements for isolated culture and stable re-isolation are more for taxonomic and evolutionary concerns. As long as we have a proper name (and record the reference sequence used) for microorganisms identified by sequencing, cultured isolates are not relevant for causal reasoning, yet would again be useful for mechanistic and therapeutic studies. The re-isolation also addresses an alternative hypothesis that some unintended microorganism caused the same disease in the animals that were innoculated with the microorganism in question (not shown); so the disease needs to be well characterized and the laboratory animals should be under standard care to ensure reproducibility. Credit: Huijue Jia.

Box 6.1 Ockham's razor, or Newton's simplicity rule

Professor Edwin Thompson Jaynes explained in Chapter 20—"Model comparison" of his posthumous book, *Probability theory: the logic of science* [5], that Ockham's razor (Occam's razor) is intrinsic to probability theory. "Do not introduce details that do not contribute to the quality of your inferences." Formulated by the Franciscan monk William of Ockham in 1330, Occam's razor states that "Plurality should not be posited without necessity." Also known as the law of economy or law of parsimony ("It is futile to do with more what can be done with fewer").

Ockham's factor penalizes a model, essentially by considering prior information [5]. Orthodox statistical theory compares models entirely in terms of "sampling distributions." With Bayes' theorem, if the data are highly informative compared with the prior information, the relative merit of two models is determined by how high a likelihood can be attained on the respective parameter spaces, and how much prior probability is concentrated in their respective high-likelihood regions [5]. For a reasonably informative experiment, we expect the likelihood to be concentrated in small subregions, and a "simpler" model would occupy a smaller volume of parameter space, thus favored as a more plausible model [5].

Besides, Prof. Jaynes demonstrated in Chapter 7 of his book that, if we do not have any information, the Gaussian distribution is the most plausible estimate. But when we do have prior information, we should not pretend that we do not [5].

Sir Issac Newton summarized four rules of reasoning in his 1687 book *Mathematical principles of natural philosophy*. The first one—"No more causes of natural things should be admitted than are both true and sufficient to explain their phenomena."—also emphasizes simplicity and does not allow unnecessarily complicities (parameters) that do not improve the likelihood. This was also what Albert Einstein did in formulating the theory of relativity, to incorporate evidence that was not known in Newton's time.

than the old ones (Box 6.1). Of course, some people are going to be more refractory than others [5]. Back to Koch's postulates, what alternative explanations are ruled out by each experiment? Are we more and more confident with the conclusion that *Bacillus anthracis* caused anthrax (Figs. 6.1 and 6.2)?

Finding alternative explanations (hypotheses) and assigning probabilities for each explanation can depend on our own prior knowledge. For example, an analysis of vaginal microbiota community types in the Human Microbiome Project (HMP) from the United States found an association with whether the volunteers have a college degree or not (Lactobacilli and non-Lactobacilli vaginal community types) [7]. What would be the likely explanations here? (More on the cervicovaginal microbiome in Chapter 8).

"Confounders" is also a vague term from statistics that have to be abandoned [5,6]. Although there may be too many unknown factors at the beginning of a study, each scenario would still need to be clearly drawn as path diagrams in order to be mathematically workable. To try to control for (keep constant) every factor, would lead to erroneous negative associations when colliders are controlled for, and loss of

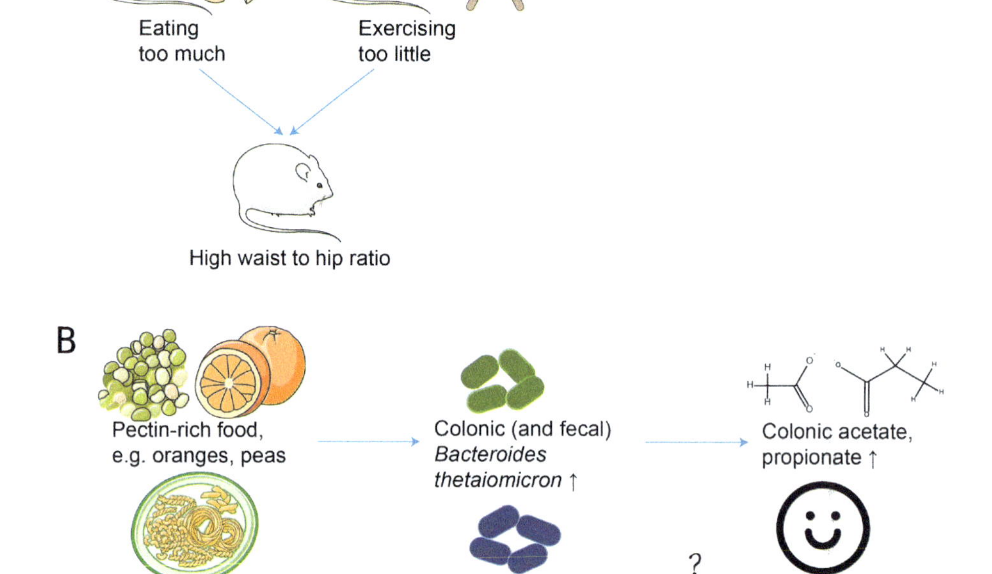

Fig. 6.3 Examples of path diagrams which would be distorted by controlling of variables, i.e., holding constant. (A) Collider. (B) On the same path. The relationship between *Faecalibacterium* and mental health is currently an association [8]. Credit: Huijue Jia, Yanzheng Meng.

effect when something in the same causal pathway is controlled for [6]. For example, if we control for fecal abundance of *Faecalibacterium* spp., we will be searching for other gut bacteria that could mediate the effect of carbohydrate (e.g., noodles) intake on a happy mood (Fig. 6.3); Or to control for tobacco use when studying lung cancer, also disabling the causal path. A more relevant example is given for a collider (Fig. 6.3), other than the famous examples of "handsome but mean guys"—apparent negative association because the mean and not handsome guys had zero opportunity, or talents and good look in actors and actresses [6]. If we all agree that eating too much or exercising too little can both make one fat (waist-hip ratio (WHR) here, Fig. 6.3), controlling for WHR or BMI (Body mass index, weight/height^2) in analyses would create a negative association between eating too much and exercising too little, i.e., an apparent positive association between eating and exercising because controlling for a collider (constant waistline or weight) opens the door there. So we are happily in the loop of eating, exercising, and more eating, while trying to maintain the same shape.

Path diagrams provide a transparent common ground for subsequent investigations, by outlining the working hypotheses [5,6]. For example, are we measuring the necessary hormones when we talk about differences in the microbiome of infants delivered by Caesarean section versus vaginal delivery, or preterm versus term birth (Fig. 6.4; More on infants in Chapter 8)? In addition to a role in the development

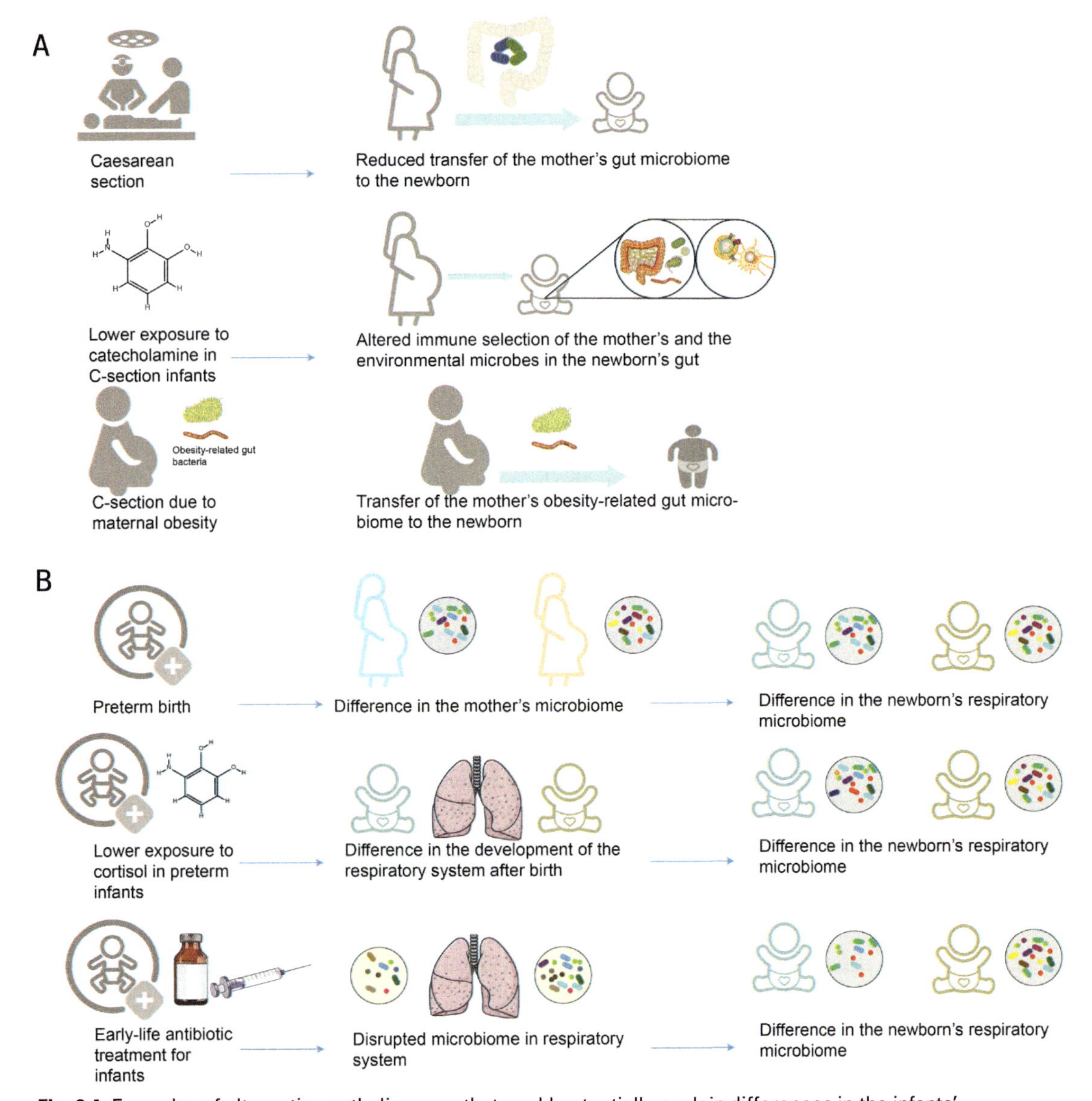

Fig. 6.4 Examples of alternative path diagrams that could potentially explain differences in the infants' microbiome. (A) A few examples for C-section. (B) A few examples of preterm birth. Inspired by [9]. Credit: Huijue Jia, Yanzheng Meng.

Worked sample 6.1

A recent publication from the American Gut project (16S rRNA gene amplicon sequencing) controlled for alcohol consumption as well as dietary information, and lost most of the fecal microbiome signal for T2D [12]. Dietary information comes down to ingredients such as amino acids, vitamins, fibers [13], sugars, lipids, and additives (e.g., Ref. [14]). Alcohol is epidemiologically known as a protective factor for rheumatoid arthritis [15]. MHC class I expression is upregulated in peripheral blood lymphocytes during acute ethanol intoxication [16]. Ethanol and its metabolite acetate (among the SCFAs, Fig. 6.5 in Section 6.3.1), potently modulate the function of T follicular help cells (T_{FH}) [18], and acetate induces IgA (immunoglobulin A) production [19].
Without getting lost in the details, how would you draw a path diagram here to interpret the chain of events? Which path do you think is more important to a T2D population you are more familiar with? What other evidence would you look for?

of the respiratory system [9], corticosteroids from the mother's milk have been shown in rats and mice to prime the hypothalamic-pituitary adrenocortical (HPA) axis [10], which would be central to stress responses [11]. New paths can always be added for a more comprehensive or more precise picture. When sailors only knew that lemons prevent scurvy, the lemons could be boiled (which destroys vitamin C) and people would no longer believe that the lemons have an effect [6].

6.2 Levels of existing evidence for the human microbiome and diseases

Numerous associations have been reported between taxa in the human microbiome and diseases, and between taxa in the microbiome and circulating molecules that are relevant to diseases, e.g., cytokines, lipids, amino acids. Phylum level claims tend to be affected by the compositional nature of relative abundance data [20]. For example, Firmicutes and Bacteroidetes almost add up to 1 in the mouse gut, so they appear negatively correlated, and whichever disease that enriched for Firmicutes would also show up as depleted for Bacteroidetes. When there are hundreds of taxa in a more even distribution (e.g., not like the vaginal samples with $>$ 90% Lactobacilli), this is less of a problem [20,21].

Besides the associations, some of the microbes' roles in diseases are reaching Level 2 evidence—(Randomized) intervention (Box 6.2), which adds the exposure to see an effect, or Level 3 evidence—Counterfactual, which removes the exposure in one's mind to assess causality (Table 6.1), something AI (artificial intelligence) cannot do.

The readers are encouraged to try to read the most layman book from Prof. Judea Pearl, *The book of why* [6], with plenty of examples from other disciplines [5,6]. Without a randomized controlled trial (RCT), mendelian randomization (MR) could also provide Level 2 causal evidence (Box 6.1) [22,26], using human genetic information (summary statistics, quantities for a population that can be analyzed without access to individual genetic data) from large cross-sectional cohorts, or having all the measurements in the same cohort (one-sample MR versus two-sample MR). Intervention experiments on mice are usually not randomized, but unlike human cohorts, we have no specific reason to believe that the assignment of this mouse and that mouse into experimental groups is affected by some phenotype of the mice (e.g., a

Box 6.2 Mendelian randomization (MR)

A common problem with interventions is that other factors may be affecting the exposure X, when we try to look at the effect of X on the outcome Y. For example, taking yogurt every day may say something about one's job, education, economic status, which may also correlate with going to the gym on a regular basis. This is not to discourage longitudinal cohorts, which have their unique values if painstakingly followed for decades, and would help with counterfactuals.

Randomized controlled trials (RCTs) assign the value of X in a random manner, thereby severing its link with other factors, e.g., taking yogurt or not have nothing to do with the other things mentioned above, and now we can more confidently see whether there is a beneficial effect of yogurt on cardiovascular health, gastrointestinal health, etc.

MR could have the same effect of removing X from other factors, when looking at the effect of X on Y. Here, the randomization is based on the naturally random assortment of parental alleles into daughter cells during meiosis. This is typically analyzed with multiple SNPs (single-nucleotide polymorphisms). The SNPs (referred to as an instrumental variable) have to associate with the exposure X, to together make an effective instrumental variable that explains a considerable portion of the variance in X (e.g., > 20%), otherwise it may be no surprise that we could not see an effect of X on Y. The instrument variable should not influence the outcome Y without going through the exposure X. And the instrument variable should not associate with another factor that influences both X and Y [22], which might then be an actual cause, instead of a "confounder."

MR can be very powerful in discovering causal relationships, including where RCTs are ethically impossible. For example, through two-sample MR combining the SNPs-microbiome association from the 4D-SZ cohort from China and the SNPs-diseases association from Biobank Japan (BBJ), gut *Streptococcus parasanguinis* has been shown to contribute to the heart problem of posterior wall thickness, as well as colorectal cancer, raising the level of causal evidence above the MWAS associations [23–25] (Table 6.1). Higher BMI (body mass index), smoking, and coffee consumption were shown by MR to increase rheumatoid arthritis (RA), while iron, linoleic acid (a major polyunsaturated fatty acid (PUFA) that could stimulate testosterone synthesis) and years of education were protective [26].

When interpreting results from large cohorts, it is important to bear in mind that we are limited by the phenotypes and questionnaires collected, so a causal signal could still mean another unasked question that strongly correlates with the available one, at least potentially on the same path.

Table 6.1 Examples of causal evidence for the microbiome.

Level of causal evidence		Evidence	Reference
1	Association	Fecal *Escherichia coli* associated with GLP-1 level; Fecal *Escherichia coli* enriched in prediabetic patients compared to controls	[27,28]
1	Association	Fecal *Escherichia coli* enriched in atherosclerotic cardiovascular disease patients compared to controls, and negatively associated with hand grip strength, a known epidemiological factor for cardiovascular events	[24,29]
1	Association	Fecal *Bacteroides caccae* enriched in T2D patients[a]	[28,30]
1	Association	Fecal *Eggerthella* enriched in (pre)diabetic patients; and associated with early frailty	[28,31,32]
1	Association	Fecal *Ruminococcus torques* enriched in ulcerative colitis patients; and associated with loose stool according to BSS	[33,34]
1	Association	Smoking associated with oral *Veillonella* and *Prevotella*	Section 4.4.1
1	Association	Oral *Lachnoanaerobaculum umeaense* and two *Oribacterium* species associated with serum urate level, and with SNP in the uric acid transporter gene *SLC2A9*	[35]
2	Intervention (Randomized controlled trials (RCT))	A multicenter RCT for *Lactobacillus* and *Bifidobacterium* (a 9-strain mix of *B. longum*, *B. breve*, *L. casei*, *L. crispatus*, *L. fermentum*, *L. plantarum*, *L. rhamnosus*, *L. salivarius*, and *L. gasseri*) together with berberine treatment (after the antibiotics gentamycin sulfate) did not result in better blood glucose level in T2D patients, yet showed some effects on lipids	[36]
2	Intervention (RCT)	RCT for *L. crispatus* CTV-05 after metronidazole treatment decreased recurrence of bacterial vaginosis within 12 weeks (46/152 recurrence with treatment versus 34/76 recurrence with placebo)	[37]
2	Intervention (Mendelian randomization (MR))	A fecal Lachnospiraceae species showed bidirectional increase with serum urate level; Fecal microbiome pectin degradation module (e.g., from *Bacteroides*, *Fusobacterium*) appeared to increase serum urate level	[25]
2	Intervention (MR)	Fecal *Escherichia coli* could lead to T2D, heart failure, colorectal cancer, etc.	[25]
2	Intervention (MR)	Fecal *Streptococcus parasanguinis* could lead to a thicker posterior wall in the heart, and colorectal cancer	[25]
2	Intervention (MR)	Fecal Saccharibacteria (TM7) could decrease serum creatinine and increase the estimated glomerular filtration rate (eGFR)	[38]

Table 6.1 Examples of causal evidence for the microbiome—cont'd

Level of causal evidence		Evidence	Reference
3	Counterfactual	If there is no HPV (human papillomavirus), there will be no cervical cancer	Free insurance can be provided with every test
3	Counterfactual	If there is no *Helicobacter pylori*, will there be no gastric cancer? Will there be no gastric ulcer?	
3	Counterfactual	If there is no *Porphyromonas gingivitis*, *Treponema denticola*, *Tannerella forsythia*, and *Prevotella intermedia*, will there be no periodontitis?	

This is not meant to be an exhaustive list, and the associations are especially numerous. It serves to illustrate the three levels of causal evidence as was summarized by Pearl and Mackenzie [6]. GLP-1, glucagon-like peptide 1 (see also Section 6.3.3). BSS, Bristol stool scale, is mentioned in Chapter 3. SNP, single-nucleotide polymorphism.
[a]Other Bacteroides species reported in other studies, e.g., Refs. [31,39]; see also Section 6.3.
Credit: Huijue Jia.

fatter mouse is more cooperative to an intervention?). However, germ-free mice are not normal in metabolic, immunological, and neurological states [40–42]. The microbiome is also much different between SPF (specific-pathogen free) mice and humans, and between mice facilities [43,44], both regarding the species and the interactions.

Intuitively, microbial associations involving well-established molecules that contribute to disease are considered as stronger evidence as mere association with disease cases compared to healthy controls (Table 6.1). Having more associations along the same line indeed greatly reduces the likelihood that we have just caught one spurious association when we do so many statistical tests (Professor Jaynes's book [5] may be mathematically intimidating for many readers, but one can safely skip some equations). M-GWAS (Metagenome-genome-wide association studies) associations with human genes are regarded as perhaps more than Level 1 association (Table 6.1), as the established link between a gene and a disease adds to the credibility of a new association between a microbe and a disease, narrowing down the parameter space and pointing at experiments to do. The beauty of Guilt-by-association is, we are not arbitrarily following a hypothesis that may turn out to be a small question in the big picture, and to have an association is much better than nothing.

Infection by the gastric pathogen *Helicobacter pylori* is typically eradicated in developed countries, while *H. pylori* continues to colonize about 50% of the world's population [45]. Only 1%–3% of *H. pylori*-infected individuals would develop gastric cancer, depending on genetic features of the *H. pylori*, possibly other microbes (e.g., *Lactobacillus* may be protective) [46], as well as the site of *H. pylori* colonization, host genetics and immune responses [45,47]. As the strongest risk factor for duodenal and gastric ulcer, and gastric cancer, *H. pylori* has not reached Level 3 causal evidence, not because of the less than 100% disease manifestation (Koch's postulates, Fig. 6.2) which just means that other factors are also part of the path diagram, but because we cannot confidently make a counterfactual (Table 6.1), and tell people that they would not get gastric cancer within the next few years. It would still be a counterfactual if it turns out that we have to exclude some rare form of gastric cancer that is not caused by *H. pylori*. Gastric cancer patients are not routinely examined for *H. pylori*; It is likely that for some patients, eradication or spontaneous elimination (e.g., by increased Lactobacilli with coffee consumption? This is rather anecdotal at this point.) of *H. pylori* was too late, and no additional intervention has been developed to stop further deterioration of the gastric epithelium. This would need to be better worked out.

The periodontitis-promoting bacteria are usually dislodged mechanically or killed chemically (e.g., mouthwash targeting *Porphyromonas gingivalis*). If the answer is yes to the counterfactual question regarding their causal role in periodontitis (Table 6.1), what would be the consensus for best practice? Do the bacteria have some good functions for host immunity (e.g., in defense of a pathogen that is no longer of concern) or for host metabolism that we need to keep an eye on (e.g., salivary *Porphyromonas* correlated with a high copy number of the amylase gene in the human genome [48])? Or do the chemicals used for oral hygiene have side effects from the microbiome point-of-view? In addition to periodontitis and dental caries, what would be the recommendation if there are too many colorectal cancer, colitis, or liver disease-promoting bacteria in the oral microbiome (e.g., Refs. [49–51]), although not yet in the feces?

As a community of microbes, more convoluted scenarios can take place. Some microbes modify the habitat or provide a common good (nonspecific cross-feeding), thereby affecting many other microbes

Worked sample 6.2

What complications do you think multicenter RCTs intend to control for, that is of concern in single-center RCTs?

If the participants tend to toss away what they are given (e.g., 50% chance of placebo, 50% chance of the bacterial formulation being trialed), can you collect metagenomic samples to double-check that they are taking the formulation?

For how long can the participants be followed after taking the bacterial formulation? Do you expect cell phone applications or other new technologies to help? (Also for Chapter 8).

(Chapter 2). For example, *Lactobacillus* and *Bifidobacterium* can reduce nitrite into nitric oxide, which impacts blood flow and motility, while ridding the host of the potential carcinogen, nitrite [52] (worth considering for the comprehensive picture of *H. pylori* and gastric cancer). *Ruminococcus gnavus*, a common bacterium enriched in Crohn's diseases, can hydrolyze the blood group B glycan into blood group O [53], degrade mucin glycans in a way different from other commensals [54], and secrete a complex glucorhamnan polysaccharide that induces inflammation through TLR4 (Toll-like receptor 4) [55]. In gnotobiotic mice, it has been shown that sialidase from *Bacteroides thetaiotaomicron* liberates sialic acid from the gut mucosa, which is required for the expansion of pathogens such as *Salmonella typhimurium* and *Peptoclostridium difficile* (renamed from *Clostridium difficile*) [56]. Vaginal *Gardnerella vaginalis* can express sialidase; *Prevotella bivia* encode sulfatase and sialidase that all damage the healthy mucus layer and promote a dysbiotic (imbalanced) microbiome on the way to bacterial vaginosis, and with more steps, preterm birth [57,58]. Intestinal *Candida albicans* is a major fungal inducer of T helper 17 (Th17) cell response, which would cross-react to fight against other fungi, such as airway inflammation due to *Aspergillus fumigatus* [59]. In animal models, the dose of *Staphylococcus aureus* pathogenesis was lowered by the co-presence of *Staphylococcus epidermis*, *Micrococcus luteus* or *M. luteus* cell wall peptidoglycan, due to reduced oxidative bursts of liver Kuffer cells in the presence of the commensals [60]. Some of the currently single-microbe conclusions may also turn out to be path diagrams with more branches. We will eventually understand each path, the probability in each path, and make informed decisions for each patient (Chapter 7).

6.3 From microbes to molecules

In analogy to SARS-CoV-2 causing COVID-19, a systematic understanding of the pathogenicity would need to include everything from human and microbial genetics to etiology of the complex disease. Multiple molecules from one microbe could all contribute to its pathogenic or beneficial role in disease.

6.3.1 Multiple effective molecules from *Akkermansia muciniphila*

Akkermansia muciniphila inversely correlated with body weight [61], fecal salinity [62], increased in mice model of Roux-en-Y gastric bypass (RYGB) [63], and in T2D patients treated with metformin [64]. An outer membrane protein Amuc_1100 from *A. muciniphila* signaled through Toll-like receptor 2 (TLR2), and was partly responsible for the beneficial effect of pastuerized (heat-killed) *A. muciniphila* [65]. A single-center RCT has been completed on 32 volunteers using pastuerized *A. muciniphila*, showing significantly better insulin sensitivity, and some difference in body weight, fat mass, and hip circumference [66]. On the other hand, exposure to cold temperature decreased *A. muciniphila* relative to some of the *Firmicutes* [67], while *A. muciniphila* increased thermogenesis [68]. This function has recently led to the discovery that the bacterium could secrete a protein called P9, which induced GLP-1 (glucagon-like peptide-1) in circulation to promote glucose homeostasis, and induced uncoupling protein 1 in brown adipose tissue for thermogenesis [68]. *Akkermansia* and many other bacteria produce acetate (a major SCFA, Fig. 6.5), which just as vinegar does, can increase appetite [69], as well as promoting T follicular help cells (T_{FH}) [18], and inducing more IgA [19]. Fecal propionate, however, was reported by one MR study to promote Type 2 diabetes [70]; yet this might reflect pleiotropic effects of diabetes-promoting *Bacteroides* and *Prevotella*, which produce propionate (please continue reading to Sections 6.3.2 and 6.3.3), or less succinate production by *Bacteroides* and *Prevotella* for intestinal gluconeogenesis [71]. In chicken, duodenal *A. muciniphila* abundance associated with feed efficiency; so did cecal abundance of *Parabacteroides*, *Lactobacillus*, *Corynebacterium*, etc. [72].

As mentioned in Chapter 2, *Akkermansia muciniphila* has also been associated with diseases such as colorectal cancer [73,74], atherosclerotic cardiovascular disease [24], Alzheimer's disease, and schizophrenia [75,76]. The disease associations may involve some different genomic features of the bacterium [77,78], access to some different host molecules (e.g., mucin proteins other than Muc2 expressed in cases of infection and cancer [79]), circadian rhythm (Chapter 2, Box 2.5) or some other mechanisms.

Back to the membrane proteins, T cell-interacting peptides from both Amuc_RS03735 and Amuc_RS03740 of *Akkermansia muciniphila* specifically induced IgG1 (immunoglobulin G1), instead of the more common IgA in the intestine [80]. Such immune modulation raises the stakes of supplementing the bacterium for more effective PD-1 checkpoint immunotherapy.

6.3.2 Branched chain amino acids for muscles and diabetes

All clades of the *Prevotella copri* complex produce branched chain amino acids (BCAAs) [81]. In this case, even though bacteria other than *Prevotella copri* also contribute to the production and metabolism of BCAAs, *Prevotella copri* makes a causal contribution to BCAAs, and consequently susceptibility to diseases that have been established for BCAAs.

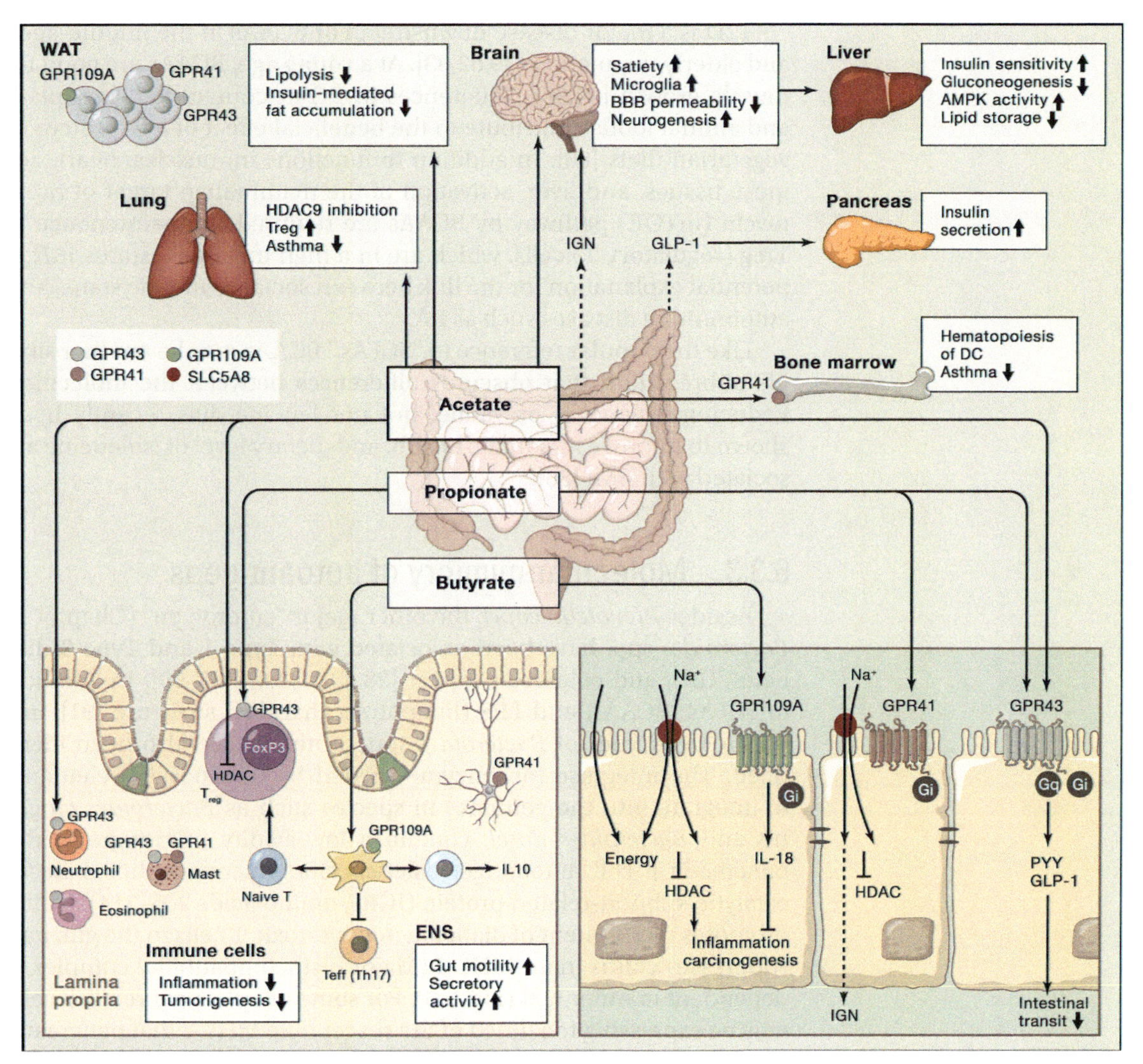

Fig. 6.5 A summary of physiological functions of microbially produced acetate, propionate, and butyrate (SCFAs), including the human receptors. Fermentation of dietary fiber leads to the production of SCFAs via various biochemical pathways. The size of the letters symbolizes the ratio of SCFAs present. In the distal gut, SCFAs can enter the cells through diffusion or SLC5A8-mediated transport and act as an energy source or an HDAC inhibitor. Luminal acetate or propionate sensed by GPR41 and GPR43 releases PYY and GLP-1, affecting satiety and intestinal transit. Luminal butyrate exerts antiinflammatory effects via GPR109A and HDAC inhibition. Furthermore, propionate can be converted into glucose by IGN, leading to satiety and decreased hepatic glucose production. SCFAs can also act on other sites in the gut, like the ENS (enteric neural system), where they stimulate motility and secretory activity, or the immune cells in the lamina propria, where they reduce inflammation and tumorigenesis. Small amounts of SCFAs (mostly acetate and possibly propionate) reach the circulation and can also directly affect the adipose tissue, brain, and liver, inducing overall beneficial metabolic effects. Solid arrows indicate the direct action of each SCFA, and dashed arrows from the gut are indirect effects. For diseases, also see [17]. Credit: Fig. 4 of Koh A, De Vadder F, Kovatcheva-Datchary P, Bäckhed F. From dietary fiber to host physiology: short-chain fatty acids as key bacterial metabolites. Cell 2016;165:1332–345. https://doi.org/10.1016/j.cell.2016.05.041.

T2D is a major disease downstream of BCAAs in the middle-aged and elderly population [39,82,83]. At a young age, BCAAs are good for muscle growth [84,85]. Differences in BCAA content between plant and animal foods contribute to the beneficial effect of low-protein or vegetarian diets [86]. In addition to functions in muscles, heart, adipose tissues, and liver, activation of the mammalian target of rapamycin (mTOR) pathway by BCAAs are required for maintenance of Treg (regulatory T) cells, which are in a high metabolic status [87], a potential explanation for the link between socio-economic status and autoimmune diseases such as RA.

Like the popular reference to "SCFAs," BCAAs may be another simple abbreviation that obscures differences between the molecules. Reducing isoleucine or valine, but not leucine, has recently been shown to improve metabolic health, and dietary level of isoleucine associated with BMI [86].

6.3.3 Molecular mimicry of autoantigens

Besides *Prevotella copri*, the other major "enterotype" (Chapter 2) *Bacteroides* spp. have been associated with Type 1 and Type 2 diabetes, IBD, and colorectal cancer [28,30,31,39,74,88–90]. In addition to SCFAs, BCAAs, and LPS (lipopolysaccharides) structure [91], antigenic properties of *Bacteroides* spp. proteins have also been identified. The integrase (an enzyme carried by transposable elements to integrate into the genome) in species such as *Bacteroides vulgatus* and *Bacteroides dorei*, contain a low-avidity mimotope of the pancreatic β cell autoantigen islet-specific glucose-6-phosphatase-catalytic-subunit-related protein (IGRP, amino acids 206–214), which promotes recruitment of diabetogenic cytotoxic T cells to the gut, and suppresses colitis in an MHC-I (Major histocompatibility complex I) dependent manner [92] (Fig. 6.6). For some people, pancreatic β cells may be expressing too much of the susceptible MHC-I that pancreatic β cells, instead of autoantigen-loaded dendritic cells (DC) are killed in the pancreatic lymph node (PLN).

Another peptide from an outer membrane polysaccharide utilization (PUL) protein of *Bacteroides thetaiotaomicron* (BT4295, amino acids 541–554) was recognized by T-cell receptors (TCRs), without known mimicry, and also contribute to protection against colitis through Treg cells in a mouse model [93]. BT4295 is incorporated into outer membrane vesicles (OMVs), and BT4295 expression is downregulated by dietary glucose [93]. The *Bacteroides* "enterotype" may need to be further divided into subtypes such as *Bacteroides thetaiotaomicron* versus *Bacteroides uniformis*, for potentially different susceptibility to inflammation and metabolic diseases.

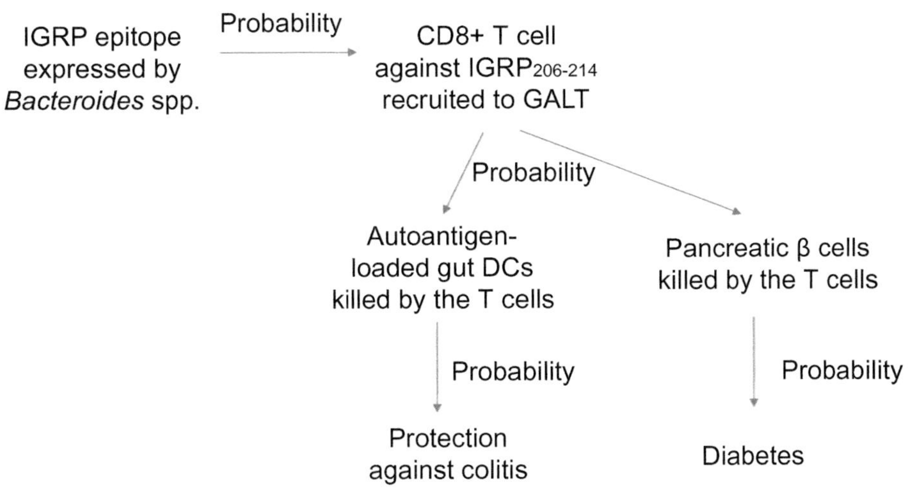

Fig. 6.6 Working model in path diagrams for the autoantigen in *Bacteroides* integrase. Summarized from [92]. There is a probability for each step, depending on the cell populations and their evolutionary history. Some people may never get the disease. Credit: Huijue Jia.

Molecular mimicry of the gut and oral microbiome to self-antigens has been tentatively explored for RA, ankylosing spondylitis, and Sjogren's syndrome [94–96]. For systemic lupus erythematosus (SLE), microbiome from multiple body sites of both healthy adults and SLE patients contain homologous sequences that map to the autoantigen Ro60 [97]. Patient T cells that react to the Ro60-mimic from skin *Propionibacterium propionicum* and from gut *Bacteroides thetaiotaomicron* showed some activity against human Ro60; Patients T cells against human Ro60 also cross-reacted with *Propionibacterium propionicum* [97].

6.3.4 Other examples, outer membrane vesicles, phages

One explanation for the enrichment of *Escherichia coli* in prediabetic individuals (Table 6.1) is its production of indole, which stimulates enteroendocrine L cells to produce glucagon-like peptide-1 (GLP-1) and therefore insulin secretion by pancreatic β cells [98]. So the increase of *Escherichia coli*, together with *Akkermansia muciniphila*, in the mice model of RYGB [63] is likely central to the effect of the surgery in treating metabolic syndrome.

This is of course a very versatile bacterium with many other functions. In pancreatic cancer, cytidine deaminase (CDD_L) encoded by *Escherichia coli* could metabolize the chemotherapeutic drug gentamicin (2′,2′-difluorodeoxycytidine) (Fig. 6.7) [99]. Such in vitro experiments after bioinformatic analyses are very useful for our understanding of individual members of the microbiome.

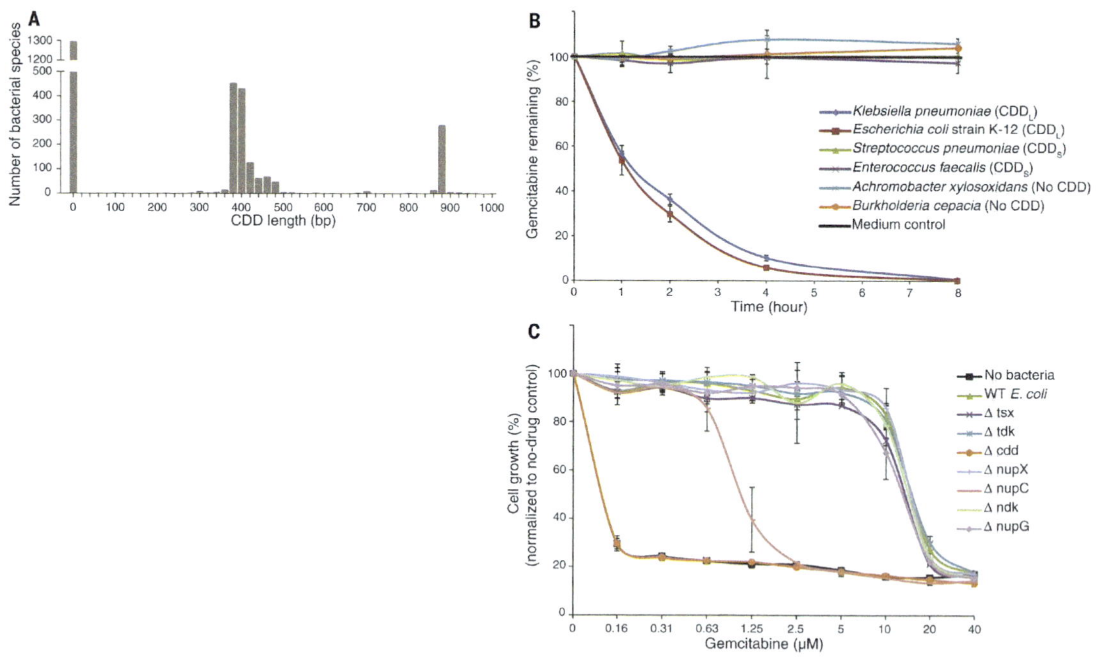

Fig. 6.7 Experiments to show that the long isoform of intestinal bacterial cytidine deaminase (CDD) mediates gemcitabine metabolism. (A) Histogram of CDD DNA sequence length across all bacteria in the KEGG database. bp, base pair. (B) Gemcitabine (4 mM) was incubated with 107 bacteria in an M9 minimal salt medium. Bacteria were filtered from the media at different time points, and the remaining gemcitabine was detected by HPLCMS/MS. Bars represent the standard deviation between two biological replicates, each containing two technical repeats. (C) WT (wild-type parental) *Escherichia coli* strain K-12 (long CDD), CDD knockout (delta) strains of the parental *Escherichia coli*, and bacteria-free media were each incubated with different gemcitabine concentrations for 4 h. Bacteria were then filtered out, and the flow-through media were added to GFP-labeled AsPC1 human pancreatic adenocarcinoma cells. The growth of AsPC1 cells after 7 days, as measured by GFP, was normalized to a no-drug control. Bars represent the standard deviation between four replicates. Credit: Fig. 2 of Geller LT, Barzily-Rokni M, Danino T, Jonas OH, Shental N, Nejman D, et al. Potential role of intratumor bacteria in mediating tumor resistance to the chemotherapeutic drug gemcitabine. Science 2017;357:1156–160. https://doi.org/10.1126/science.aah5043.

In the lung, outer membrane vesicles (OMVs) from *Bacteroides ovatus*, *Bacteroides stercoris*, and *Prevotella melaninogenica* could induce expression of the cytokine interleukin-17B (IL-17B) by alveolar macrophages and promote lung fibrosis [100].

Production of OMVs is no longer a feature of Gram-negative bacteria only; Gram-positive bacteria can also release membrane vesicles [101]. The vaginal Lactobacilli strains *L. crispatus* BC3 and *L. gasseri* BC12 produce vesicles that are over 100 nm in diameter, which decrease the adhesion of HIV-1 (Human immunodeficiency virus-1) to target cells and protect against viral entry [102]. This OMV pathway of viral defense involves host cells, in addition to better-known functions of vaginal Lactobacilli against other bacteria, through the production of lactic acid and hydrogen peroxide.

For alcoholic liver disease, patients with cytolysin in the fecal microbiome, e.g., cytolysin expressed by some *Enterococcus faecalis* strains, have been reported to show higher mortality than patients without fecal cytolysin [103]. Bacteriophages that target cytolysin-expressing *Enterococcus faecalis* were shown in mice to reduce ethanol-induced liver injury and steatosis [103].

6.4 Summary

As a data-driven field of research, metagenomic studies discover many associations, which are Level 1 evidence on the way to causality. Randomized controlled trials (RCTs) and Mendelian randomization (MR) in large cohorts are providing Level 2 evidence for some of the associations (Table 6.1). Prospective cohorts and more basic research on the microbiome will add to our confidence in Level 3 evidence—counterfactuals, which will guide public health decisions. If our understanding of members of the human microbiome and their products are as thorough as some of the traditional pathogens, there would be little doubt in a causal role, even if the molecular mechanisms and the interactions are more complicated. However, things always need to be prioritized.

References

[1] Tortora GJ, Funke BR, CL CT, editors. Chapter 14 Principles of disease and epidemiology copyright. In: Microbiol. An introd. 10th ed. Pearson; 2010.

[2] Tu A-HT. 15.2 How pathogens cause disease—microbiology | OpenStax. In: Parker N, Schneegurt M, Lister PM, Forster B, editors. Microbiology. The American Society for Microbiology Press; 2016.

[3] Dickson RP, Erb-Downward JR, Martinez FJ, Huffnagle GB. The microbiome and the respiratory tract. Annu Rev Physiol 2016;78:481–504. https://doi.org/10.1146/annurev-physiol-021115-105238.

[4] Levine R, Evers C. The slow death of spontaneous generation (1668–1859). Biotech Chronicles. https://webprojects.oit.ncsu.edu/project/bio183de/Black/cellintro/cellintro_reading/Spontaneous_Generation.html [Accessed 5 July 2021].

[5] Jaynes ET. Probability theory: the logic of science. Cambridge: Cambridge University Press; 2003.

[6] Pearl J, Mackenzie D. The book of why: the new science of cause and effect. 1st. New York: Basic Books; 2018.

[7] Ding T, Schloss PD. Dynamics and associations of microbial community types across the human body. Nature 2014;509:357–60. https://doi.org/10.1038/nature13178.

[8] Valles-Colomer M, Falony G, Darzi Y, Tigchelaar EF, Wang J, Tito RY, et al. The neuroactive potential of the human gut microbiota in quality of life and depression. Nat Microbiol 2019. https://doi.org/10.1038/s41564-018-0337-x.

[9] Ben-Ari Y. Is birth a critical period in the pathogenesis of autism spectrum disorders? Nat Rev Neurosci 2015;16:498–505. https://doi.org/10.1038/nrn3956.
[10] Apps PJ, Weldon PJ, Kramer M. Chemical signals in terrestrial vertebrates: search for design features. Nat Prod Rep 2015;32:1131–53. https://doi.org/10.1039/c5np00029g.
[11] Powell N, Walker MM, Talley NJ. The mucosal immune system: master regulator of bidirectional gut–brain communications. Nat Rev Gastroenterol Hepatol 2017;14:143–59. https://doi.org/10.1038/nrgastro.2016.191.
[12] Vujkovic-Cvijin I, Sklar J, Jiang L, Natarajan L, Knight R, Belkaid Y. Host variables confound gut microbiota studies of human disease. Nature 2020;2020:1–7. https://doi.org/10.1038/s41586-020-2881-9.
[13] Qi Q, Li J, Yu B, Moon J-Y, Chai JC, Merino J, et al. Host and gut microbial tryptophan metabolism and type 2 diabetes: an integrative analysis of host genetics, diet, gut microbiome and circulating metabolites in cohort studies. Gut 2021. https://doi.org/10.1136/gutjnl-2021-324053.
[14] Chassaing B, Koren O, Goodrich JK, Poole AC, Srinivasan S, Ley RE, et al. Dietary emulsifiers impact the mouse gut microbiota promoting colitis and metabolic syndrome. Nature 2015;519:92–6. https://doi.org/10.1038/nature14232.
[15] Turk JN, Zahavi ER, Gorman AE, Murray K, Turk MA, Veale DJ. Exploring the effect of alcohol on disease activity and outcomes in rheumatoid arthritis through systematic review and meta-analysis. Sci Rep 2021;11:10474. https://doi.org/10.1038/s41598-021-89618-1.
[16] Kolber MA, Walls RM, Hinners ML, Singer DS. Evidence of increased class I MHC expression on human peripheral blood lymphocytes during acute ethanol intoxication. Alcohol Clin Exp Res 1988;12:820–3. https://doi.org/10.1111/j.1530-0277.1988.tb01353.x.
[17] Nicolas GR, Chang PV. Deciphering the chemical lexicon of host–gut microbiota interactions. Trends Pharmacol Sci 2019;40:430–45. https://doi.org/10.1016/j.tips.2019.04.006.
[18] Azizov V, Dietel K, Steffen F, Dürholz K, Meidenbauer J, Lucas S, et al. Ethanol consumption inhibits TFH cell responses and the development of autoimmune arthritis. Nat Commun 2020;11:1998. https://doi.org/10.1038/s41467-020-15855-z.
[19] Wu W, Sun M, Chen F, Cao AT, Liu H, Zhao Y, et al. Microbiota metabolite short-chain fatty acid acetate promotes intestinal IgA response to microbiota which is mediated by GPR43. Mucosal Immunol 2017;10:946–56. https://doi.org/10.1038/mi.2016.114.
[20] Cao Y, Lin W, Li H. Large covariance estimation for compositional data via composition-adjusted thresholding. J Am Stat Assoc 2018;1–45. https://doi.org/10.1080/01621459.2018.1442340.
[21] Friedman J, Alm EJ. Inferring correlation networks from genomic survey data. PLoS Comput Biol 2012;8. https://doi.org/10.1371/journal.pcbi.1002687, e1002687.
[22] Holmes MV, Ala-Korpela M, Smith GD. Mendelian randomization in cardiometabolic disease: challenges in evaluating causality. Nat Rev Cardiol 2017;14:577–90. https://doi.org/10.1038/nrcardio.2017.78.
[23] Liu X, Tang S, Zhong H, Tong X, Jie Z, Ding Q, et al. A genome-wide association study for gut metagenome in Chinese adults illuminates complex diseases. Cell Discov 2021;7(1):9. https://doi.org/10.1038/s41421-020-00239-w.
[24] Jie Z, Xia H, Zhong S-L, Feng Q, Li S, Liang S, et al. The gut microbiome in atherosclerotic cardiovascular disease. Nat Commun 2017;8:845. https://doi.org/10.1038/s41467-017-00900-1.
[25] Liu X, Tong X, Zou Y, Lin X, Zhao H, Tian L, et al. Inter-determination of blood metabolite levels and gut microbiome supported by Mendelian randomization. BioRxiv 2020. https://doi.org/10.1101/2020.06.30.181438. 2020.06.30.

[26] Jiang X, Alfredsson L. Modifiable environmental exposure and risk of rheumatoid arthritis - current evidence from genetic studies. Arthritis Res Ther 2020;22. https://doi.org/10.1186/s13075-020-02253-5.
[27] Karlsson FH, Tremaroli V, Nookaew I, Bergström G, Behre CJ, Fagerberg B, et al. Gut metagenome in European women with normal, impaired and diabetic glucose control. Nature 2013;498:99–103. https://doi.org/10.1038/nature12198.
[28] Zhong H, Ren H, Lu Y, Fang C, Hou G, Yang Z, et al. Distinct gut metagenomics and metaproteomics signatures in prediabetics and treatment-naïve type 2 diabetics. EBioMedicine 2019. https://doi.org/10.1016/j.ebiom.2019.08.048.
[29] Jie Z, Liang S, Ding Q, Li F, Sun X, Lin Y, et al. Dairy consumption and physical fitness tests associated with fecal microbiome in a Chinese cohort. Med Microecol 2021. https://doi.org/10.1016/j.medmic.2021.100038.
[30] Schüssler-Fiorenza Rose SM, Contrepois K, Moneghetti KJ, Zhou W, Mishra T, Mataraso S, et al. A longitudinal big data approach for precision health. Nat Med 2019;25:792–804. https://doi.org/10.1038/s41591-019-0414-6.
[31] Qin J, Li Y, Cai Z, Li S, Zhu J, Zhang F, et al. A metagenome-wide association study of gut microbiota in type 2 diabetes. Nature 2012;490:55–60. https://doi.org/10.1038/nature11450.
[32] Jackson M, Jeffery IB, Beaumont M, Bell JT, Clark AG, Ley RE, et al. Signatures of early frailty in the gut microbiota. Genome Med 2016;8:8. https://doi.org/10.1186/s13073-016-0262-7.
[33] Jie Z, Liang S, Ding Q, Li F, Tang S, Wang D, et al. A transomic cohort as a reference point for promoting a healthy gut microbiome. Med Microecol 2021. https://doi.org/10.1016/j.medmic.2021.100039.
[34] Png CW, Lindén SK, Gilshenan KS, Zoetendal EG, McSweeney CS, Sly LI, et al. Mucolytic bacteria with increased prevalence in IBD mucosa augment in vitro utilization of mucin by other bacteria. Am J Gastroenterol 2010;105:2420–8. https://doi.org/10.1038/ajg.2010.281.
[35] Liu X, Tong X, Zhu J, Tian L, Jie Z, Zou Y, et al. Metagenome-genome-wide association studies reveal human genetic impact on the oral microbiome. bioRxiv 2021. https://doi.org/10.1101/2021.05.06.443017.
[36] Zhang Y, Gu Y, Ren H, Wang S, Zhong H, Zhao X, et al. Gut microbiome-related effects of berberine and probiotics on type 2 diabetes (the PREMOTE study). Nat Commun 2020;11:5015. https://doi.org/10.1038/s41467-020-18414-8.
[37] Cohen CR, Wierzbicki MR, French AL, Morris S, Newmann S, Reno H, et al. Randomized trial of lactin-V to prevent recurrence of bacterial vaginosis. N Engl J Med 2020;382:1906–15. https://doi.org/10.1056/NEJMoa1915254.
[38] Xu F, Fu Y, Sun TY, Jiang Z, Miao Z, Shuai M, et al. The interplay between host genetics and the gut microbiome reveals common and distinct microbiome features for complex human diseases. Microbiome 2020;8:145. https://doi.org/10.1186/s40168-020-00923-9.
[39] Pedersen HK, Gudmundsdottir V, Nielsen HB, Hyotylainen T, Nielsen T, Jensen BAH, et al. Human gut microbes impact host serum metabolome and insulin sensitivity. Nature 2016;535:376–81. https://doi.org/10.1038/nature18646.
[40] Mukherji A, Kobiita A, Ye T, Chambon P. Homeostasis in intestinal epithelium is orchestrated by the circadian clock and microbiota cues transduced by TLRs. Cell 2013;153:812–27. https://doi.org/10.1016/j.cell.2013.04.020.
[41] Braniste V, Al-Asmakh M, Kowal C, Anuar F, Abbaspour A, Tóth M, et al. The gut microbiota influences blood-brain barrier permeability in mice. Sci Transl Med 2014;6. https://doi.org/10.1126/scitranslmed.3009759, 263ra158.
[42] Olszak T, An D, Zeissig S, Vera MP, Richter J, Franke A, et al. Microbial exposure during early life has persistent effects on natural killer T cell function. Science 2012;336:489–93. https://doi.org/10.1126/science.1219328.

[43] Xiao L, Feng Q, Liang S, Sonne SB, Xia Z, Qiu X, et al. A catalog of the mouse gut metagenome. Nat Biotechnol 2015;33:1103–8. https://doi.org/10.1038/nbt.3353.

[44] Rausch P, Basic M, Batra A, Bischoff SC, Blaut M, Clavel T, et al. Analysis of factors contributing to variation in the C57BL/6J fecal microbiota across German animal facilities. Int J Med Microbiol 2016. https://doi.org/10.1016/j.ijmm.2016.03.004.

[45] Wroblewski LE, Peek RM, Wilson KT. Helicobacter pylori and gastric cancer: factors that modulate disease risk. Clin Microbiol Rev 2010;23:713–39. https://doi.org/10.1128/CMR.00011-10.

[46] Chen X-H, Wang A, Chu A-N, Gong Y-H, Yuan Y. Mucosa-associated microbiota in gastric cancer tissues compared with non-cancer tissues. Front Microbiol 2019;10:1261. https://doi.org/10.3389/fmicb.2019.01261.

[47] Bugaytsova JA, Björnham O, Chernov YA, Gideonsson P, Henriksson S, Mendez M, et al. *Helicobacter pylori* adapts to chronic infection and gastric disease via pH-responsive BabA-mediated adherence. Cell Host Microbe 2017;21:376–89. https://doi.org/10.1016/j.chom.2017.02.013.

[48] Poole AC, Goodrich JK, Youngblut ND, Luque GG, Ruaud A, Sutter JL, et al. Human salivary amylase gene copy number impacts oral and gut microbiomes. Cell Host Microbe 2019;25(4):553–564.e7. https://doi.org/10.1016/j.chom.2019.03.001.

[49] Zhu J, Tian L, Chen P, Han M, Song L, Tong X, et al. Over 50,000 metagenomically assembled draft genomes for the human oral microbiome reveal new taxa. Genomics Proteomics Bioinformatics 2021;18:2790. https://doi.org/10.1016/j.gpb.2021.05.001.

[50] Atarashi K, Suda W, Luo C, Kawaguchi T, Motoo I, Narushima S, et al. Ectopic colonization of oral bacteria in the intestine drives T H 1 cell induction and inflammation. Science 2017;358:359–65. https://doi.org/10.1126/science.aan4526.

[51] Bajaj JS, Matin P, White MB, Fagan A, Golob Deeb J, Acharya C, et al. Periodontal therapy favorably modulates the oral-gut-hepatic axis in cirrhosis. Am J Physiol Liver Physiol 2018. https://doi.org/10.1152/ajpgi.00230.2018. ajpgi.00230.2018.

[52] Sobko T, Reinders CI, Jansson E, Norin E, Midtvedt T, Lundberg JO. Gastrointestinal bacteria generate nitric oxide from nitrate and nitrite. Nitric Oxide 2005;13:272–8. https://doi.org/10.1016/j.niox.2005.08.002.

[53] Hata DJ, Smith DS. Blood group B degrading activity of *Ruminococcus gnavus* α-galactosidase. Artif Cells Blood Substit Immobil Biotechnol 2004;32:263–74. https://doi.org/10.1081/BIO-120037831.

[54] Tailford LE, Owen CD, Walshaw J, Crost EH, Hardy-Goddard J, Le Gall G, et al. Discovery of intramolecular trans-sialidases in human gut microbiota suggests novel mechanisms of mucosal adaptation. Nat Commun 2015;6:7624. https://doi.org/10.1038/ncomms8624.

[55] Henke MT, Kenny DJ, Cassilly CD, Vlamakis H, Xavier RJ, Clardy J. *Ruminococcus gnavus*, a member of the human gut microbiome associated with Crohn's disease, produces an inflammatory polysaccharide. Proc Natl Acad Sci U S A 2019;116:12672–7. https://doi.org/10.1073/pnas.1904099116.

[56] Ng KM, Ferreyra JA, Higginbottom SK, Lynch JB, Kashyap PC, Gopinath S, et al. Microbiota-liberated host sugars facilitate post-antibiotic expansion of enteric pathogens. Nature 2013;502:96–9. https://doi.org/10.1038/nature12503.

[57] McGregor JA, French JI, Jones W, Milligan K, McKinney PJ, Patterson E, et al. Bacterial vaginosis is associated with prematurity and vaginal fluid mucinase and sialidase: results of a controlled trial of topical clindamycin cream. Am J Obstet Gynecol 1994;170:1048–59. discussion 1059-60 https://doi.org/10.1016/s0002-9378(94)70098-2.

[58] dos Santos Santiago GL, Tency I, Verstraelen H, Verhelst R, Trog M, Temmerman M, et al. Longitudinal qPCR study of the dynamics of *L. crispatus, L. iners, A. vaginae,* (sialidase positive) *G. vaginalis,* and *P. bivia* in the vagina. PLoS One 2012;7. https://doi.org/10.1371/journal.pone.0045281, e45281.

[59] Bacher P, Hohnstein T, Beerbaum E, Röcker M, Blango MG, Kaufmann S, et al. Human anti-fungal Th17 immunity and pathology rely on cross-reactivity against *Candida albicans*. Cell 2019. https://doi.org/10.1016/j.cell.2019.01.041.

[60] Boldock E, Surewaard BGJ, Shamarina D, Na M, Fei Y, Ali A, et al. Human skin commensals augment *Staphylococcus aureus* pathogenesis. Nat Microbiol 2018. https://doi.org/10.1038/s41564-018-0198-3.

[61] Everard A, Belzer C, Geurts L, Ouwerkerk JP, Druart C, Bindels LB, et al. Cross-talk between *Akkermansia muciniphila* and intestinal epithelium controls diet-induced obesity. Proc Natl Acad Sci U S A 2013;110:9066–71. https://doi.org/10.1073/pnas.1219451110.

[62] Seck EH, Senghor B, Merhej V, Bachar D, Cadoret F, Robert C, et al. Salt in stools is associated with obesity, gut halophilic microbiota and *Akkermansia muciniphila* depletion in humans. Int J Obes (Lond) 2019;43:862–71. https://doi.org/10.1038/s41366-018-0201-3.

[63] Liou AP, Paziuk M, Luevano J-M, Machineni S, Turnbaugh PJ, Kaplan LM. Conserved shifts in the gut microbiota due to gastric bypass reduce host weight and adiposity. Sci Transl Med 2013;5:178ra41. https://doi.org/10.1126/scitranslmed.3005687.

[64] Wu H, Esteve E, Tremaroli V, Khan MT, Caesar R, Mannerås-Holm L, et al. Metformin alters the gut microbiome of individuals with treatment-naive type 2 diabetes, contributing to the therapeutic effects of the drug. Nat Med 2017;23:850–8. https://doi.org/10.1038/nm.4345.

[65] Plovier H, Everard A, Druart C, Depommier C, Van Hul M, Geurts L, et al. A purified membrane protein from *Akkermansia muciniphila* or the pasteurized bacterium improves metabolism in obese and diabetic mice. Nat Med 2017;23:107–13. https://doi.org/10.1038/nm.4236.

[66] Depommier C, Everard A, Druart C, Plovier H, Van Hul M, Vieira-Silva S, et al. Supplementation with *Akkermansia muciniphila* in overweight and obese human volunteers: a proof-of-concept exploratory study. Nat Med 2019. https://doi.org/10.1038/s41591-019-0495-2.

[67] Chevalier C, Stojanović O, Colin DJ, Suarez-Zamorano N, Tarallo V, Veyrat-Durebex C, et al. Gut microbiota orchestrates energy homeostasis during cold. Cell 2015;163:1360–74. https://doi.org/10.1016/j.cell.2015.11.004.

[68] Yoon HS, Cho CH, Yun MS, Jang SJ, You HJ, Hyeong KJ, et al. *Akkermansia muciniphila* secretes a glucagon-like peptide-1-inducing protein that improves glucose homeostasis and ameliorates metabolic disease in mice. Nat Microbiol 2021;1–11. https://doi.org/10.1038/s41564-021-00880-5.

[69] Perry RJ, Peng L, Barry NA, Cline GW, Zhang D, Cardone RL, et al. Acetate mediates a microbiome–brain–β-cell axis to promote metabolic syndrome. Nature 2016;534:213–7. https://doi.org/10.1038/nature18309.

[70] Sanna S, van Zuydam NR, Mahajan A, Kurilshikov A, Vich Vila A, Võsa U, et al. Causal relationships among the gut microbiome, short-chain fatty acids and metabolic diseases. Nat Genet 2019;51:600–5. https://doi.org/10.1038/s41588-019-0350-x.

[71] De Vadder F, Kovatcheva-Datchary P, Zitoun C, Duchampt A, Bäckhed F, Mithieux G. Microbiota-produced succinate improves glucose homeostasis via intestinal gluconeogenesis. Cell Metab 2016;24:151–7. https://doi.org/10.1016/j.cmet.2016.06.013.

[72] Wen C, Yan W, Mai C, Duan Z, Zheng J, Sun C, et al. Joint contributions of the gut microbiota and host genetics to feed efficiency in chickens. Microbiome 2021;9:126. https://doi.org/10.1186/s40168-021-01040-x.

[73] Weir TL, Manter DK, Sheflin AM, Barnett BA, Heuberger AL, Ryan EP. Stool microbiome and metabolome differences between colorectal cancer patients and healthy adults. PLoS One 2013;8. https://doi.org/10.1371/journal.pone.0070803, e70803.

[74] Feng Q, Liang S, Jia H, Stadlmayr A, Tang L, Lan Z, et al. Gut microbiome development along the colorectal adenoma–carcinoma sequence. Nat Commun 2015;6:6528. https://doi.org/10.1038/ncomms7528.

[75] Liu P, Wu L, Peng G, Han Y, Tang R, Ge J, et al. Altered microbiomes distinguish Alzheimer's disease from amnestic mild cognitive impairment and health in a Chinese cohort. Brain Behav Immun 2019. https://doi.org/10.1016/j.bbi.2019.05.008.

[76] Zhu F, Ju Y, Wang W, Wang Q, Guo R, Ma Q, et al. Metagenome-wide association of gut microbiome features for schizophrenia. Nat Commun 2020;11:1612. https://doi.org/10.1038/s41467-020-15457-9.

[77] Xie H, Guo R, Zhong H, Feng Q, Lan Z, Qin B, et al. Shotgun metagenomics of 250 adult twins reveals genetic and environmental impacts on the gut microbiome. Cell Syst 2016;3:572–584.e3. https://doi.org/10.1016/j.cels.2016.10.004.

[78] Guo X, Li S, Zhang J, Wu F, Li X, Wu D, et al. Genome sequencing of 39 *Akkermansia muciniphila* isolates reveals its population structure, genomic and functional diverisity, and global distribution in mammalian gut microbiotas. BMC Genomics 2017;18:800. https://doi.org/10.1186/s12864-017-4195-3.

[79] Johansson MEV, Hansson GC. Immunological aspects of intestinal mucus and mucins. Nat Rev Immunol 2016;16:639–49. https://doi.org/10.1038/nri.2016.88.

[80] Ansaldo E, Slayden LC, Ching KL, Koch MA, Wolf NK, Plichta DR, et al. *Akkermansia muciniphila* induces intestinal adaptive immune responses during homeostasis. Science 2019;364:1179–84. https://doi.org/10.1126/science.aaw7479.

[81] Tett A, Huang KD, Asnicar F, Fehlner-Peach H, Pasolli E, Karcher N, et al. The *Prevotella copri* complex comprises four distinct clades underrepresented in westernized populations. Cell Host Microbe 2019. https://doi.org/10.1016/j.chom.2019.08.018.

[82] Newgard CB. Interplay between lipids and branched-chain amino acids in development of insulin resistance. Cell Metab 2012;15:606–14. https://doi.org/10.1016/j.cmet.2012.01.024.

[83] Neinast MD, Jang C, Hui S, Murashige DS, Chu Q, Morscher RJ, et al. Quantitative analysis of the whole-body metabolic fate of branched-chain amino acids. Cell Metab 2019;29:417–429.e4. https://doi.org/10.1016/j.cmet.2018.10.013.

[84] Lim MT, Pan BJ, Toh DWK, Sutanto CN, Kim JE. Animal protein versus plant protein in supporting lean mass and muscle strength: a systematic review and meta-analysis of randomized controlled trials. Nutrients 2021;13. https://doi.org/10.3390/nu13020661.

[85] Duan Y, Guo Q, Wen C, Wang W, Li Y, Tan B, et al. Free amino acid profile and expression of genes implicated in protein metabolism in skeletal muscle of growing pigs fed low-protein diets supplemented with branched-chain amino acids. J Agric Food Chem 2016;64:9390–400. https://doi.org/10.1021/acs.jafc.6b03966.

[86] Yu D, Richardson NE, Green CL, Spicer AB, Murphy ME, Flores V, et al. The adverse metabolic effects of branched-chain amino acids are mediated by isoleucine and valine. Cell Metab 2021;33:905–922.e6. https://doi.org/10.1016/j.cmet.2021.03.025.

[87] Ikeda K, Kinoshita M, Kayama H, Nagamori S, Kongpracha P, Umemoto E, et al. Slc3a2 mediates branched-chain amino-acid-dependent maintenance of regulatory T cells. Cell Rep 2017;21:1824–38. https://doi.org/10.1016/j.celrep.2017.10.082.

[88] He Q, Gao Y, Jie Z, Yu X, Laursen JMJM, Xiao L, et al. Two distinct metacommunities characterize the gut microbiota in Crohn's disease patients. Gigascience 2017;6:1–11. https://doi.org/10.1093/gigascience/gix050.

[89] de Groot P, Nikolic T, Pellegrini S, Sordi V, Imangaliyev S, Rampanelli E, et al. Faecal microbiota transplantation halts progression of human new-onset type 1 diabetes in a randomised controlled trial. Gut 2020. https://doi.org/10.1136/gutjnl-2020-322630. gutjnl-2020-322630.

[90] Paun A, Yau C, Meshkibaf S, Daigneault MC, Marandi L, Mortin-Toth S, et al. Association of HLA-dependent islet autoimmunity with systemic antibody responses to intestinal commensal bacteria in children. Sci Immunol 2019;4. https://doi.org/10.1126/sciimmunol.aau8125, eaau8125.

[91] Vatanen T, Kostic AD, D'Hennezel E, Siljander H, Franzosa EA, Yassour M, et al. Variation in microbiome LPS immunogenicity contributes to autoimmunity in humans. Cell 2016;165:842–53. https://doi.org/10.1016/j.cell.2016.04.007.

[92] Hebbandi Nanjundappa R, Ronchi F, Wang J, Clemente-Casares X, Yamanouchi J, Sokke Umeshappa C, et al. A gut microbial mimic that hijacks diabetogenic autoreactivity to suppress colitis. Cell 2017;171:655–667.e17. https://doi.org/10.1016/j.cell.2017.09.022.

[93] Wegorzewska MM, Glowacki RWP, Hsieh SA, Donermeyer DL, Hickey CA, Horvath SC, et al. Diet modulates colonic T cell responses by regulating the expression of a *Bacteroides thetaiotaomicron* antigen. Sci Immunol 2019;4:eaau9079. https://doi.org/10.1126/sciimmunol.aau9079.

[94] Zhang X, Zhang D, Jia H, Feng Q, Wang D, Di Liang D, et al. The oral and gut microbiomes are perturbed in rheumatoid arthritis and partly normalized after treatment. Nat Med 2015;21:895–905. https://doi.org/10.1038/nm.3914.

[95] Zhou C, Zhao H, Xiao X, Chen B, Guo R, Wang Q, et al. Metagenomic profiling of the pro-inflammatory gut microbiota in ankylosing spondylitis. J Autoimmun 2020;107:102360. https://doi.org/10.1016/j.jaut.2019.102360.

[96] Szymula A, Rosenthal J, Szczerba BM, Bagavant H, Fu SM, Deshmukh US. T cell epitope mimicry between Sjögren's syndrome antigen a (SSA)/Ro60 and oral, gut, skin and vaginal bacteria. Clin Immunol 2014;152:1–9. https://doi.org/10.1016/j.clim.2014.02.004.

[97] Greiling TM, Dehner C, Chen X, Hughes K, Iñiguez AJ, Boccitto M, et al. Commensal orthologs of the human autoantigen Ro60 as triggers of autoimmunity in lupus. Sci Transl Med 2018;10. https://doi.org/10.1126/scitranslmed.aan2306, eaan2306.

[98] Agus A, Planchais J, Sokol H. Gut microbiota regulation of tryptophan metabolism in health and disease. Cell Host Microbe 2018;23:716–24. https://doi.org/10.1016/j.chom.2018.05.003.

[99] Geller LT, Barzily-Rokni M, Danino T, Jonas OH, Shental N, Nejman D, et al. Potential role of intratumor bacteria in mediating tumor resistance to the chemotherapeutic drug gemcitabine. Science 2017;357:1156–60. https://doi.org/10.1126/science.aah5043.

[100] Yang D, Chen X, Wang J, Lou Q, Lou Y, Li L, et al. Dysregulated lung commensal bacteria drive interleukin-17B production to promote pulmonary fibrosis through their outer membrane vesicles. Immunity 2019;50:692–706.e7. https://doi.org/10.1016/j.immuni.2019.02.001.

[101] Toyofuku M, Nomura N, Eberl L. Types and origins of bacterial membrane vesicles. Nat Rev Microbiol 2019;17:13–24. https://doi.org/10.1038/s41579-018-0112-2.

[102] Ñahui Palomino RA, Vanpouille C, Laghi L, Parolin C, Melikov K, Backlund P, et al. Extracellular vesicles from symbiotic vaginal lactobacilli inhibit HIV-1 infection of human tissues. Nat Commun 2019;10:5656. https://doi.org/10.1038/s41467-019-13468-9.

[103] Duan Y, Llorente C, Lang S, Brandl K, Chu H, Jiang L, et al. Bacteriophage targeting of gut bacterium attenuates alcoholic liver disease. Nature 2019. https://doi.org/10.1038/s41586-019-1742-x.

7

Metagenomics from bench to bedside and from bedside to bench

7.1 Metagenomics for decision-making in diagnosis and treatment

7.1.1 Metagenomics for disease screening

As we can reasonably believe based on the previous chapters, the human microbiome at various body sites could contribute to and respond to the shifting trends of human diseases in the last few decades (Fig. 7.1). When trying to improve living conditions, nutrition, and hygiene practices in an underdeveloped region, it would also probably be better to beware of the changes to expect in the human microbiome and disease prevalence.

While scientists are always excited about new technologies, in order for metagenomic sequencing to be routinely used in hospitals, it will be important to have a clear view of what is really needed for the disease in question (Table 7.1). For a patient with liver disease that is about to enter a coma? For a child with leukemia that is about to receive a bone marrow transplant? Adult data show that a very low-diversity fecal microbiome before and during the transplant could cost one's life [1–3]; Similar evidence is also emerging for pediatric patients, including fecal, oral, and nasal microbiome before and after the bone marrow transplant [4,5].

With vaccination against Hepatitis B Virus (HBV), reduced exposure to aflatoxin, and medication to treat Hepatitis C Virus (HCV), the trend for liver diseases has shifted. Nonalcoholic fatty liver disease (NAFLD), following the global increase in obesity, is by far the most prevalent liver disease on the way to acute hospitalizations or hepatocellular carcinoma (HCC) [6]. NAFLD includes a spectrum of conditions from hepatic steatosis (fatty liver), nonalcoholic steatohepatitis (NASH), and cirrhosis. Fatty liver is quite common from ultrasound examinations, and levels of liver enzymes such as ALT (alanine aminotransferase) and GGT (γ-glutamyl transpeptidase) can also be high

Investigating Human Diseases with the Microbiome: Metagenomics Bench to Bedside.
https://doi.org/10.1016/B978-0-323-91369-0.00006-6

Females

Leading causes 1990	Leading causes 2007	Mean percentage change in number of prevalent cases, 1990–2007	Mean percentage change in all-age prevalence rate, 1990–2007	Mean percentage change in age-standardised prevalence rate, 1990–2007	Leading causes 2017	Mean percentage change in number of prevalent cases, 2007–17	Mean percentage change in all-age prevalence rate, 2007–17	Mean percentage change in age-standardised prevalence rate, 2007–17
1 Oral disorders	1 Oral disorders	23·1	−2·0	−3·8	1 Oral disorders	13·5	0·3	−1·3
2 Headache disorders	2 Headache disorders	31·5	4·7	−0·4	2 Headache disorders	14·5	1·2	0·3
3 Haemoglobinopathies	3 Haemoglobinopathies	29·9	3·3	4·2	3 Haemoglobinopathies	13·4	0·2	0·8
4 Tuberculosis	4 Tuberculosis	27·7	1·6	−2·2	4 Tuberculosis	1·2	−10·6	−11·7
5 Intestinal nematode	5 Gynaecological diseases	34·0	6·6	−2·3	5 Gynaecological diseases	13·3	0·1	−0·5
6 Dietary iron deficiency	6 STIs	40·2	11·6	1·7	6 STIs	17·7	4·0	0·7
7 Gynaecological diseases	7 Dietary iron deficiency	7·2	−14·7	−14·5	7 Blindness and vision impairment	24·1	9·7	0·7
8 STIs	8 Blindness and vision impairment	43·4	14·1	0·9	8 Age-related hearing loss	26·1	11·4	0·9
9 Blindness and vision impairment	9 Intestinal nematode	−20·7	−36·9	−34·9	9 Dietary iron deficiency	6·4	−6·0	−4·9
10 Cirrhosis	10 Age-related hearing loss	45·4	15·7	1·2	10 Cirrhosis	23·5	9·2	4·6
11 Age-related hearing loss	11 Cirrhosis	40·8	12·0	5·0	11 Intestinal nematode	−15·7	−25·5	−23·4
12 Vitamin A deficiency	12 Vitamin A deficiency	11·4	−11·3	−5·2	12 Upper digestive diseases	21·1	7·0	1·5
13 Fungal skin diseases	13 Upper digestive diseases	37·1	9·1	−1·2	13 Chronic kidney disease	28·2	13·3	3·0
14 Upper digestive diseases	14 Fungal skin diseases	23·0	−2·1	−3·0	14 Vitamin A deficiency	5·9	−6·4	−4·0
15 Chronic kidney disease	15 Chronic kidney disease	43·2	14·0	−1·3	15 Fungal skin diseases	12·5	−0·6	−4·0
16 Low back pain	16 Low back pain	29·6	3·2	−7·7	16 Low back pain	17·4	3·8	−2·7
17 Other skin diseases	17 Other skin diseases	44·2	14·8	5·7	17 Other skin diseases	25·4	10·8	3·9
18 Interpersonal violence	18 Diabetes	70·2	35·4	17·6	18 Diabetes	29·8	14·7	3·8
19 Iodine deficiency	19 Interpersonal violence	28·1	1·9	−2·3	19 Interpersonal violence	14·7	1·4	1·1
20 Anxiety disorders	20 Anxiety disorders	33·1	5·9	0·3	20 Other musculoskeletal	21·6	7·5	0·9
26 Diabetes	21 Other musculoskeletal				23 Anxiety disorders			
27 Other musculoskeletal	33 Iodine deficiency				35 Iodine deficiency			

Males

Leading causes 1990	Leading causes 2007	Mean percentage change in number of prevalent cases, 1990–2007	Mean percentage change in all-age prevalence rate, 1990–2007	Mean percentage change in age-standardised prevalence rate, 1990–2007	Leading causes 2017	Mean percentage change in number of prevalent cases, 2007–17	Mean percentage change in all-age prevalence rate, 2007–17	Mean percentage change in age-standardised prevalence rate, 2007–17
1 Oral disorders	1 Oral disorders	21·6	−2·9	−4·3	1 Oral disorders	12·5	−0·2	−1·6
2 Headache disorders	2 Headache disorders	31·3	4·8	0·0	2 Headache disorders	14·3	1·5	0·7
3 Tuberculosis	3 Tuberculosis	26·2	0·7	−3·1	3 Tuberculosis	1·1	−10·2	−11·5
4 Intestinal nematode	4 Cirrhosis	42·5	13·8	6·5	4 Cirrhosis	22·8	9·0	4·6
5 Cirrhosis	5 Haemoglobinopathies	29·0	3·0	3·6	5 Age-related hearing loss	24·3	10·3	0·0
6 Dietary iron deficiency	6 Intestinal nematode	−21·4	−37·3	−35·7	6 Haemoglobinopathies	12·7	0·1	0·7
7 Haemoglobinopathies	7 Age-related hearing loss	44·6	15·4	0·4	7 Blindness and vision impairment	23·1	9·3	−0·4
8 Age-related hearing loss	8 Dietary iron deficiency	6·0	−15·4	−14·3	8 Dietary iron deficiency	5·8	−6·1	−5·2
9 Vitamin A deficiency	9 Blindness and vision impairment	39·5	11·3	−2·2	9 STIs	19·7	6·3	1·9
10 Blindness and vision impairment	10 Vitamin A deficiency	9·7	−12·4	−7·1	10 Intestinal nematode	−16·7	−26·0	−24·2
11 Fungal skin diseases	11 STIs	38·9	10·9	0·7	11 Vitamin A deficiency	5·6	−6·3	−4·0
12 STIs	12 Fungal skin diseases	20·8	−3·5	−3·5	12 Upper digestive diseases	20·3	6·8	1·3
13 Upper digestive diseases	13 Upper digestive diseases	36·5	9·0	−1·3	13 Fungal skin diseases	10·2	−2·2	−4·6
14 Low back pain	14 Chronic kidney disease	45·6	16·2	−0·1	14 Chronic kidney disease	25·4	11·4	1·1
15 Chronic kidney disease	15 Low back pain	30·3	4·0	−6·8	15 Other skin diseases	26·4	12·2	4·7
16 Other skin diseases	16 Other skin diseases	46·5	16·9	7·1	16 Low back pain	18·0	4·7	−1·3
17 Falls	17 Diabetes	77·6	41·8	21·5	17 Diabetes	29·3	14·8	4·0
18 Diabetes	18 Falls	26·4	0·9	−9·8	18 Falls	26·8	12·6	4·1
19 Asthma	19 Other musculoskeletal	41·6	13·0	0·8	19 Other musculoskeletal	16·7	3·6	−2·9
20 Dermatitis	20 COPD	31·5	4·9	−10·6	20 COPD	15·6	2·6	−10·1
21 COPD	21 Dermatitis				22 Dermatitis			
22 Other musculoskeletal	22 Asthma				23 Asthma			

Communicable, maternal, neonatal, and nutritional diseases
Non-communicable diseases
Injuries

Fig. 7.1 Leading 20 Level 3 causes of global prevalence for 1990, 2007, and 2017, with the percentage change in number of cases and all-age and age-standardized rates for each sex. Level 1 contains three broad cause groups: communicable, maternal, neonatal, and nutritional diseases; noncommunicable diseases; and injuries. For nonfatal health estimates, there are 22 Level 2 causes, 167 Level 3 causes, and 288 Level 4 causes. Causes are connected by lines between time periods; solid lines are increases and dashed lines are decreases. For the time periods 1990–2007 and 2007–17, three measures of change are shown: percentage change in the number of cases, the percentage change in the all-age prevalence rate, and percentage change in the age-standardized prevalence rate. Communicable, maternal, neonatal, and nutritional diseases are shown in *red*; noncommunicable causes in *blue*; and injuries in *green*. Statistically significant changes are shown in bold. *COPD*, chronic obstructive pulmonary disease; *STIs*, sexually transmitted infections. Credit: Fig. 7 of GBD 2017 Disease and Injury Incidence and Prevalence Collaborators. Global, regional, and national incidence, prevalence, and years lived with disability for 354 diseases and injuries for 195 countries and territories, 1990–2017: a systematic analysis for the Global Burden of Disease Study 2017. Lancet 2018;392:1789–858. https://doi.org/10.1016/S0140-6736(18)32279-7.

Table 7.1 Considering microbiome tests for clinical practice.

Considerations	Available technology
How soon would the results be needed?	
How sensitive and how accurate do the results need to be?	
Do we need another technology to work in combination?	
What is the current gold-standard practice that could still follow the microbiome test (to further decrease false negatives, and to confirm the positive diagnosis)?	Or a panel of experts
How much would the test cost, and who is paying?	

Please go through above questions and answers for the particular situation you would like to help improve.
Credit: Huijue Jia.

from routine health examinations [7]. The fecal and oral microbiome might help predict who is more likely to progress into liver cirrhosis, and from cirrhosis to carcinoma. Sex differences in the levels of secondary bile acids processed from the gut microbiome offer an explanation for the higher incidence of liver carcinoma in men [8]. Ethanol production by some *Klebsiella pneumoniae* strains in NAFLD is one of the possible mechanisms for NAFLD [9]. The microbiome may be no less important in acute events. Acetaminophen, commonly used for pains and colds, is a leading cause of acute liver failure and causes a more severe liver injury at night. Gut microbial metabolites such as 1-phenyl-1,2-propanedione have been found in a mice study to explain such diurnal difference in acetaminophen toxicity, through depletion of liver glutathione [10].

Colonoscopy is now common practice in many developed countries. Metagenomic shotgun sequencing, or qPCR for just a few bacteria, could be a more convenient and robust technology than FOBT (fecal occult blood test) and qPCR for methylation of host genes. The incentive for governments would be to save costs on unnecessary colonoscopy. The incentive for individuals would be no fear of microbiome disturbance due to bowel cleansing, and in the case of metagenomic sequencing, to potentially know ones' risks for other diseases and options for treatment (more in Chapter 8). For lung cancer, sending in sputum samples for metagenomics [11] may help reduce the waiting line for high-resolution computed tomography (CT) scans and might work in combination with cell-free DNA (cfDNA) (Box 7.1) to improve the utility of both technologies in screening for new patients and in watching out for relapses. Oral microbiome biomarkers have also been reported for diseases such as pancreatic cancer

[12–14], awaiting further validation. Fecal *Achromobacter* appeared to promote biliary tract cancer, according to results from Mendelian Randomization (MR, Chapter 6, Box 6.2) [15]. More generally, the microbiome is probably a key factor in the different disease incidences in different populations around the world (Chapter 8, Fig. 8.1).

For cardiovascular diseases (Chapter 4) and mental disorders, oral or fecal metagenomic samples could be sent from patients' home on a regular basis, to capture early signs of a relapse. This may also be a valid approach for detecting early signs of relapse in cancer patients. For body fluid samples with a high proportion of human sequences, cell-free DNA or RNA would also be an option to detect pathogens (Box. 7.1) [16,17], although losing intracellular and adherent microbes.

For diagnostic purposes, one has to be aware of the false positives and false negatives of the method as well as the model (Table 7.1). Life-threatening diseases require diagnosis as early as possible, so the cutoff is typically not at the largest area under the curve (AUC), but tends to minimize false negatives (e.g., can 1 miss in 10,000 people be tolerated?). Insurance can be combined with the test to prepare for rare incidences. False positives could be checked with another existing method, yet one also needs to estimate the number of such further tests for the cutoff value used. Trying a multicancer blood

Box 7.1 Cell-free DNA or RNA in human body fluids

Prenatal diagnosis of diseases using cell-free DNA (cfDNA) from pregnant mothers has safeguarded childbirth in many countries. Screening for tumors and predicting recurrence, using cfDNA or cell-free RNA (cfRNA) from plasma samples, is also proceeding into clinics [18–20]. Although the coverage is typically low, and the data may be intrinsically fragmental [21], nonhuman reads in such plasma cfDNA or cfRNA could map to viruses, bacteria, and fungi. For all these microbes, rapid sequencing and bioinformatic analyses could allow the doctor to know the taxa that are enriched in the patient, along with drug resistance genes, virulence, and lineage according to the metagenomically derived microbial genomes (Chapter 5), before trying for a few days to culture all the possible microbes. The pathogen database and the population baseline for such efforts are by no means perfect at this early stage of application.

cfDNA in umbilical cord blood identified bacteria that enriched in cases of suspected chorioamnionitis (infection of the membranes that surround the fetus and the amniotic fluid) compared to healthy controls [22]. For invasive fungi infections, e.g., during chronic immune suppression after organ transplant, fungi identified by plasma cfDNA have shown good agreement with plate culture experiments or targeted sequencing results in 7 out of 9 patients and could relieve the need for tissue biopsy [23]. The strength of metagenomics lies more in unbiased detection of microbes, even in culture-negative or (targeted) PCR-negative samples (Chapter 1, Fig. 1.2) [16,17], and would be the first test to see a shifting trend in pathogens for the same apparent symptoms.

test (human cfDNA and protein markers) in 10,006 women between 65 and 75 years old in the United States led to the detection of 26 cancer cases, followed by PET-CT (positron emission tomography-computed tomography) imaging, while conventional methods detected another 24 cases [18]. Actually, individuals who tested positive but do not yet have clinical symptoms should still be followed in subsequent years. So the tests are not necessarily false but reflect individual differences in the time course and severity of clinical manifestation. Ethically and economically, each population-wide screen should be carefully designed to minimize unnecessary anxiety and costs. Most screens are performed on older people, because the disease incidence is too low in young people and the tests would result in too many false positives (a small fraction of a big number of healthy individuals is still a big number). Young people, however, could be interested in participating in longitudinal studies without a simple answer (Chapter 8).

7.1.2 Metagenomics for personalized treatment

Microbiome composition could predict response to cancer immunotherapy and chemotherapy (Table 7.2), response to medication for rheumatoid arthritis, type 2 diabetes, etc. [32,33]. For example, it remains to be validated with more patients that *Veillonella* sp. in the saliva of rheumatoid arthritis patients was better reduced after treatment with methotrexate plus *Tripterygium wilfordi* (thunder god vine) glycosides or *T. wilfordi* glycosides alone, compared to methotrexate alone [32]. Methotrexate treatment has been found to negatively associate with pulmonary fibrosis in early rheumatoid arthritis patients [34], without studying the respiratory or the oral microbiome. For malignant glioma, clinical trials are being performed for immune checkpoint inhibitors, peptide vaccines, dendritic cell vaccines, etc. [35], and the gut, oral, and potentially cerebrospinal fluid (CSF) microbiome might all influence the outcome. Gut microbial tryptophan metabolism has been implicated in metabolic, autoimmune, neuropsychiatric diseases and cancer [36–41], and clinical trials targeting the tryptophan pathways should probably take into account the microbiome (Tables 7.3 and 7.4), which would potentially impact the dose, toxicity and efficacy of the drug (Fig. 7.2).

Mechanistically, metabolism of a drug by the microbiome could activate, deactivate or toxify the drug (Fig. 7.2, Table 7.5), and could work together with human genetic variations (Table 7.6). Note that mice experiments are typically performed on male mice, to decrease variability due to the female mice's estrus cycle, so the difference between animal results and human cohorts could have

Table 7.2 Microbiome and response to cancer therapies.

Cancer	Treatment	Microbes	Reference
Mice with subcutaneous injection of fibrosarcoma, melanoma, or mastocytoma cell lines	CpG-oligonucleotide (CpG-ODN) immunotherapy; platinum chemotherapy (oxaliplatin)	For CpG-ODN, *Alistipes shahii*, *Ruminococcus* sp. positively correlated with intratumoral TNF (tumor necrosis factor) expression; *Lactobacillus* spp. negatively correlated with TNF. Effects of *Alistipes shahii* and *L. fermentum* were verified experimentally	[24]
Mice with subcutaneous injection of lymphoma, melanoma or colon carcinoma cell lines	Cyclophosphamide (CTX)	CTX induced translocation of *Lactobacillus* spp. (e.g., *L. johnsonii*) and *Enterococcus hirae* into secondary lymphoid organs. Such Gram-positive bacteria were necessary for the induction of Th17 cells that mediate the effects of CTX	[25]
Mice with subcutaneous injection of fibrosarcoma or colon carcinoma cell lines; patients with advanced lung cancer or ovarian cancer	CTX	*Enterococcus hirae* translocated from the small intestine to secondary lymphoid organs and increased the intratumoral CD8/Treg ratio; *Barnesiella intestinihominis* accumulated in the colon and promoted the infiltration of IFN-γ-producing $\gamma\delta$T cells in cancer lesions. Both were restrained by Nod2. *E. hirae* and *B. intestinihominis* specific-memory Th1 cell immune responses predicted longer progression-free survival in advanced lung and ovarian cancer patients treated with chemo-immunotherapy	[26]
Mice model of melanoma	Anticytotoxic T lymphocyte antigen (CTLA-4)	T-cell responses for *Bacteroides thetaiotaomicron* or *Bacteroides fragilis* associated with the efficacy of CTLA-4 blockade; Experimentally verified effects of *B. thetaiotaomicron*, *B. fragilis* alone, *B. fragilis* together with *Burkholderia cepacia*, *B. fragilis* polysacharides, and *B. fragilis*-specific T-cells	[27]
Mice model of melanoma	Antiprogrammed cell death ligand 1 (PD-L1)	*Bifidobacterium* spp. (*B. breve*, *B. longum*) enhanced the efficacy of anti-PD-L1 therapy	[28]

Table 7.2 Microbiome and response to cancer therapies—cont'd

Cancer	Treatment	Microbes	Reference
Nonsmall cell lung cancer (NSCLC) patients, renal cell carcinoma (RCC) patients	Antiprogrammed cell death 1 (PD-1)	Higher relative abundance of *Akkermansia muciniphila* in patients who responded to PD-1 treatment; Verified by supplementing *Akkermansia muciniphila* into nonresponder feces to show response in mice model	[29]
Melanoma patients	PD-1	Higher α-diversity, higher relative abundance of the Ruminococcaceae family in patients who responded to PD-1 treatment	[30]

Existing research has focused on the fecal microbiome. A number of clinical trials are being performed [31].
Credit: Huijue Jia.

been increased partly due to the microbiome and immunological differences between sexes, in addition to the "enterotype" difference (Chapter 2).

Testing for the microbes before treatment saves critical time and helps recommend the effective medication and other therapies to prescribe (Figs. 7.2–7.4). The optimal dose for each patient depends on both the human genome and the microbiome, which could activate, inactivate or convert a diverse range of molecules, including medication (Figs. 7.2 and 7.3; Tables 7.5 and 7.6) [91,92]. Some of the disease-associated microbes remain unaffected by standard medication [32,93], calling for combinatorial treatment, or further development of new drugs. Microbiome information could also be leveraged to predict or manage side effects, e.g., during cancer immunotherapy [94]. If patients tend to spontaneously discontinue the use of a medication, regular tests of the microbiome showing the trend of improvement toward a healthy state might also help them stay on. Microbiome tests after treatment may also inform decisions to continue, stop, or switch to other treatments.

High-intensity interval training has been tried in prediabetic patients, and patients with Alzheimer's disease [95–98]. Given the associations between fecal, oral, vaginal microbiomes and physical activity, measures of muscle strength, lung capacity, etc. [99–101], microbiome tests could also help guide more personalized physical exercises. In addition, exercises may be a proxy for sweating, which secretes nonessential or toxic trace metals in addition to sodium chloride, urea, lactate, creatinine, etc. [102], and might also reduce the disease risks.

Table 7.3 List of currently investigated IDO1 (indoleamine 2,3-dioxygenase 1) inhibitors.

Molecule	Structure and properties	Investigations	Published studies	Active or recruiting studies
1-MT-L-Trp (1-methyl-L-tryptophan)	Analog of L-Trp; Nonspecific competitive inhibitor of IDO1; Increases the effectiveness of anticancer drugs and increases KYNA in vivo and ex vivo regardless of IDO	Fundamental research [42]	Advanced malignancies: well tolerated (monotherapy) [43]	Phase I/II: breast (NCT01042535, NCT01792050), pancreatic (NCT02077881), prostate (NCT01560923), nonsmall cell lung cancer (NCT02460367), solid (NCT00567931, NCT01191216), brain tumors (NCT04049669, NCT02052648, NCT02502708), leukemia (NCT02835729), and melanoma (NCT03301636, NCT02073123)
1-MT-D-Trp (1-methyl-D-tryptophan. indoximod)	Low in vitro activity but effective in vivo, preferentially inhibit IDO2; May promote tumor growth by off-target effect; Prodrug: NLG802	Cancers (alone or in combination) [44,45]		
Epacadostat INCB024360 OH HN N H N Br H_2N S H N N N-O F O O	Selective reversible competitive inhibitor of IDO1; Antitumoral (decreases Tregs, increases the synthesis of IFNγ by T cells) but lack of activity as a monotherapy; Metabolized by the intestinal microbiota and the enzyme UGT1A9 (AhR target)	Cancers (only in combination) [46,47]	Ovarian cancer: no benefit [48] Tumors: well tolerated and had encouraging antitumor activity [49] Metastatic melanoma: no benefit [50]	Phase I/II: thymic carcinoma (NCT02364076), naso-pharyngeal (NCT04231864), gastric (NCT03196232), gastrointestinal (NCT03291054), pancreatic (NCT03006302), urothelial bladder (NCT03832673), nonsmall cell lung (NCT03322566, NCT03322540), and rectal (NCT03516708) cancers, melanoma (NCT01961115), sarcoma (NCT03414229), metastatic solid tumors (NCT03347123) Phase III: urothelial (NCT03361865, NCT03374488) and renal carcinoma (NCT03260894), head and neck carcinoma (NCT03358472)

Linrodostat BMS-986205 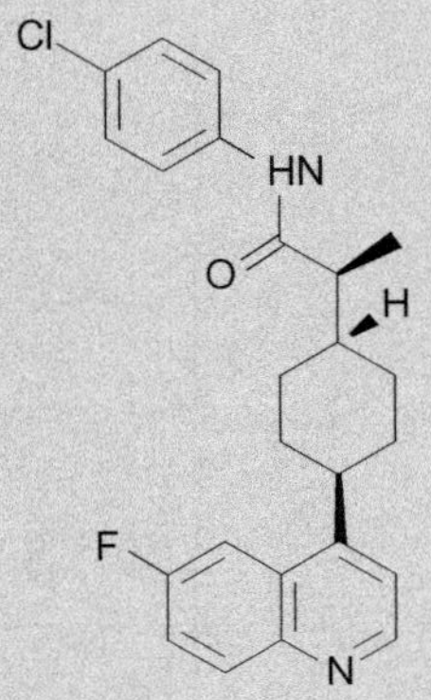	Potent, selective, and irreversible IDO1 inhibitor, restores T-cell proliferation and reduces intratumoral L-kyn up to 90%	Cancers [51–53]	Tumors: well tolerated (± nivolumab), need further investigations for efficacy [52]	Phase I/II: pharmacokinetics (NCT03378310, NCT03312426) and safety (NCT03192943), endometrial (NCT04106414), liver (NCT03695250), gastric (NCT02935634) head and neck (NCT03854032) and bladder (NCT03519256) cancers, solid tumors (NCT03792750, NCT03459222, NCT02658890) glioblastoma (NCT04047706) Phase III: bladder cancer (NCT03661320, NCT03661320), melanoma (NCT03329846)
EOS200271 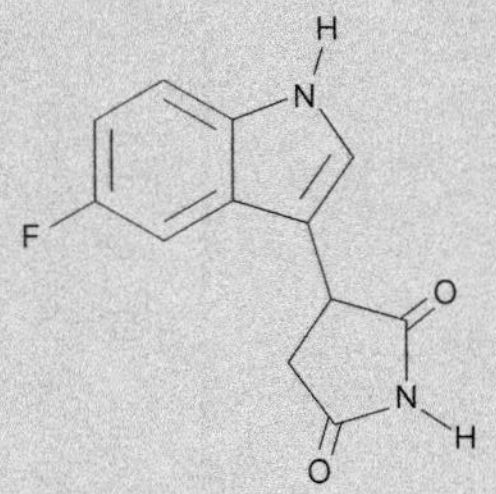	IDO1 specific noncompetitive inhibitor; Oral use Brain permeable	Glioma Association with PD-L1 inhibitors [54,55]	Malignant glioma: well tolerated [55]	
Navoximod, GDC-0919, or NLG-919 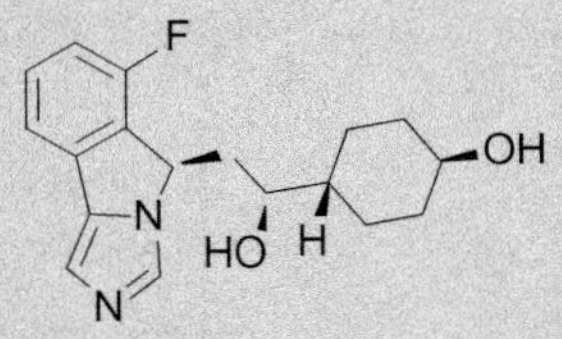	Moderately selective noncompetitive reversible inhibitor; Dose-dependent activation and proliferation of effector T cells; Regression of large established tumors; Synergy with indoximod; Increases survival (± chemotherapy) currently optimized by prodrug formulation	Cancers [56]	Recurrent advances solid tumors: well tolerated and reduced plasmatic L-kyn [57]	Phase I/II: solid tumors (NCT02471846, NCT02048709)

IFN, interferon; *KYNA*, kynurenic acid; *L-kyn*, L-kynurenine; *Treg*, regulatory T cell. Clinical trials can be accessed at https://www.clinicaltrials.gov.
Credit: Table 1 of Modoux M, Rolhion N, Mani S, Sokol H. Tryptophan metabolism as a pharmacological target. Trends Pharmacol Sci 2021;42:60–73. https://doi.org/10.1016/j.tips.2020.11.006.

Table 7.4 List of currently investigated AhR (aryl hydrocarbon receptor) agonists and antagonists.

Molecule	Structure and properties	Investigations	Published studies	Active or recruiting studies
AhR agonists				
Laquinimod 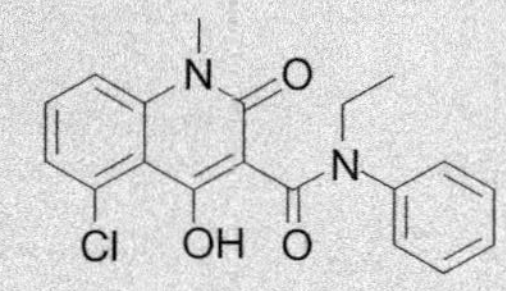	Quinoline 3-carboxamide structural similar to KYNA; AhR-dependent effects on encephalomyelitis; Mixed results (Phase II and III clinical trials—multiple sclerosis); Allows remyelination	Huntington's Multiple sclerosis Crohn's disease [58,59]	Multiple sclerosis: well tolerated, significant reduction in brain atrophy [60,61] Crohn's disease: well tolerated, promising effects [62]	Phase I/II: efficacity and safety in relapsing multiple sclerosis (NCT01047319), Huntington's disease (NCT02215616), lupus arthritis (NCT01085084), lupus nephritis (NCT01085097), Crohn's disease (NCT00737932), relapsing multiple sclerosis (NCT01975298)
Tranilast 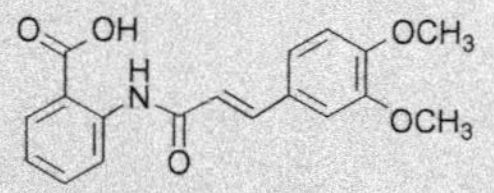	Synthetic analog of ANA	Asthma (marketed) Rheumatoid arthritis Multiple sclerosis Hyperuricemia Cancer [63]	Prostate cancer: benefit on prognosis [63]	Phase I/II: mucinoses (NCT03490708), scleredema diabeticorum (NCT03512873), sarcoidosis (NCT03528070), cryopyrin-associated periodic syndrome (NCT03923140), pterygium (NCT01003613), hyperuricemia (NCT00995618, NCT01052987), gout (NCT01109121), rheumatoid arthritis (NCT00882024)
Tapinarof (benvitimod) 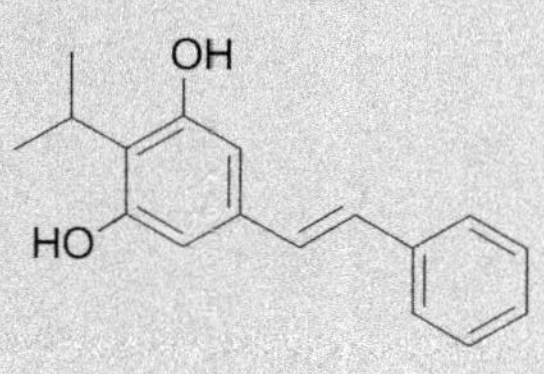	Bacterial stilbene; Free radical scavenger; Dermal application	Psoriasis atopic dermatitis [64]	Psoriasis and atopic dermatitis: well tolerated [65,66]	Phase I/II: safety, tolerability, and pharmacokinetics of tapinarof cream, 1% (extensive plaque psoriasis) (NCT04042103) Phase III: efficacy and safety of topical tapinarof cream, 1% (plaque psoriasis) (NCT03956355)

AhR antagonists				
CH223191 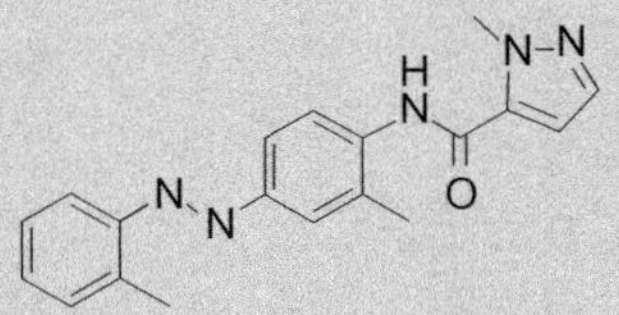	Competitive selective antagonist; No antagonistic activity with non-HAH ligands	Fundamental research but may be a promising effect in pancreatic cancer [67]	No active clinical trials	
CB7993113 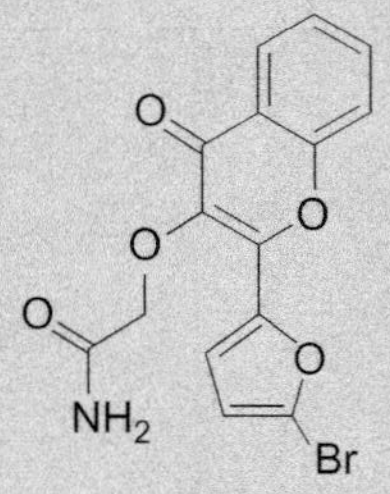	Good oral bioavailability; Blocks tumor cell migration and reduces the invasive phenotype of ER-/PR-/ HER2- breast cancer cells in vitro	[68,69]		
StemRegenin-1	Ex vivo application; Expand CD34 + cells	Stem cell transplantation Neutropenia Thrombocytopenia	CD34 + cell expansion [68,70]	Malignant hemopathies (NCT01474681 and NCT01930162) Neutropenia and thrombocytopenia (NCT03406962)

Non-HAH ligands (halogenated aromatic hydrocarbons) include polycyclic aromatic hydrocarbons (PAHs) as well as endogenous L-Trp ligands. HAHs are distinguished from PAHS and endogenous ligands by very slow metabolism and a prolonged effect on the AhR receptor. *ANA*, Anthranilic acid; *ER*, estrogen receptor; *HER*, human epidermal growth factor receptor 2; *KYNA*, kynurenic acid; *PR*, progesterone receptor. Clinical trials can be accessed at https://www.clinicaltrials.gov/.

Credit: Table 3 of Modoux M, Rolhion N, Mani S, Sokol H. Tryptophan metabolism as a pharmacological target. Trends Pharmacol Sci 2021;42:60–73. https://doi.org/10.1016/j.tips.2020.11.006.

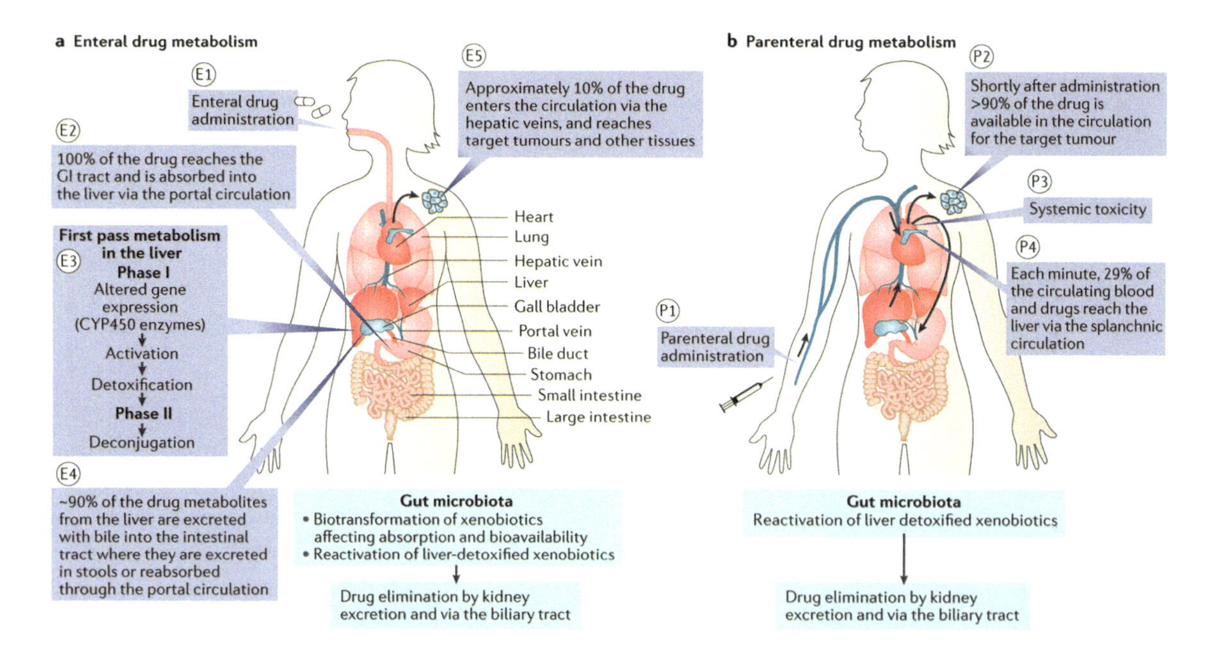

Fig. 7.2 Major pathways of drug metabolism and the role of microbiota following enteral (e.g., oral) or parenteral (e.g., intravenous) administration. (A) Enteral drug metabolism. Orally administered drugs (E1) sit in the stomach for 30–45 min before reaching the intestine and being absorbed into the liver by the portal circulation (E2). In the intestine, host and microbial enzymes induce metabolic alterations to the drug that together with direct binding to bacterial products and segregation control intestinal absorption. In the liver, following phase I and phase II processing (first pass metabolism; E3), approximately 90% of the oral drug is metabolized and destroyed or eliminated through biliary secretion (E4). The drugs secreted into the intestine via the biliary duct can be reabsorbed via portal circulation or excreted in stools. As a consequence, only 10% of the oral drug enters the circulation through the hepatic veins and is available to reach the target tumors and other tissues (E5). Phase I and phase II processing are also affected by the gut microbiota through the regulation of the level of host enzymes involved in drug processing. (B) Parenteral drug metabolism. Following intravenous administration (P1) close to 100% of the drug enters the circulation and is available to reach the target tumors (P2); however, the drug is also distributed systemically, inducing adverse toxic reactions (P3). Any remaining drug not retained in tissues can be rapidly excreted by the kidney. Each minute 29% of the circulating drug is transported via the splanchnic circulation (hepatic, mesenteric, and splenic arteries) to the liver (P4), where the drug is processed similarly to enterally administered drugs. The detoxified drugs that are secreted from the liver to the intestine through the biliary excretion route can be reactivated by bacterial enzymes, inducing intestinal toxicity. *CYP450*, cytochrome P450; *GI*, gastrointestinal. Credit: Fig. 2 of Roy S, Trinchieri G. Microbiota: a key orchestrator of cancer therapy. Nat Rev Cancer 2017. https://doi.org/10.1038/nrc.2017.13.

Table 7.5 Selected drug modifications made by human gut microbiome.

Phenotypic effect	Microbial modification	Subclass: drugs	Outcome	Host effect	Reference
Activation and reactivation	Reduction	Azoreduction: sulfasalazine (SSZ), balsalazide, ipsalazide, olsalazine	Prodrug activation: local 5-ASA release	Antiinflammatory treatment	[71]
		Azoreduction: prontosil, neoprontosil	Antibiotic activation	Bacterial killing	[72]
	Dealkylation	*N*-dealkylation: amiodarone	Increased bioavailability of active metabolite	Increased half-life, possible drug interactions	[71]
	Deconjugation	Deglucuronidation: morphine, codeine	Reformation of active metabolite	Increased AUC, enterohepatic circulation	[71]
	Other	Desulfation: sodium picosulfate	Solubility increase	Activation of laxative effect	[72]
Inactivation	Reduction	Nitroreduction: benzodiazepines: nitrazepam, clonazepam, bromazepam	Change to inactive metabolite	Inactivation of drug, a possible overdose intervention	[71,72]
		Lactone ring reduction: digoxin	Change to inactive metabolite	Narrow therapeutic window	[71]
	Dealkylation	*N*-demethylation: methamphetamine	Change to inactive metabolite	Decreases therapeutic effect	[71]
	Dehydroxylation	*P*-dehydroxylation: L-dopa	Decrease in L-dopa absorption, caused by *Helicobacter pylori*	Decreases therapeutic effect	[71,72]
	Proteolysis	Insulin, calcitonin	Breakdown of therapeutic protein	Decreases therapeutic effect	[72]
	Acetylation	*N*-acetylation: 5-ASA	Change to inactive metabolite	Less efficacy, possible pancreatic toxicity	[71]

Continued

Table 7.5 Selected drug modifications made by human gut microbiome—cont'd

Phenotypic effect	Microbial modification	Subclass: drugs	Outcome	Host effect	Reference
Toxification	Reduction	Nitroreduction: chloramphenicol	*p*-Aminophenyl-2-morphine-glucuronide amino-1,3-propanediol generation (speculated)	Bone marrow toxicity	[72]
		Nitroreduction: benzodiazepines: nitrazepam, clonazepam, bromazepam	Amino-metabolite generation, Inactivation	Teratogenicity	[71,72]
	Dealkylation	*N*-dealkylation: brivudine, sorivudine	Generation of additional bromovinyluracil, drug AUC decrease, interaction with 5-fluorouracil (5-FU)	*Bacteroides*-mediated hepatotoxicity, potentially fatal 5-FU accumulation	[73]
	Deconjugation	Deglucuronidation: irinotecan, diclofenac, ketoprofen, indomethacin	Reformation of cytotoxic drug	Diarrhea, bowel distress, GI lesions	[71,72]

5-ASA, 5-aminosalicylate. AUC, area under the curve, which shows plasma concentration of a drug over time, so a higher AUC means more drug in the body.
Credit: Table 1 of Hitchings R, Kelly L. Predicting and understanding the human Microbiome's impact on pharmacology. Trends Pharmacol Sci 2019;40:495–505. https://doi.org/10.1016/j.tips.2019.04.014.

Table 7.6 Drugs with potential human and bacterial sources of variance.

Drug	Human pharmacogene	Effect of polymorphism	Microbiome-associated metabolism	Effect of microbiome metabolism	References
Warfarin	*CYP2C9*	Altered activity of drug	Vitamin K production	Microbiomes produce variable concentrations of vitamin K. Alterations in vitamin K production by microbiome may alter warfarin metabolism	[74–76]
Irinotecan	*UGT1A1**28 "Gilbert's syndrome"	Defect in glucuronidation, increased toxicity	Deglucuronidation of excreted SN-38G metabolite	Reformation of cytotoxic Irinotecan	[77,78]
Codeine	*CYP2D6*	Variant alleles may cause absent, decreased, or increased rate of biotransformation to morphine	Deglucuronidation of excreted morphine-glucuronide metabolite	Reformation of morphine, higher morphine AUC due to enterohepatic circulation	[79,80]
Morphine	*SLC22A1, OCT1*	Decreased clearance of morphine	Deglucuronidation of excreted morphine-glucuronide metabolite	Reformation of morphine, higher morphine AUC due to enterohepatic circulation Induces virulence in some strains of *Pseudomonas aeruginosa*	[79,81,82]
Acetaminophen	*UGT1A, SULT1A3*	Increased rate of glucuronidation and decreased risk of liver failure due to unintentional overdose, decreased sulfation	Sulfonation	Increase in sulfonated metabolite, may be competitively inhibited by p-cresol sulfonation	[81,83,84]
Simvastatin	*SLCO1B1*	221% increase in simvastatin AUC for homozygotes	Unknown	Increased efficacy hypothesized to be due to microbial alteration of primary bile acids	[75,85–87]
Digoxin	*ABCB1*	Increased AUC may increase toxicity	Lactone ring reduction	Decreased AUC, narrow therapeutic window	[88,89]
Brivudine and sorivudine	*DYPD*	Increased drug-drug interactions with pyrimidine analogs	Generation of additional bromovinyluracil	Hepatotoxicity, bromovinyluracil prevents clearance of 5-FU	[73,90]

Vitamin K is also known as menaquinone, which often shows up in KEGG (Kyoto Encyclopedia of Genes and Genomes) analyses of the microbiome (e.g., Ref. [32]). AUC, area under the curve, which shows plasma concentration of a drug over time, so a higher AUC means more drug in the body; 5-FU, 5-fluorouracil.
Credit: Table 2 of Hitchings R, Kelly L. Predicting and understanding the human Microbiome's impact on pharmacology. Trends Pharmacol Sci 2019;40:495–505. https://doi.org/10.1016/j.tips.2019.04.014.

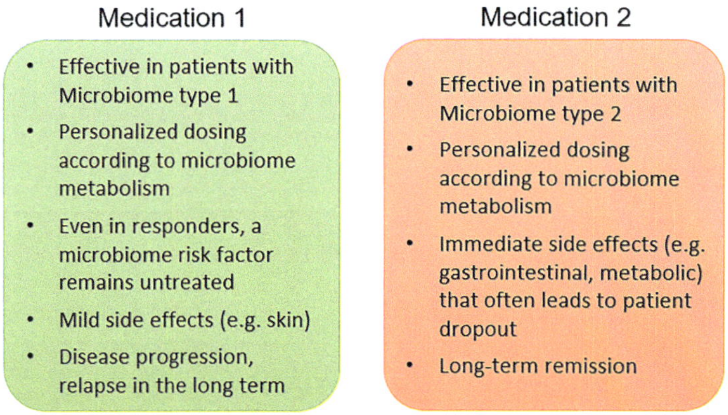

Medication 1

- Effective in patients with Microbiome type 1
- Personalized dosing according to microbiome metabolism
- Even in responders, a microbiome risk factor remains untreated
- Mild side effects (e.g. skin)
- Disease progression, relapse in the long term

Medication 2

- Effective in patients with Microbiome type 2
- Personalized dosing according to microbiome metabolism
- Immediate side effects (e.g. gastrointestinal, metabolic) that often leads to patient dropout
- Long-term remission

Fig. 7.3 A hypothetical example of how microbiome information could facilitate more personalized choice of the types and dose of medication, better compliance despite side effects, and more effective long-term management of diseases. Credit: Huijue Jia.

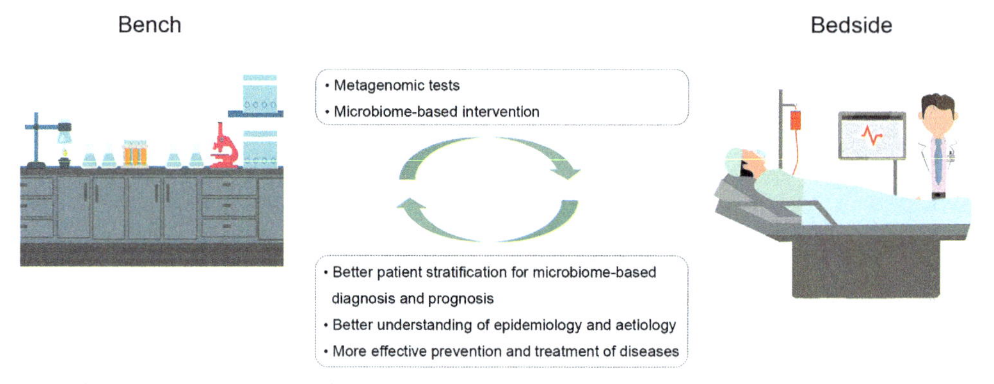

Fig. 7.4 From bench to bedside and from bedside to bench. Credit: Huijue Jia, Yanmei Ju, BGI-Shenzhen.

7.2 Further research to be inspired by clinical practice

New medication, new vaccines, new foods or additives, new materials and devices to be inserted into the body. The human microbiome might again play important roles. As discussed in Chapter 4, doctors are uniquely positioned to find out the full loop of events that takes place in the human body. Noninvasive tests of the oral and fecal microbiome can be performed on individuals at-risk for pancreatic cancer (Fig. 7.5, the very deadly pancreatic ductal adenocarcinoma, PDAC), while tissue samples may be available before treatment, followed by more noninvasive tests during long-term management. Metastasis may also carry microbes from the original site, which may be sequenced to trace their evolution, and treated locally if necessary. A lot of the drug and food metabolites end up in the urine, and it is currently unknown how these may influence the microbiome there.

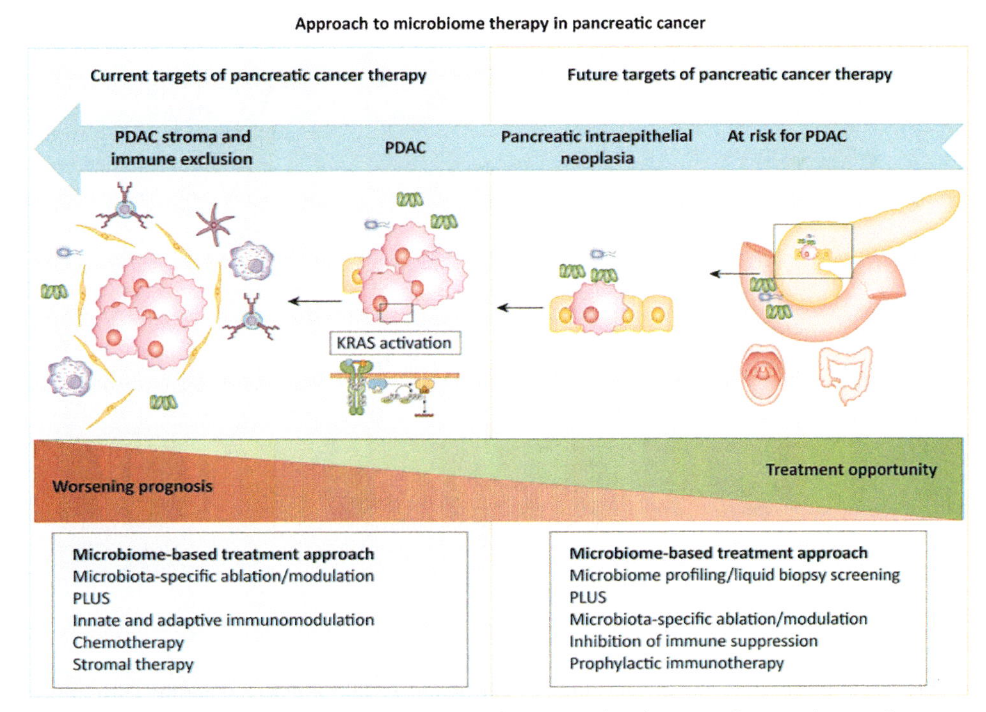

Fig. 7.5 Graphic illustration of the stages of pancreatic cancer development. Current therapy for pancreatic cancer is focused on early and advanced PDAC *(light blue box)*, which generally harbors a poor prognosis. At this stage, microbiota-specific ablation and immunomodulation have the potential to improve pancreatic cancer outcomes, but the therapeutic effect may be limited due to additional oncogenic factors including KRAS activation and immune cell exclusion in the tumor microenvironment. Instead, microbiome modulation may prove more impactful at the earliest stages of pancreatic cancer development *(light orange box)*, when microbiota directly contributes to tumor oncogenesis in the absence of an unfavorable tumor microenvironment. Microbiome profiling, screening, and augmentation may also lead to earlier PDAC diagnosis and open more therapeutic opportunities. Abbreviation: *PDAC*, pancreatic ductal adenocarcinoma. Credit: Fig. 1 of Vitiello GA, Cohen DJ, Miller G. Harnessing the microbiome for pancreatic cancer immunotherapy. Trends Cancer 2019;5:670–76. https://doi.org/10.1016/j.trecan.2019.10.005.

If metagenomic tests for semen, endometrium/cervical, urine, and fecal samples can enter fertility clinics, there will be more work to do regarding the various types of male and female fertility [103–107]. Other than getting a baby, long-term effects on health should also be an important consideration (Chapter 8).

Lung infections can evolve over time (Fig. 7.6). Metagenomically assembled genomes could complement traditional methodology, and improve the database for faster analyses and action in the future. The metagenomic associations among different members of the microbiome would be informative for the accurate prediction of outcomes in each patient. It has been shown in mice that the lung microbiome is

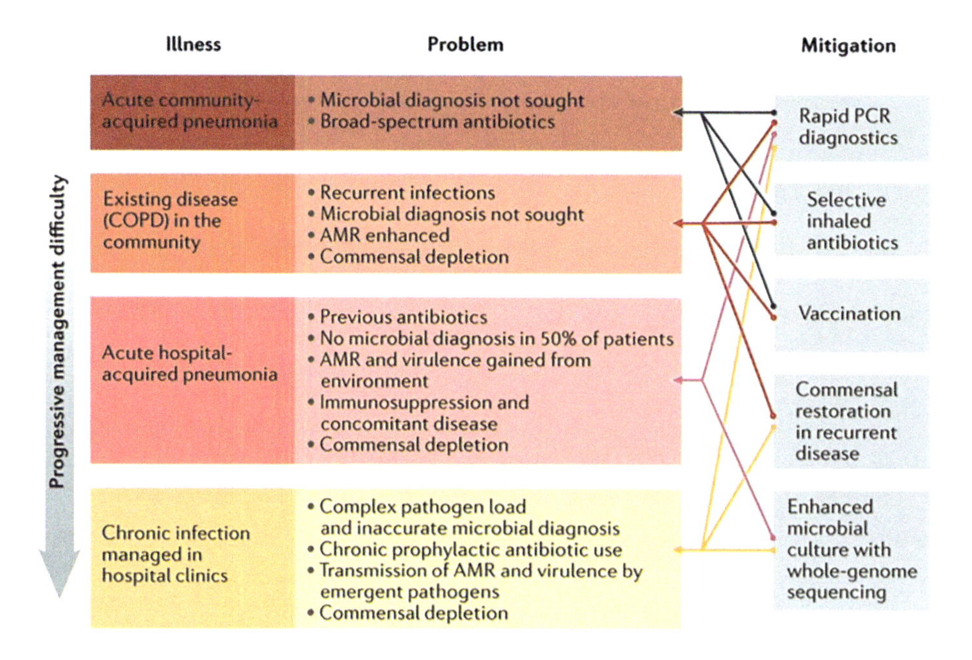

Fig. 7.6 Management of acute and chronic lung infections. PCR, if designed for the correct pathogen, is still more real-time than metagenomic shotgun sequencing. Detailed analyses of the microbiome and the evolution of members of the microbiome could lead to better long-term care. *AMR*, antimicrobial resistance; *COPD*, chronic obstructive pulmonary disease. Credit: Fig. 2 of Cookson WOCM, Cox MJ, Moffatt MF. New opportunities for managing acute and chronic lung infections. Nat Rev Microbiol 2017. https://doi.org/10.1038/nrmicro.2017.122.

required for the progression of lung adenocarcinoma through γδT cells and neutrophils [108]. Lung adenocarcinoma would also reorganize the circadian metabolic clock in the liver [109]. Epidemiologically, dietary fiber, cruciferous vegetables, and probiotics are associated with a reduced risk of lung cancer, while a high intake of coffee in men interacted with smoking and showed a higher risk [110–112].

Skin infections also tend to be refractory. How the different fungi *Malassezia* species, the lack of *Dermacoccus*, and strain-level evolution (Chapter 5) in the skin microbiome predicts atopic dermatitis flares warrant further investigation [113]. The different presentations of autoimmune disorders might be matched with different skin, mucosal, and circulating microbiome.

A cocktail of three bacteriophages against antibiotic-resistant *Mycobacterium abscessus* was used to treat multiple skin lesions on a 15-year-old cystic fibrous patient following a lung transplant. After the intravenous phage treatment which was generally effective and only elicited weak immune reactions, *M. abscessus* could still be cultured from slowly resolving skin nodules [114]. Antibodies against the *M. abscessus* phages were also detected in an 81-year-old patient with bronchiectasis, in which case the phages became ineffective after two months [115].

For the major types of inflammatory bowel diseases (IBD), feces of Crohn's diseases (CD) patients might have more *Ruminococcus gnavus*, while feces of Ulcerative colitis (UC) patients might have more *R. torques* [116], and the decrease in *Bacteroides* spp. is usually accompanied with overgrown *Enterobactericeae* [117,118]. The subtypes and sequence of events need to be better worked out in patients. Some *R. gnavus* strains encode superantigens that stimulate a potent IgA response [119], which is expected to impact the gut microbiome (Chapter 2). For some people, the more *R. torques*, blood group B, and loose stool at 30 years old [99] may never manifest as UC at an older age. Epidemic strains of *Peptoclostridium difficile* (formerly *Clostridium difficile*) that emerged in North America in the early 2000s grow fast in the presence of trehalose [120], but people with a functional gut microbiome do not need to be too worried about trehalose consumption. Fecal Microbiome Transplant (FMT) for IBD is not nearly as effective as FMT for *P. difficile* infections [121–124], and replacement of patients' strains with donors' strains using FMT was more difficult for CD than for UC [125]. The fecal microbiome may help predict immune markers [125,126], the oral microbiome is also a reservoir for immune derangement [127,128], and more effective treatment can potentially be selected for each patient.

For dentists, will we one day have enough data to be able to predict which teeth are more likely to fall off, and keep the other ones for longer? Orthodontal practices may also change the aeration in the mouth, and the protective layer of saliva on teeth. Given the association between the oral microbiome and all kinds of diseases, how can hospitals foster more collaborations between different departments?

7.3 Potential to modify existing categorization of diseases with knowledge of the microbiome

Naming of diseases is perhaps no less historical as the naming of microbes. With the key layer of information provided by the microbiome, some grouping, regrouping, and dividing of disease categories might be warranted.

Colorectal cancer without a strong genetic cause (e.g., Lynch syndrome) is referred to as sporadic. But we now know that on top of the dietary and obesity risk factors, a few bacteria could be the culprits, and a patient does not have to have all of them. If more evidence becomes available regarding the prognosis, and the optimal treatment, for the different fecal or mucosal bacteria enriched singly or in combination, they may well be named as subtypes of colorectal cancer and adenomas [129–136]. Presumably, the mutation and immune subtypes [137,138] result from the long-term interaction between gene, microbiome, and environment. *Fusobacterium* spp., especially *F. nucleatum*, is most studied for colorectal cancer in recent years [139,140]. In addition to being a biomarker for adenomas and carcinomas, a higher amount of tissue *F. nucleatum* DNA was associated with tumor location in the proximal colon, higher pT stage (deeper invasion), poor tumor differentiation, Microsatellite Instability (MSI)-high, *MLH1* hypermethylation, CpG island methylator phenotype (CIMP)-high, and *BRAF* mutation [141]. *F. nucleatum* has also been implicated with recurrence after chemotherapy [142].

Such updates for the nomenclature of complex diseases may also be needed for autoimmune diseases, in combination with the underlying genetics (Fig. 7.7) [143]. For example, none of the bacteria implicated in rheumatoid arthritis (Chapter 4) is 100% prevalent, just like none of the autoimmune antibodies is 100% prevalent. Which of the patients are more likely to have faster bone erosion and may require more aggressive/expensive treatments to begin with, instead of beginning with methotrexate alone and waiting for an unsatisfactory response [144]?

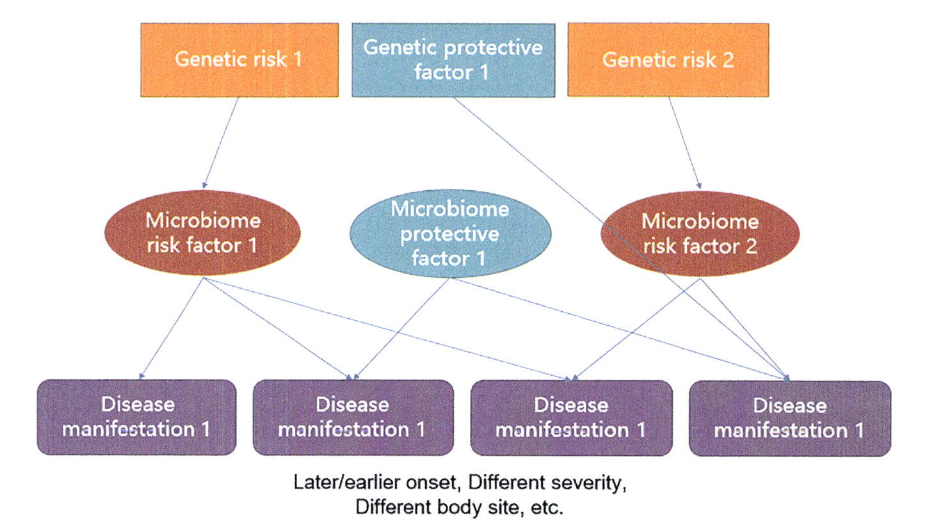

Fig. 7.7 An oversimplified illustration of genetic and microbiome factors that could lead to disease subtypes. Other clinically available data, e.g., autoantibodies, affected lymph nodes, should also be incorporated into the classification. Credit: Huijue Jia.

Worked sample 7.1

For patients with colorectal cancer, how would you look for clinical differences between those with high *Parvimonas micra*, *Peptostreptococcus stomatis* (or *Peptostreptococcus anaerobius* [130,145]), *Porphyromonas asaccharolytica*, or *Escherichia coli* in the fecal microbiome? Or, starting with existing subtypes, do you see certain subtypes to be more frequent in a particular group of people?

What kind of samples and other information could you collect before, during, and after treatment?

Besides surgical removal of the tumor, how do you think the treatment could be more targeted?

7.4 Summary

With all the knowledge about members of the microbiome that contribute to or prevent diseases, it is high time that we apply this knowledge to clinical practices wherever necessary. Healthcare professionals would have to decide whether to collect surgical samples for investigation, and whether to prescribe metagenomic tests before or after a treatment to see whether the medication works for a particular patient. The microbiome heterogeneity among patients is also an important consideration, after the human genomes, for the rational design of clinical trials in the development of effective new drugs. While the biomarkers for various diseases, in combination with current best practice, would enable population-scale screening. It would also be important to keep the clinical investigations going, and continue to refine the microbiome models for diagnosis and treatment, and reach a better understanding of many complex diseases (Figs. 7.4 and 7.7).

References

[1] Liao C, Taylor BP, Ceccarani C, Fontana E, Amoretti LA, Wright RJ, et al. Compilation of longitudinal microbiota data and hospitalome from hematopoietic cell transplantation patients. Sci Data 2021;8:71. https://doi.org/10.1038/s41597-021-00860-8.

[2] Khan N, Lindner S, Gomes ALC, Devlin SM, Shah GL, Sung AD, et al. Fecal microbiota diversity disruption and clinical outcomes after auto-HCT: a multicenter observational study. Blood 2021;137:1527–37. https://doi.org/10.1182/blood.2020006923.

[3] Peled JU, Gomes ALC, Devlin SM, Littmann ER, Taur Y, Sung AD, et al. Microbiota as predictor of mortality in allogeneic hematopoietic-cell transplantation. N Engl J Med 2020;382:822–34. https://doi.org/10.1056/NEJMoa1900623.

[4] Căcilia Ingham A, Kielsen K, Mordhorst H, Ifversen M, Gottlob Müller K, Johanna Pamp S. Microbiota long-term dynamics and prediction of acute graft-versus-host-disease in pediatric allogeneic stem cell transplantation. MedRxiv 2021. https://doi.org/10.1101/2021.02.19.21252040. 2021.02.19.

[5] Elgarten CW, Tanes C, Lee J-J, Danziger-Isakov LA, Grimley MS, Green M, et al. TITLE: early microbiome and metabolome signatures in pediatric patients undergoing allogeneic hematopoietic cell transplantation. MedRxiv 2021. https://doi.org/10.1101/2021.06.08.21258499. 2021.06.08.21258499.
[6] Moon AM, Singal AG, Tapper EB. Contemporary epidemiology of chronic liver disease and cirrhosis. Clin Gastroenterol Hepatol 2020;18:2650–66. https://doi.org/10.1016/j.cgh.2019.07.060.
[7] Newsome PN, Cramb R, Davison SM, Dillon JF, Foulerton M, Godfrey EM, et al. Guidelines on the management of abnormal liver blood tests. Gut 2018;67:6–19. https://doi.org/10.1136/gutjnl-2017-314924.
[8] Xie G, Wang X, Zhao A, Yan J, Chen W, Jiang R, et al. Sex-dependent effects on gut microbiota regulate hepatic carcinogenic outcomes. Sci Rep 2017;7:45232. https://doi.org/10.1038/srep45232.
[9] Yuan J, Chen C, Cui J, Lu J, Yan C, Wei X, et al. Fatty liver disease caused by high-alcohol-producing *Klebsiella pneumoniae*. Cell Metab 2019. https://doi.org/10.1016/j.cmet.2019.08.018.
[10] Gong S, Lan T, Zeng L, Luo H, Yang X, Li N, et al. Gut microbiota mediates diurnal variation of acetaminophen induced acute liver injury in mice. J Hepatol 2018;69:51–9. https://doi.org/10.1016/j.jhep.2018.02.024.
[11] Ramírez-Labrada AG, Isla D, Artal A, Arias M, Rezusta A, Pardo J, et al. The influence of lung microbiota on lung carcinogenesis, immunity, and immunotherapy. Trends Cancer 2020;6:86–97. https://doi.org/10.1016/j.trecan.2019.12.007.
[12] Farrell JJ, Zhang L, Zhou H, Chia D, Elashoff D, Akin D, et al. Variations of oral microbiota are associated with pancreatic diseases including pancreatic cancer. Gut 2012;61:582–8. https://doi.org/10.1136/gutjnl-2011-300784.
[13] Fan X, Alekseyenko AV, Wu J, Peters BA, Jacobs EJ, Gapstur SM, et al. Human oral microbiome and prospective risk for pancreatic cancer: a population-based nested case-control study. Gut 2018;67:120–7. https://doi.org/10.1136/gutjnl-2016-312580.
[14] Zhang X, Hoffman KL, Wei P, Elhor Gbito KY, Joseph R, Li F, et al. Baseline oral microbiome and all-cancer incidence in a cohort of nonsmoking Mexican American women. Cancer Prev Res (Phila) 2021;14:383–92. https://doi.org/10.1158/1940-6207.CAPR-20-0405.
[15] Liu X, Tong X, Zou Y, Lin X, Zhao H, Tian L, et al. Inter-determination of blood metabolite levels and gut microbiome supported by Mendelian randomization. BioRxiv 2020. https://doi.org/10.1101/2020.06.30.181438. 2020.06.30.
[16] Chiu CY, Miller SA. Clinical metagenomics. Nat Rev Genet 2019. https://doi.org/10.1038/s41576-019-0113-7.
[17] Gu W, Deng X, Lee M, Sucu YD, Arevalo S, Stryke D, et al. Rapid pathogen detection by metagenomic next-generation sequencing of infected body fluids. Nat Med 2021;27:115–24. https://doi.org/10.1038/s41591-020-1105-z.
[18] Lennon AM, Buchanan AH, Kinde I, Warren A, Honushefsky A, Cohain AT, et al. Feasibility of blood testing combined with PET-CT to screen for cancer and guide intervention. Science 2020;369. https://doi.org/10.1126/science.abb9601, eabb9601.
[19] Tie J, Cohen JD, Wang Y, Christie M, Simons K, Lee M, et al. Circulating tumor DNA analyses as markers of recurrence risk and benefit of adjuvant therapy for stage III colon cancer. JAMA Oncol 2019. https://doi.org/10.1001/jamaoncol.2019.3616.
[20] Nakamura Y, Taniguchi H, Ikeda M, Bando H, Kato K, Morizane C, et al. Clinical utility of circulating tumor DNA sequencing in advanced gastrointestinal cancer: SCRUM-Japan GI-SCREEN and GOZILA studies. Nat Med 2020. https://doi.org/10.1038/s41591-020-1063-5.

[21] Lo YMD, Han DSC, Jiang P, Chiu RWK. Epigenetics, fragmentomics, and topology of cell-free DNA in liquid biopsies. Science 2021;372. https://doi.org/10.1126/science.aaw3616, eaaw3616.

[22] Witt RG, Blair L, Frascoli M, Rosen MJ, Nguyen QH, Bercovici S, et al. Detection of microbial cell-free DNA in maternal and umbilical cord plasma in patients with chorioamnionitis using next generation sequencing. PLoS One 2020;15. https://doi.org/10.1371/journal.pone.0231239.

[23] Hong DK, Blauwkamp TA, Kertesz M, Bercovici S, Truong C, Banaei N. Liquid biopsy for infectious diseases: sequencing of cell-free plasma to detect pathogen DNA in patients with invasive fungal disease. Diagn Microbiol Infect Dis 2018;92:210–3. https://doi.org/10.1016/j.diagmicrobio.2018.06.009.

[24] Iida N, Dzutsev A, Stewart CA, Smith L, Bouladoux N, Weingarten RA, et al. Commensal bacteria control cancer response to therapy by modulating the tumor microenvironment. Science 2013;342:967–70. https://doi.org/10.1126/science.1240527.

[25] Viaud S, Saccheri F, Mignot G, Yamazaki T, Daillère R, Hannani D, et al. The intestinal microbiota modulates the anticancer immune effects of cyclophosphamide. Science 2013;342:971–6.

[26] Daillère R, Vétizou M, Waldschmitt N, Yamazaki T, Isnard C, Poirier-Colame V, et al. *Enterococcus hirae* and *Barnesiella intestinihominis* facilitate cyclophosphamide-induced therapeutic immunomodulatory effects. Immunity 2016;45:931–43. https://doi.org/10.1016/j.immuni.2016.09.009.

[27] Vétizou M, Pitt JM, Daillère R, Lepage P, Waldschmitt N, Flament C, et al. Anticancer immunotherapy by CTLA-4 blockade relies on the gut microbiota. Science 2015;350:1079–84. https://doi.org/10.1126/science.aad1329.

[28] Sivan A, Corrales L, Hubert N, Williams JB, Aquino-Michaels K, Earley ZM, et al. Commensal Bifidobacterium promotes antitumor immunity and facilitates anti-PD-L1 efficacy. Science 2015;350:1084–9. https://doi.org/10.1126/science.aac4255.

[29] Routy B, Le Chatelier E, Derosa L, Duong CPM, Alou MT, Daillère R, et al. Gut microbiome influences efficacy of PD-1–based immunotherapy against epithelial tumors. Science 2017. https://doi.org/10.1126/science.aan3706, eaan3706.

[30] Gopalakrishnan V, Spencer CN, Nezi L, Reuben A, Andrews MC, Karpinets TV, et al. Gut microbiome modulates response to anti–PD-1 immunotherapy in melanoma patients. Science 2017. https://doi.org/10.1126/science.aan4236, eaan4236.

[31] Zhou C-B, Zhou Y-L, Fang J-Y. Gut microbiota in cancer immune response and immunotherapy. Trends Cancer 2021;7:647–60. https://doi.org/10.1016/j.trecan.2021.01.010.

[32] Zhang X, Zhang D, Jia H, Feng Q, Wang D, Di Liang D, et al. The oral and gut microbiomes are perturbed in rheumatoid arthritis and partly normalized after treatment. Nat Med 2015;21:895–905. https://doi.org/10.1038/nm.3914.

[33] Gu Y, Wang X, Li J, Zhang Y, Zhong H, Liu R, et al. Analyses of gut microbiota and plasma bile acids enable stratification of patients for antidiabetic treatment. Nat Commun 2017;8:1785. https://doi.org/10.1038/s41467-017-01682-2.

[34] Jönsson E, Ljung L, Norrman E, Freyhult E, Ärlestig L, Dahlqvist J, et al. Pulmonary fibrosis in relation to genetic loci in an inception cohort of patients with early rheumatoid arthritis from northern Sweden. Rheumatology (Oxford) 2021. https://doi.org/10.1093/rheumatology/keab441.

[35] Wang H, Xu T, Huang Q, Jin W, Chen J. Immunotherapy for malignant glioma: current status and future directions. Trends Pharmacol Sci 2020;41:123–38. https://doi.org/10.1016/j.tips.2019.12.003.

[36] Li G, Young KD. Indole production by the tryptophanase TnaA in *Escherichia coli* is determined by the amount of exogenous tryptophan. Microbiology 2013;159:402–10. https://doi.org/10.1099/mic.0.064139-0.
[37] Ye L, Bae M, Cassilly CD, Jabba SV, Thorpe DW, Martin AM, et al. Enteroendocrine cells sense bacterial tryptophan catabolites to activate enteric and vagal neuronal pathways. Cell Host Microbe 2020;29:1–18. https://doi.org/10.1101/2020.06.09.142133.
[38] Qi Q, Li J, Yu B, Moon J-Y, Chai JC, Merino J, et al. Host and gut microbial tryptophan metabolism and type 2 diabetes: an integrative analysis of host genetics, diet, gut microbiome and circulating metabolites in cohort studies. Gut 2021. https://doi.org/10.1136/gutjnl-2021-324053.
[39] Zhu F, Guo R, Wang W, Ju Y, Wang Q, Ma Q, et al. Transplantation of microbiota from drug-free patients with schizophrenia causes schizophrenia-like abnormal behaviors and dysregulated kynurenine metabolism in mice. Mol Psychiatry 2019. https://doi.org/10.1038/s41380-019-0475-4.
[40] Wlodarska M, Luo C, Kolde R, D'Hennezel E, Annand JW, Heim CE, et al. Indoleacrylic acid produced by commensal peptostreptococcus species suppresses inflammation. Cell Host Microbe 2017;22:25–37.e6. https://doi.org/10.1016/j.chom.2017.06.007.
[41] Dvořák Z, Sokol H, Mani S. Drug mimicry: promiscuous receptors PXR and AhR, and microbial metabolite interactions in the intestine. Trends Pharmacol Sci 2020;41:900–8. https://doi.org/10.1016/j.tips.2020.09.013.
[42] Wirthgen E, Leonard AK, Scharf C, Domanska G. The immunomodulator 1-methyltryptophan drives tryptophan catabolism toward the kynurenic acid branch. Front Immunol 2020;11. https://doi.org/10.3389/fimmu.2020.00313.
[43] Masab M, Saif MW. Telotristat ethyl: proof of principle and the first oral agent in the management of well-differentiated metastatic neuroendocrine tumor and carcinoid syndrome diarrhea. Cancer Chemother Pharmacol 2017;80:1055–62. https://doi.org/10.1007/s00280-017-3462-y.
[44] Soliman HH, Minton SE, Han HS, Ismail-Khan R, Neuger A, Khambati F, et al. A phase I study of indoximod in patients with advanced malignancies. Oncotarget 2016;7:22928–38. https://doi.org/10.18632/oncotarget.8216.
[45] Kumar S, Jaipuri FA, Waldo JP, Potturi H, Marcinowicz A, Adams J, et al. Discovery of indoximod prodrugs and characterization of clinical candidate NLG802. Eur J Med Chem 2020;198:112373. https://doi.org/10.1016/j.ejmech.2020.112373.
[46] Boer J, Young-Sciame R, Lee F, Bowman KJ, Yang X, Shi JG, et al. Roles of UGT, P450, and gut microbiota in the metabolism of epacadostat in humans. Drug Metab Dispos 2016;44:1668–74. https://doi.org/10.1124/dmd.116.070680.
[47] Lewis-Ballester A, Pham KN, Batabyal D, Karkashon S, Bonanno JB, Poulos TL, et al. Structural insights into substrate and inhibitor binding sites in human indoleamine 2,3-dioxygenase 1. Nat Commun 2017;8:1693. https://doi.org/10.1038/s41467-017-01725-8.
[48] Kristeleit R, Davidenko I, Shirinkin V, El-Khouly F, Bondarenko I, Goodheart MJ, et al. A randomised, open-label, phase 2 study of the IDO1 inhibitor epacadostat (INCB024360) versus tamoxifen as therapy for biochemically recurrent (CA-125 relapse)-only epithelial ovarian cancer, primary peritoneal carcinoma, or fallopian tube cancer. Gynecol Oncol 2017;146:484–90. https://doi.org/10.1016/j.ygyno.2017.07.005.
[49] Mitchell TC, Hamid O, Smith DC, Bauer TM, Wasser JS, Olszanski AJ, et al. Epacadostat plus pembrolizumab in patients with advanced solid tumors: phase I results from a multicenter, open-label phase I/II trial (ECHO-202/KEYNOTE-037). J Clin Oncol 2018;36:3223–30. https://doi.org/10.1200/JCO.2018.78.9602.

[50] Long GV, Dummer R, Hamid O, Gajewski TF, Caglevic C, Dalle S, et al. Epacadostat plus pembrolizumab versus placebo plus pembrolizumab in patients with unresectable or metastatic melanoma (ECHO-301/KEYNOTE-252): a phase 3, randomised, double-blind study. Lancet Oncol 2019;20:1083–97. https://doi.org/10.1016/S1470-2045(19)30274-8.

[51] Hunt JT, Balog A, Huang C, Lin T-A, Lin T-A, Maley D, et al. Abstract 4964: Structure, in vitro biology and in vivo pharmacodynamic characterization of a novel clinical IDO1 inhibitor. Exp Mol Ther 2017;4964. https://doi.org/10.1158/1538-7445.AM2017-4964. American Association for Cancer Research.

[52] Luke JJ, Tabernero J, Joshua A, Desai J, Varga AI, Moreno V, et al. BMS-986205, an indoleamine 2, 3-dioxygenase 1 inhibitor (IDO1i), in combination with nivolumab (nivo): updated safety across all tumor cohorts and efficacy in advanced bladder cancer (advBC). J Clin Oncol 2019;37:358. https://doi.org/10.1200/JCO.2019.37.7_suppl.358.

[53] Siu LL, Gelmon K, Chu Q, Pachynski R, Alese O, Basciano P, et al. Abstract CT116: BMS-986205, an optimized indoleamine 2,3-dioxygenase 1 (IDO1) inhibitor, is well tolerated with potent pharmacodynamic (PD) activity, alone and in combination with nivolumab (nivo) in advanced cancers in a phase 1/2a trial. Clin Trials 2017;CT116. https://doi.org/10.1158/1538-7445.AM2017-CT116. American Association for Cancer Research.

[54] Crosignani S, Bingham P, Bottemanne P, Cannelle H, Cauwenberghs S, Cordonnier M, et al. Discovery of a novel and selective indoleamine 2,3-dioxygenase (IDO-1) inhibitor 3-(5-fluoro-1 H -indol-3-yl)pyrrolidine-2,5-dione (EOS200271/PF-06840003) and its characterization as a potential clinical candidate. J Med Chem 2017;60:9617–29. https://doi.org/10.1021/acs.jmedchem.7b00974.

[55] Reardon DA, Desjardins A, Rixe O, Cloughesy T, Alekar S, Williams JH, et al. A phase 1 study of PF-06840003, an oral indoleamine 2,3-dioxygenase 1 (IDO1) inhibitor in patients with recurrent malignant glioma. Invest New Drugs 2020;38:1784–95. https://doi.org/10.1007/s10637-020-00950-1.

[56] Sun J, Chen Y, Huang Y, Zhao W, Liu Y, Venkataramanan R, et al. Programmable co-delivery of the immune checkpoint inhibitor NLG919 and chemotherapeutic doxorubicin via a redox-responsive immunostimulatory polymeric prodrug carrier. Acta Pharmacol Sin 2017;38:823–34. https://doi.org/10.1038/aps.2017.44.

[57] Nayak-Kapoor A, Hao Z, Sadek R, Dobbins R, Marshall L, Vahanian NN, et al. Phase Ia study of the indoleamine 2,3-dioxygenase 1 (IDO1) inhibitor navoximod (GDC-0919) in patients with recurrent advanced solid tumors. J Immunother Cancer 2018;6:61. https://doi.org/10.1186/s40425-018-0351-9.

[58] Kaye J, Piryatinsky V, Birnberg T, Hingaly T, Raymond E, Kashi R, et al. Laquinimod arrests experimental autoimmune encephalomyelitis by activating the aryl hydrocarbon receptor. Proc Natl Acad Sci 2016;113:E6145–52. https://doi.org/10.1073/pnas.1607843113.

[59] Nyamoya S, Steinle J, Chrzanowski U, Kaye J, Schmitz C, Beyer C, et al. Laquinimod supports remyelination in non-supportive environments. Cell 2019;8:1363. https://doi.org/10.3390/cells8111363.

[60] Ziemssen T, Tumani H, Sehr T, Thomas K, Paul F, Richter N, et al. Safety and in vivo immune assessment of escalating doses of oral laquinimod in patients with RRMS. J Neuroinflammation 2017;14:172. https://doi.org/10.1186/s12974-017-0945-z.

[61] Vollmer TL, Sorensen PS, Selmaj K, Zipp F, Havrdova E, Cohen JA, et al. A randomized placebo-controlled phase III trial of oral laquinimod for multiple sclerosis. J Neurol 2014;261:773–83. https://doi.org/10.1007/s00415-014-7264-4.

[62] D'Haens G, Sandborn WJ, Colombel JF, Rutgeerts P, Brown K, Barkay H, et al. A phase II study of laquinimod in Crohn's disease. Gut 2015;64:1227–35. https://doi.org/10.1136/gutjnl-2014-307118.
[63] Darakhshan S, Pour AB. Tranilast: a review of its therapeutic applications. Pharmacol Res 2015;91:15–28. https://doi.org/10.1016/j.phrs.2014.10.009.
[64] Smith SH, Jayawickreme C, Rickard DJ, Nicodeme E, Bui T, Simmons C, et al. Tapinarof is a natural AhR agonist that resolves skin inflammation in mice and humans. J Invest Dermatol 2017;137:2110–9. https://doi.org/10.1016/j.jid.2017.05.004.
[65] Robbins K, Bissonnette R, Maeda-Chubachi T, Ye L, Peppers J, Gallagher K, et al. Phase 2, randomized dose-finding study of tapinarof (GSK2894512 cream) for the treatment of plaque psoriasis. J Am Acad Dermatol 2019;80:714–21. https://doi.org/10.1016/j.jaad.2018.10.037.
[66] Peppers J, Paller AS, Maeda-Chubachi T, Wu S, Robbins K, Gallagher K, et al. A phase 2, randomized dose-finding study of tapinarof (GSK2894512 cream) for the treatment of atopic dermatitis. J Am Acad Dermatol 2019;80:89–98.e3. https://doi.org/10.1016/j.jaad.2018.06.047.
[67] Leja-Szpak A, Góralska M, Link-Lenczowski P, Czech U, Nawrot-Porąbka K, Bonior J, et al. The opposite effect of L-kynurenine and Ahr inhibitor Ch223191 on apoptotic protein expression in pancreatic carcinoma cells (Panc-1). Anticancer Agents Med Chem 2020;19:2079–90. https://doi.org/10.2174/1871520619666190415165212.
[68] Parks AJ, Pollastri MP, Hahn ME, Stanford EA, Novikov O, Franks DG, et al. In silico identification of an aryl hydrocarbon receptor antagonist with biological activity in vitro and in vivo. Mol Pharmacol 2014;86:593–608. https://doi.org/10.1124/mol.114.093369.
[69] Cheong JE, Sun L. Targeting the IDO1/TDO2-KYN-AhR pathway for cancer immunotherapy—challenges and opportunities. Trends Pharmacol Sci 2018;39:307–25. https://doi.org/10.1016/j.tips.2017.11.007.
[70] Boitano AE, Wang J, Romeo R, Bouchez LC, Parker AE, Sutton SE, et al. Aryl hydrocarbon receptor antagonists promote the expansion of human hematopoietic stem cells. Science 2010;329:1345–8. https://doi.org/10.1126/science.1191536.
[71] Zhang JJ, Zhang JJ, Wang R. Gut microbiota modulates drug pharmacokinetics. Drug Metab Rev 2018;50:357–68. https://doi.org/10.1080/03602532.2018.1497647.
[72] Wilson ID, Nicholson JK. Gut microbiome interactions with drug metabolism, efficacy, and toxicity. Transl Res 2017;179:204–22. https://doi.org/10.1016/j.trsl.2016.08.002.
[73] Zimmermann M, Zimmermann-Kogadeeva M, Wegmann R, Goodman AL. Separating host and microbiome contributions to drug pharmacokinetics and toxicity. Science 2019;363. https://doi.org/10.1126/science.aat9931, eaat9931.
[74] Violi F, Lip GY, Pignatelli P, Pastori D. Interaction between dietary vitamin K intake and anticoagulation by vitamin K antagonists. Medicine (Baltimore) 2016;95. https://doi.org/10.1097/MD.0000000000002895, e2895.
[75] Ong FS, Deignan JL, Kuo JZ, Bernstein KE, Rotter JI, Grody WW, et al. Clinical utility of pharmacogenetic biomarkers in cardiovascular therapeutics: a challenge for clinical implementation. Pharmacogenomics 2012;13:465–75. https://doi.org/10.2217/pgs.12.2.
[76] Sconce EA, Kamali F. Appraisal of current vitamin K dosing algorithms for the reversal of over-anticoagulation with warfarin: the need for a more tailored dosing regimen. Eur J Haematol 2006;77:457–62. https://doi.org/10.1111/j.0902-4441.2006.t01-1-EJH2957.x.
[77] Guthrie L, Gupta S, Daily J, Kelly L. Human microbiome signatures of differential colorectal cancer drug metabolism. Npj Biofilms Microbiomes 2017;3:27. https://doi.org/10.1038/s41522-017-0034-1.

[78] Lankisch TO, Schulz C, Zwingers T, Erichsen TJ, Manns MP, Heinemann V, et al. Gilbert's syndrome and irinotecan toxicity: combination with UDP-glucuronosyltransferase 1A7 variants increases risk. Cancer Epidemiol Biomarkers Prev 2008;17:695–701. https://doi.org/10.1158/1055-9965.EPI-07-2517.

[79] Sistonen J, Madadi P, Ross CJ, Yazdanpanah M, Lee JW, Landsmeer MLA, et al. Prediction of codeine toxicity in infants and their mothers using a novel combination of maternal genetic markers. Clin Pharmacol Ther 2012;91:692–9. https://doi.org/10.1038/clpt.2011.280.

[80] Kirchheiner J, Schmidt H, Tzvetkov M, Keulen J-T, Lötsch J, Roots I, et al. Pharmacokinetics of codeine and its metabolite morphine in ultra-rapid metabolizers due to CYP2D6 duplication. Pharmacogenomics J 2007;7:257–65. https://doi.org/10.1038/sj.tpj.6500406.

[81] Bairam AF, Rasool MI, Alherz FA, Abunnaja MS, El Daibani AA, Kurogi K, et al. Effects of human SULT1A3/SULT1A4 genetic polymorphisms on the sulfation of acetaminophen and opioid drugs by the cytosolic sulfotransferase SULT1A3. Arch Biochem Biophys 2018;648:44–52. https://doi.org/10.1016/j.abb.2018.04.019.

[82] Tzvetkov MV, dos Santos Pereira JN, Meineke I, Saadatmand AR, Stingl JC, Brockmöller J. Morphine is a substrate of the organic cation transporter OCT1 and polymorphisms in OCT1 gene affect morphine pharmacokinetics after codeine administration. Biochem Pharmacol 2013;86:666–78. https://doi.org/10.1016/j.bcp.2013.06.019.

[83] Clayton TA, Baker D, Lindon JC, Everett JR, Nicholson JK. Pharmacometabonomic identification of a significant host-microbiome metabolic interaction affecting human drug metabolism. Proc Natl Acad Sci 2009;106:14728–33. https://doi.org/10.1073/pnas.0904489106.

[84] Court MH, Freytsis M, Wang X, Peter I, Guillemette C, Hazarika S, et al. The UDP-glucuronosyltransferase (UGT) 1A polymorphism c.2042C>G (rs8330) is associated with increased human liver acetaminophen glucuronidation, increased UGT1A Exon 5a/5b splice variant mRNA ratio, and decreased risk of unintentional acetaminophen-ind. J Pharmacol Exp Ther 2013;345:297–307. https://doi.org/10.1124/jpet.112.202010.

[85] SLCO1B1. Variants and statin-induced myopathy—a genomewide study. N Engl J Med 2008;359:789–99. https://doi.org/10.1056/NEJMoa0801936.

[86] Voora D, Shah SH, Spasojevic I, Ali S, Reed CR, Salisbury BA, et al. The SLCO1B1*5 genetic variant is associated with statin-induced side effects. J Am Coll Cardiol 2009;54:1609–16. https://doi.org/10.1016/j.jacc.2009.04.053.

[87] Ramsey LB, Johnson SG, Caudle KE, Haidar CE, Voora D, Wilke RA, et al. The clinical pharmacogenetics implementation consortium guideline for SLCO1B1 and simvastatin-induced myopathy: 2014 update. Clin Pharmacol Ther 2014;96:423–8. https://doi.org/10.1038/clpt.2014.125.

[88] Haiser HJ, Seim KL, Balskus EP, Turnbaugh PJ. Mechanistic insight into digoxin inactivation by *Eggerthella lenta* augments our understanding of its pharmacokinetics. Gut Microbes 2014;5:233–8. https://doi.org/10.4161/gmic.27915.

[89] Aarnoudse A-JLHJ, Dieleman JP, Visser LE, Arp PP, van der Heiden IP, van Schaik RHN, et al. Common ATP-binding cassette B1 variants are associated with increased digoxin serum concentration. Pharmacogenet Genomics 2008;18:299–305. https://doi.org/10.1097/FPC.0b013e3282f70458.

[90] Diasio RB. Sorivudine and 5-fluorouracil; a clinically significant drug-drug interaction due to inhibition of dihydropyrimidine dehydrogenase. Br J Clin Pharmacol 1998;46:1–4. https://doi.org/10.1046/j.1365-2125.1998.00050.x.

[91] Spanogiannopoulos P, Bess EN, Carmody RN, Turnbaugh PJ. The microbial pharmacists within us: a metagenomic view of xenobiotic metabolism. Nat Rev Microbiol 2016;14:273–87. https://doi.org/10.1038/nrmicro.2016.17.

[92] Hitchings R, Kelly L. Predicting and understanding the human Microbiome's impact on pharmacology. Trends Pharmacol Sci 2019;40:495–505. https://doi.org/10.1016/j.tips.2019.04.014.

[93] Zhu F, Ju Y, Wang W, Wang Q, Guo R, Ma Q, et al. Metagenome-wide association of gut microbiome features for schizophrenia. Nat Commun 2020;11:1612. https://doi.org/10.1038/s41467-020-15457-9.

[94] Chang AE, Golob JL, Schmidt TM, Peltier DC, Lao CD, Tewari M. Targeting the gut microbiome to mitigate immunotherapy-induced colitis in cancer. Trends Cancer 2021;7:583–93. https://doi.org/10.1016/j.trecan.2021.02.005.

[95] da Silva DE, Grande AJ, Roever L, Tse G, Liu T, Biondi-Zoccai G, et al. High-intensity interval training in patients with type 2 diabetes mellitus: a systematic review. Curr Atheroscler Rep 2019;21:8. https://doi.org/10.1007/s11883-019-0767-9.

[96] Asle Mohammadi Zadeh M, Kargarfard M, Marandi SM, Habibi A. Diets along with interval training regimes improves inflammatory & anti-inflammatory condition in obesity with type 2 diabetes subjects. J Diabetes Metab Disord 2018;17:253–67. https://doi.org/10.1007/s40200-018-0368-0.

[97] Jensen CS, Bahl JM, Østergaard LB, Høgh P, Wermuth L, Heslegrave A, et al. Exercise as a potential modulator of inflammation in patients with Alzheimer's disease measured in cerebrospinal fluid and plasma. Exp Gerontol 2019;121:91–8. https://doi.org/10.1016/j.exger.2019.04.003.

[98] Fiuza-Luces C, Santos-Lozano A, Joyner M, Carrera-Bastos P, Picazo O, Zugaza JL, et al. Exercise benefits in cardiovascular disease: beyond attenuation of traditional risk factors. Nat Rev Cardiol 2018;15:731–43. https://doi.org/10.1038/s41569-018-0065-1.

[99] Jie Z, Liang S, Ding Q, Li F, Tang S, Wang D, et al. A transomic cohort as a reference point for promoting a healthy gut microbiome. Med Microecol 2021. https://doi.org/10.1016/j.medmic.2021.100039.

[100] Jie Z, Chen C, Hao L, Li F, Song L, Zhang X, et al. Life history recorded in the vagino-cervical microbiome along with multi-omics. Genomics Proteomics Bioinformatics 2021. https://doi.org/10.1016/j.gpb.2021.01.005.

[101] Wilmanski T, Rappaport N, Earls JC, Magis AT, Manor O, Lovejoy J, et al. Blood metabolome predicts gut microbiome α-diversity in humans. Nat Biotechnol 2019. https://doi.org/10.1038/s41587-019-0233-9.

[102] Cohn JR, Emmett EA. The excretion of trace metals in human sweat. Ann Clin Lab Sci 1978;8:270–5.

[103] Lundy SD, Sangwan N, Parekh NV, Selvam MKP, Gupta S, McCaffrey P, et al. Functional and taxonomic dysbiosis of the gut, urine, and semen microbiomes in male infertility. Eur Urol 2021;79:826–36. https://doi.org/10.1016/j.eururo.2021.01.014.

[104] Chen H, Luo T, Chen T, Wang G. Seminal bacterial composition in patients with obstructive and non-obstructive azoospermia. Exp Ther Med 2018. https://doi.org/10.3892/etm.2018.5778.

[105] Chen C, Song X, Wei W, Zhong H, Dai J, Lan Z, et al. The microbiota continuum along the female reproductive tract and its relation to uterine-related diseases. Nat Commun 2017;8:875. https://doi.org/10.1038/s41467-017-00901-0.

[106] Koedooder R, Singer M, Schoenmakers S, Savelkoul PHM, Morré SA, de Jonge JD, et al. The vaginal microbiome as a predictor for outcome of in vitro fertilization with or without intracytoplasmic sperm injection: a prospective study. Hum Reprod 2019;34:1042–54. https://doi.org/10.1093/humrep/dez065.

[107] Farquhar CM, Bhattacharya S, Repping S, Mastenbroek S, Kamath MS, Marjoribanks J, et al. Female subfertility. Nat Rev Dis Primers 2019;5:1–21. https://doi.org/10.1038/s41572-018-0058-8.

[108] Jin C, Lagoudas GK, Zhao C, Bullman S, Bhutkar A, Hu B, et al. Commensal microbiota promote lung cancer development via γδ T cells. Cell 2019;176:998–1013.e16. https://doi.org/10.1016/j.cell.2018.12.040.

[109] Masri S, Papagiannakopoulos T, Kinouchi K, Liu Y, Cervantes M, Baldi P, et al. Lung adenocarcinoma distally rewires hepatic circadian homeostasis. Cell 2016;165:896–909. https://doi.org/10.1016/j.cell.2016.04.039.
[110] Yang JJ, Yu D, Xiang Y-B, Blot W, White E, Robien K, et al. Association of dietary fiber and yogurt consumption with lung cancer risk. JAMA Oncol 2019. https://doi.org/10.1001/jamaoncol.2019.4107.
[111] Xie Y, Qin J, Nan G, Huang S, Wang Z, Su Y. Coffee consumption and the risk of lung cancer: an updated meta-analysis of epidemiological studies. Eur J Clin Nutr 2016;70:199–206. https://doi.org/10.1038/ejcn.2015.96.
[112] Vang O. Chemopreventive potential of compounds in cruciferous vegetables. CRC Press; 2005.
[113] Chng KR, Tay ASL, Li C, Ng AHQ, Wang J, Suri BK, et al. Whole metagenome profiling reveals skin microbiome-dependent susceptibility to atopic dermatitis flare. Nat Microbiol 2016;1:16106. https://doi.org/10.1038/nmicrobiol.2016.106.
[114] Dedrick RM, Guerrero-Bustamante CA, Garlena RA, Russell DA, Ford K, Harris K, et al. Engineered bacteriophages for treatment of a patient with a disseminated drug-resistant mycobacterium abscessus. Nat Med 2019;25:730–3. https://doi.org/10.1038/s41591-019-0437-z.
[115] Dedrick RM, Freeman KG, Nguyen JA, Bahadirli-Talbott A, Smith BE, Wu AE, et al. Potent antibody-mediated neutralization limits bacteriophage treatment of a pulmonary *Mycobacterium abscessus* infection. Nat Med 2021;27(8):1357–61. https://doi.org/10.1038/s41591-021-01403-9.
[116] Png CW, Lindén SK, Gilshenan KS, Zoetendal EG, McSweeney CS, Sly LI, et al. Mucolytic bacteria with increased prevalence in IBD mucosa augment in vitro utilization of mucin by other bacteria. Am J Gastroenterol 2010;105:2420–8. https://doi.org/10.1038/ajg.2010.281.
[117] Hebbandi Nanjundappa R, Ronchi F, Wang J, Clemente-Casares X, Yamanouchi J, Sokke Umeshappa C, et al. A gut microbial mimic that hijacks diabetogenic autoreactivity to suppress colitis. Cell 2017;171:655–667.e17. https://doi.org/10.1016/j.cell.2017.09.022.
[118] He Q, Gao Y, Jie Z, Yu X, Laursen JMJM, Xiao L, et al. Two distinct metacommunities characterize the gut microbiota in Crohn's disease patients. Gigascience 2017;6:1–11. https://doi.org/10.1093/gigascience/gix050.
[119] Bunker JJ, Drees C, Watson AR, Plunkett CH, Nagler CR, Schneewind O, et al. B cell superantigens in the human intestinal microbiota. Sci Transl Med 2019;11. https://doi.org/10.1126/scitranslmed.aau9356.
[120] Collins J, Robinson C, Danhof H, Knetsch CW, van Leeuwen HC, Lawley TD, et al. Dietary trehalose enhances virulence of epidemic *Clostridium difficile*. Nature 2018;553(7688):291–4. https://doi.org/10.1038/nature25178.
[121] Moayyedi P, Surette MG, Kim PT, Libertucci J, Wolfe M, Onischi C, et al. Fecal microbiota transplantation induces remission in patients with active ulcerative colitis in a randomized controlled trial. Gastroenterology 2015;149:102–109.e6. https://doi.org/10.1053/j.gastro.2015.04.001.
[122] Colman RJ, Rubin DT. Fecal microbiota transplantation as therapy for inflammatory bowel disease: a systematic review and meta-analysis. J Crohns Colitis 2014. https://doi.org/10.1016/j.crohns.2014.08.006.
[123] Kelly CP. Fecal microbiota transplantation—an old therapy comes of age. N Engl J Med 2013;368:474–5. https://doi.org/10.1056/NEJMe1214816.
[124] Rossen NG. Fecal microbiota transplantation as novel therapy in gastroenterology: a systematic review. World J Gastroenterol 2015;21:5359. https://doi.org/10.3748/wjg.v21.i17.5359.
[125] Zou M, Jie Z, Cui B, Wang H, Feng Q, Zou Y, et al. Fecal microbiota transplantation results in bacterial strain displacement in patients with inflammatory bowel diseases. FEBS Open Bio 2019. https://doi.org/10.1002/2211-5463.12744.

[126] Schirmer M, Smeekens SP, Vlamakis H, Jaeger M, Oosting M, Franzosa EA, et al. Linking the human gut microbiome to inflammatory cytokine production capacity. Cell 2016;167:1125–1136.e8. https://doi.org/10.1016/j.cell.2016.10.020.

[127] Atarashi K, Suda W, Luo C, Kawaguchi T, Motoo I, Narushima S, et al. Ectopic colonization of oral bacteria in the intestine drives T H 1 cell induction and inflammation. Science 2017;358:359–65. https://doi.org/10.1126/science.aan4526.

[128] Williams DW, Greenwell-Wild T, Brenchley L, Dutzan N, Overmiller A, Sawaya AP, et al. Human oral mucosa cell atlas reveals a stromal-neutrophil axis regulating tissue immunity. Cell 2021. https://doi.org/10.1016/j.cell.2021.05.013.

[129] Feng Q, Liang S, Jia H, Stadlmayr A, Tang L, Lan Z, et al. Gut microbiome development along the colorectal adenoma–carcinoma sequence. Nat Commun 2015;6:6528. https://doi.org/10.1038/ncomms7528.

[130] Yu J, Feng Q, Wong SHSH, Zhang D, Liang QY, Qin Y, et al. Metagenomic analysis of faecal microbiome as a tool towards targeted non-invasive biomarkers for colorectal cancer. Gut 2017;66:70–8. https://doi.org/10.1136/gutjnl-2015-309800.

[131] Zeller G, Tap J, Voigt AY, Sunagawa S, Kultima JR, Costea PI, et al. Potential of fecal microbiota for early-stage detection of colorectal cancer. Mol Syst Biol 2014;10:766. https://doi.org/10.15252/msb.20145645.

[132] Thomas AM, Manghi P, Asnicar F, Pasolli E, Armanini F, Zolfo M, et al. Metagenomic analysis of colorectal cancer datasets identifies cross-cohort microbial diagnostic signatures and a link with choline degradation. Nat Med 2019;25:667–78. https://doi.org/10.1038/s41591-019-0405-7.

[133] Kasai C, Sugimoto K, Moritani I, Tanaka J, Oya Y, Inoue H, et al. Comparison of human gut microbiota in control subjects and patients with colorectal carcinoma in adenoma: terminal restriction fragment length polymorphism and next-generation sequencing analyses. Oncol Rep 2016;35:325–33. https://doi.org/10.3892/or.2015.4398.

[134] Yachida S, Mizutani S, Shiroma H, Shiba S, Nakajima T, Sakamoto T, et al. Metagenomic and metabolomic analyses reveal distinct stage-specific phenotypes of the gut microbiota in colorectal cancer. Nat Med 2019;25:968–76. https://doi.org/10.1038/s41591-019-0458-7.

[135] Wirbel J, Pyl PT, Kartal E, Zych K, Kashani A, Milanese A, et al. Meta-analysis of fecal metagenomes reveals global microbial signatures that are specific for colorectal cancer. Nat Med 2019;25:679–89. https://doi.org/10.1038/s41591-019-0406-6.

[136] Osman MA, Neoh H-M, Ab Mutalib N-S, Chin S-F, Mazlan L, Raja Ali RA, et al. Parvimonas micra, *Peptostreptococcus stomatis*, *Fusobacterium nucleatum* and *Akkermansia muciniphila* as a four-bacteria biomarker panel of colorectal cancer. Sci Rep 2021;11:2925. https://doi.org/10.1038/s41598-021-82465-0.

[137] Soldevilla B, Carretero-Puche C, Gomez-Lopez G, Al-Shahrour F, Riesco MC, Gil-Calderon B, et al. The correlation between immune subtypes and consensus molecular subtypes in colorectal cancer identifies novel tumour microenvironment profiles, with prognostic and therapeutic implications. Eur J Cancer 2019;123:118–29. https://doi.org/10.1016/j.ejca.2019.09.008.

[138] Komor MA, Bosch LJ, Bounova G, Bolijn AS, Delis-van Diemen PM, Rausch C, et al. Consensus molecular subtype classification of colorectal adenomas. J Pathol 2018;246:266–76. https://doi.org/10.1002/path.5129.

[139] Alexander JL, Scott AJ, Pouncey AL, Marchesi J, Kinross J, Teare J. Colorectal carcinogenesis: an archetype of gut microbiota-host interaction. Ecancermedicalscience 2018;12:865. https://doi.org/10.3332/ecancer.2018.865.

[140] Slade DJ. New roles for *Fusobacterium nucleatum* in cancer: target the bacteria, host, or both? Trends Cancer 2021;7:185–7. https://doi.org/10.1016/j.trecan.2020.11.006.

[141] Mima K, Nishihara R, Qian ZR, Cao Y, Sukawa Y, Nowak JA, et al. *Fusobacterium nucleatum* in colorectal carcinoma tissue and patient prognosis. Gut 2016;65:1973–80. https://doi.org/10.1136/gutjnl-2015-310101.
[142] Yu T, Guo F, Yu Y, Sun T, Ma D, Han J, et al. *Fusobacterium nucleatum* promotes chemoresistance to colorectal cancer by modulating autophagy. Cell 2017;170:548–563.e16. https://doi.org/10.1016/j.cell.2017.07.008.
[143] Giacomelli R, Afeltra A, Bartoloni E, Berardicurti O, Bombardieri M, Bortoluzzi A, et al. The growing role of precision medicine for the treatment of autoimmune diseases; results of a systematic review of literature and experts' consensus. Autoimmun Rev 2021;20:102738. https://doi.org/10.1016/j.autrev.2020.102738.
[144] Smolen JS, Landewé RBM, Bijlsma JWJ, Burmester GR, Dougados M, Kerschbaumer A, et al. EULAR recommendations for the management of rheumatoid arthritis with synthetic and biological disease-modifying antirheumatic drugs: 2019 update. Ann Rheum Dis 2020;79:685–99. https://doi.org/10.1136/annrheumdis-2019-216655.
[145] Long X, Wong CC, Tong L, Chu ESH, Ho Szeto C, Go MYY, et al. *Peptostreptococcus anaerobius* promotes colorectal carcinogenesis and modulates tumour immunity. Nat Microbiol 2019;4:2319–30. https://doi.org/10.1038/s41564-019-0541-3.

8

A microbiome record for life

8.1 Proactive sampling of the microbiome at important time periods

From birth to old age, the microbiome in various body sites contains important information that could predict disease risks in the future (Fig. 8.1). Colonization of microbes in early life leads to trafficking of microbial antigens to the thymus by antigen-presenting cells in the intestine, skin, and probably other mucosal sites, and the T-cells induced can then protect the host against related pathogens, and go awry at times [1–3].

8.1.1 A microbiome record from birth

Hospital records are really impressive in some countries already. If an infant is born preterm or full term, will there be microbiome samples for the feces, the oropharynx, and the skin? Were there antibiotics, oxytocin (to stimulate labor) [4], or other medication given to the mother before birth? Should the infants' microbiome be followed afterward (Table 8.1)? In extremely preterm infants, the difference between delivery modes was no longer a big deal. Infants born through vaginal delivery were more colonized by *Bacteroides* spp. and *Bifidobacterium* spp. from the mother [19,28,29], but what about the father and the grandparents, and the natural playgrounds that are also contributing some good bacteria [30]?

Vaginal microbes from the mother, including bacterial vaginosis (BV)-related *Gardnerella vaginalis* (now formally in the *Bifidobacterium* genus, as *Bifidobacterium vaginalis* [31]) and *Atopobium vaginae* (renamed as *Fannyhessea vaginae*) could be detected in the infant's gut in the first week after delivery [32]. Emerging from the amniotic fluid into a dry environment, does the ambient temperature influence the maturation of the newborn's respiratory microbiome, skin microbiome, and gut microbiome? Are there pets in the house [33,34]? How many hours did the baby sleep, before and after a stable rhythm has been established [35] (Table 8.1)?

Newborns have a continuous airway optimized for sucking and breathing, with the larynx high up in the mouth like that in chimpanzees.

Investigating Human Diseases with the Microbiome: Metagenomics Bench to Bedside.
https://doi.org/10.1016/B978-0-323-91369-0.00005-4

Fig. 8.1 Monitoring the microbiome in different age groups. Credit: Huijue Jia, Xin Tong, Fei Li.

The larynx gradually moves down to an adult-like position (Chapter 3, Fig. 3.5), swallowing more solid food instead of breastmilk (swallowing interrupts breathing), and beginning to speak [36,37]. We do not know yet, how this series of remodeling affect the oral, respiratory, and gastrointestinal microbiome. As the infant begins to produce its own antibodies, instead of depending on the mother's antibodies through the placenta and in the milk (and sometimes from the mouth) [38,39], how do the antibodies and the microbes coevolve into an appropriate affinity (weak binding help retain the commensals, Chapter 2) that could sufficiently protect the infant against pathogens? The infant liver shifts from the hematopoietic function in embryos to the (circadian) metabolic and immune functions more similar to adults [40,41]. Besides the human genome-encoded amylase that gives the sweet taste when we chew on starchy food such as rice, do the oral and gastrointestinal microbiome influence taste and food preference? Pectin-degrading *Bacteroides thetaiotaomicron*, for example, might contribute to a preference for oranges and peas [42]?

The baseline for a healthy infant microbiome may need to be established for each ethnic group [43,44], depending on the local habits. As mentioned in Chapter 2, human genetics contribute to the microbiome composition. The alleles associated with more fecal *Prevotella* spp. instead of *Bacteroides* spp. are more prevalent in African and East Asian populations compared to European populations [45,46]. *Prevotella* spp. were more prevalent in the airway of children without

Table 8.1 A metagenomic record from birth.

Age	Event	Oral	Fecal	Airway	Skin	Conditions to watch for	Reference
0	Exposure to antibiotics at delivery or during infancy	?	Y		?	Childhood obesity; impaired antibody induction after vaccination, but enhanced T cell responses	[5–7]
0	Exposure to oxytocin at delivery	?	Y		?	Supplements the endogenous oxytocin to reduce anxiety and autism risk, or reduce the risk of over-feeding and obesity?	[8–10]
1–2 months	Preterm birth, C-section, antibiotics, lack of elder siblings, lack of dog, etc.		Y	Y	?	Risk of asthma in the following years	[11–14]
Any age	Vaccination	?	Y			Prediction of vaccination efficacy depending on the microbiome, and use of additional ways if possible	[14]
2–6 months	Hospitalized for bronchiolitis with more nasal *Moraxella* and *Streptococcus*	Y	?	Y		Recurrent wheezing by 3 years old	[15]
3–6 months	Milk allergy	?	Y		?	Prediction of resolved allergy in childhood	[16–18]
0–12 months	Lack of breastfeeding	Y	Y	?		Lack of *Bifidobacterium* from the mother, and consequently difference in immune development	[19–22]
4–11 months	Introduction of solid food	Y	Y		?	Early exposure to solid food before cessation of breasting-feeding, to prevent food allergy	[23,24]
4–12 months	More regular sleep and feeding	?	Y			Establishment of a diurnal gut microbiome rhythm?	[19], Chapter 2, Box 2.5
Childhood to adolescence	Rural environment	Y	Y	Y	Y	Difference in microbiome composition with potential long-lasting consequences	[25–27]

The associations are not necessarily causal (Chapter 6) and would need to be further elucidated. This is not meant to be an exhaustive list, but it serves to illustrate the range of conditions and samples to consider. Y, samples strongly recommended; ?, evidence needed.
Credit: Huijue Jia.

asthma and in healthy adults [47]. *Bifidobacterium* spp., for example, are not so prevalent in the gut microbiome of infants in some Asian, African, or Latin American regions [48,49]; multiple alleles other than the lactase *LCT* also associate with *Bifidobacterium* [45,46]. Proteobacteria such as *Escherichia coli* and *Klebsiella* sp., for example, encodes enzymes that could metabolize sulphoquinovose (6-deoxy-6-sulphoglucose), which takes up at least 10 mg/g of plant leaf dry weight [50,51]. The local grain, soil, and water are important sources of micronutrients and heavy metals that could influence the microbiome [46,52–54] (Fig. 8.1). Water in developed countries such as the United States are not necessarily all at the healthy standards [55].

8.1.2 Immediate and historical events for a wholesome microbiome

For the mother, many kinds of precious samples could also be collected (Table 8.2). Factors that shape the vaginal microbiome span from puberty to postmenopausal years. Vaccination would not be sufficient to prevent all infections and tumors. Microbiome samples collected during the first pregnancy could also take better care of things the next time. But the mother's microbiome is going to be different for the subsequent children, with less *Lactobacillus crispatus* in the vagina [54], more *Prevotella copri* in the gut (and possibly the amniotic fluid [69]). *Prevotella copri* in the pregnant mother's fecal microbiome associated with a lower risk of food allergy for the kids [70]. A lower BMI (body mass index) before pregnancy relates to more *Bifidobacterium* in breast milk [71]. Fecal microbiome markers for gestational diabetes are partly similar to those for T2D [69,72], consistent with a family history of T2D and heightened risk for subsequent T2D in women who had gestational diabetes during pregnancy [73,74]. Women with menstrual pains (dysmenorrhea) showed more Pseudomonadales, *Acinetobacter,* and *Moraxellaceae* in the cervicovaginal microbiome, while lower in plasma level of histidine; The relative abundances of *Acinetobacter* and *Moraxellaceae* were indeed lower in married women [54]. In vitro fertilization (IVF) is rather common nowadays. In addition to contributing to a higher success rate of consumption and term pregnancy (Table 8.2), the vaginal, oral and gut microbiome could also contain information for disease risks much later in life. For example, can we predict risk for breast cancer according to the fecal and the milk microbiome (Chapter 2, Table 2.1; Chapter 4, Fig. 4.2)? Recovery of the fecal, cervical, urinary, and other microbiome after each pregnancy may be more difficult in some people than others, which necessitates additional effort. Hip fractures due to osteoporosis is a life-threatening incident in postmenopausal women, and the fecal microbiome markers for bone mineral density

Table 8.2 A metagenomic record for women.

Event	Sample					Microbes	Reference
	Vaginocervical	Oral	Fecal	Urine	Breastmilk		
Menstrual cycle	Y					More *Lactobacillus iners* and bacterial vaginosis-related bacteria during menses; More *L. crispatus* in the secretory phase	[54,56]
		Y				More bacterial cells before menses, bad breath; *Streptococcus*?	
Sexual debut	Y					More susceptible to bacterial vaginosis right after menses; Older age at sexual debut associated with *Bifidobacterium breve*	[54,57]
Contraception	Y					More *G. vaginalis* for no contraception; Oral contraceptives appeared to associate with cervical *L. iners*, *Ureaplasma parvum* and *Comamonas*, and some associations in the fecal microbiome	[54]
Pregnancy	Y			?		More Lactobacilli during pregnancy	[58]
In vitro fertilization	Y			?		Endometrium (or fallopian fluid) samples to test for Lactobacilli and other bacteria, in correlation with successful pregnancy	[59–61]
Spontaneous abortion	Y			?		*L. iners* negatively associated with spontaneous abortion	[54]
		Y	Y			Mouse gavaged with *Fusobacterium* showed preterm birth	[62]
Preterm birth	Y			?		A diverse vaginal microbiome lacking Lactobacilli	Many studies

Continued

Table 8.2 A metagenomic record for women—cont'd

Event	Sample					Microbes	Reference
	Vaginocervical	Oral	Fecal	Urine	Breastmilk		
Delivery	Y					Less *L. crispatus* for mothers who had given birth in the past; *Streptococcus anginosis* (GBS), *Ureaplasma*, etc. screened to prevent infection of newborns	[54,56]
			Y			Likely more obesity-related bacteria for mothers who deliver by C-section; Risks for diseases such as (gestational) diabetes and hypertension monitored during and after pregnancy	Follow up studies needed
Breastfeeding	Y					Recovery to a Lactobacilli-dominated microbiota, perhaps shifting to *L. iners*	[54,56]
			Y		Y	Monitor risk of breast cancer in the years to come; *Bifidobacterium* and other bacteria match those in the baby?	[63–66]
Menopause	Y	Y	Y	Y		Maintaining a healthy bone mineral density, metabolic health and urinary function, free of HPV and other pathogens	Prospective cohorts needed
Hysteromyoma (uterine fibroids)	Y			Y		More *L. iners* instead of *L. crispatus*?	[67,68]
Rheumatoid arthritis	Y	Y	Y	Y		Fungal infection in synovial fluid matches that in the vagina? Bacteria in synovial fluid matched that in the mouth and in the gut? More effective medication	Chapter 4

The associations are not necessarily causal (Chapter 6) and would need to be further elucidated. This is not meant to be an exhaustive list, but it serves to illustrate the range of conditions and samples to consider. Y, samples strongly recommended. *HPV*, human papillomavirus.
Credit: Huijue Jia.

(BMD) should be a target for early intervention through measures such as probiotics and tea (Intervention evidence needed) [75–78]. Like the skin [79], the vaginal metabolome also include compounds that likely come from cosmetic products [80].

In developed countries such as the U.S., a lot of food allergy cases start in adulthood (Table 8.3), and more commonly in women [81]. The comorbidities such as asthma, allergic rhinitis, are also known to involve the gut microbiome and the respiratory microbiome.

We do not know as much for men. Infant boys already differed in gut microbiome composition and functional capacity compared to girls [19]. An allele between *NEGR1* and *LINC01360* that has been implicated in autism spectrum disorder (ASD) and schizophrenia showed a significant M-GWAS (Metagenome-genome-wide association study) association with the gut bacterium *Acidaminococcus* (e.g., *Acidaminococcus intestinalis*) only in males [45,82]. Hyperuricemia (high level of blood uric acid) and gout are more prevalent in men, and the gut and oral microbiome are heavily involved [45,83,84]. *Anaerococcus* and *Prevotella* in semen are associated with low sperm quality, while *Pseudomonas* correlated with total sperm count [85,86]. There were microbial differences between subtypes of male infertility, e.g., nonobstructive and obstructive azoospermia (no sperm) [87], with and without varicocele (bulging blood vessels) [86]. *Staphylococcus* has been reported to be enriched in prostate cancer tissues [88]; both *Staphylococcus* and the more abundant *Cutibacterium acnes* (renamed from *Propionibacterium acnes*) implicated in chronic prostatitis showed a reduced level in the seminal fluid of *Estrogen Receptor-α* (*ESR1*) knockout mice [89]. A wild speculation would be to modulate hormone levels with physical fitness training and see how the microbiome improves throughout the body.

We mentioned biological aging in Chapter 2. To better understand the process, keeping a good record of the microbiome at each stage (Fig. 8.1), along with other omics such as hormone levels, trace metals, physical activity, would also be useful for the general population [53,90]. Handgrip strength lower than people of the same age group, a known epidemiological factor for cardiovascular events, associated with fecal relative abundance of *E. coli* [83], which was shown by Mendelian Randomization (MR, Chapter 6) to promote Type 2 diabetes, congestive heart failure, colorectal cancer [46] and could mediate the disease risks for the years to come. Sleeping for longer than average (> 9 or 10 h) is epidemiologically linked to a reduced lifespan [91][92], and it remains to be seen whether that has something to do with lack of saliva secretion while asleep, hypoxia, and the accumulation of oral and lung microbes. Effects of intervention, e.g., dietary fibers (Table 8.4) [42,94,95], dairy products, vitamins, high-intensity interval training, going to bed earlier, could be quickly assayed with

Table 8.3 Overall and age-specific prevalence of specific food allergies among all US adults.

Specific food allergy	Prevalence, % (95% confidence interval (CI))					
	All ages	18–29 years	30–39 years	40–49 years	50–59 years	≥60 years
Any food allergy	10.8 (10.4–11.1)	11.3 (10.5–12.2)	12.7 (11.8–13.7)	10.0 (9.2–10.9)	11.9 (11.0–12.8)	8.8 (8.2–9.4)
Peanut	1.8 (1.7–1.9)	2.5 (2.2–2.8)	2.9 (2.5–3.3)	1.8 (1.5–2.1)	1.4 (1.1–1.7)	0.8 (0.7–1.0)
Tree nut	1.2 (1.1–1.3)	1.6 (1.3–1.9)	1.7 (1.4–2.1)	1.1 (0.9–1.4)	1.2 (0.9–1.5)	0.6 (0.4–0.7)
Walnut	0.6 (0.6–0.7)	0.8 (0.7–1.1)	0.9 (0.7–1.3)	0.6 (0.5–0.8)	0.7 (0.5–0.9)	0.3 (0.2–0.4)
Almond	0.7 (0.6–0.8)	0.9 (0.7–1.2)	1.0 (0.7–1.3)	0.7 (0.6–1.0)	0.7 (0.5–0.9)	0.3 (0.2–0.4)
Hazelnut	0.6 (0.5–0.7)	0.7 (0.5–0.9)	0.9 (0.6–1.2)	0.6 (0.4–0.8)	0.6 (0.4–0.8)	0.3 (0.2–0.4)
Pecan	0.5 (0.5–0.6)	0.6 (0.5–0.8)	0.8 (0.5–1.1)	0.6 (0.5–0.8)	0.5 (0.4–0.8)	0.5 (0.4–0.8)
Cashew	0.5 (0.5–0.6)	0.8 (0.6–1.0)	0.8 (0.6–1.1)	0.5 (0.4–0.7)	0.5 (0.3–0.7)	0.2 (0.1–0.3)
Pistachio	0.4 (0.3–0.5)	0.6 (0.4–0.8)	0.6 (0.4–0.8)	0.5 (0.3–0.6)	0.4 (0.3–0.6)	0.1 (0.1–0.2)
Other tree nut	0.2 (0.1–0.2)	0.1 (0.1–0.2)	0.1 (0.0–0.2)	0.3 (0.2–0.6)	0.2 (0.1–0.5)	0.1 (0.1–0.2)
Milk	1.9 (1.8–2.1)	2.4 (2.0–2.9)	2.3 (1.9–2.8)	2.0 (1.6–2.4)	1.9 (1.6–2.2)	1.9 (1.6–2.2)
Shellfish	2.9 (2.7–3.1)	2.8 (2.4–3.2)	3.6 (3.1–4.2)	2.5 (2.2–3.0)	3.3 (2.8–3.8)	2.6 (2.2–3.0)
Shrimp	1.9 (1.8–2.1)	1.8 (1.5–2.1)	2.5 (2.1–3.0)	1.8 (1.4–2.1)	2.2 (1.8–2.6)	1.6 (1.3–1.9)
Lobster	1.3 (1.2–1.4)	1.2 (1.0–1.5)	1.6 (1.3–2.0)	1.3 (1.0–1.5)	1.4 (1.1–1.7)	1.1 (0.9–1.3)
Crab	1.3 (1.2–1.5)	1.2 (1.0–1.5)	1.6 (1.3–2.0)	1.3 (1.0–1.6)	1.6 (1.3–2.0)	1.1 (0.9–1.4)
Mollusk	1.6 (1.4–1.7)	1.6 (1.3–2.0)	2.0 (1.7–2.5)	1.3 (1.1–1.7)	1.7 (1.4–2.0)	1.2 (1.0–1.5)
Other shellfish	0.3 (0.2–0.3)	0.3 (0.1–0.5)	0.1 (0.1–0.2)	0.3 (0.2–0.4)	0.3 (0.2–0.5)	0.3 (0.2–0.4)
Egg	0.8 (0.7–0.9)	1.1 (0.7–1.5)	1.1 (0.9–1.3)	0.7 (0.5–0.9)	0.8 (0.6–1.1)	0.5 (0.3–0.7)
Fin fish	0.9 (0.8–1.0)	1.1 (0.9–1.4)	1.0 (0.8–1.2)	0.8 (0.6–1.1)	1.0 (0.7–1.3)	0.6 (0.4–0.7)
Wheat	0.8 (0.7–0.9)	1.0 (0.7–1.3)	1.0 (0.8–1.3)	0.8 (0.6–1.0)	0.7 (0.5–0.9)	0.6 (0.4–0.8)
Soy	0.6 (0.5–0.7)	0.7 (0.5–0.9)	0.8 (0.6–1.0)	0.6 (0.5–0.8)	0.7 (0.5–0.9)	0.4 (0.3–0.6)
Sesame	0.2 (0.2–0.3)	0.3 (0.2–0.4)	0.3 (0.2–0.5)	0.2 (0.1–0.4)	0.3 (0.2–0.5)	0.1 (0.0–0.2)

Credit: Table 2 of Gupta RS, Warren CM, Smith BM, Jiang J, Blumenstock JA, Davis MM, et al. Prevalence and severity of food allergies among US adults. JAMA Netw Open 2019;2:e185630. https://doi.org/10.1001/jamanetworkopen.2018.5630.

Table 8.4 Naturally occurring fibers according to solubility and fermentation properties.

Fiber type	Chain length	Sources	Potential benefits for IBS[a]	Potential risks for IBS[a]
Soluble highly fermentable oligosaccharides (includes FOS,GOS)	Short-chain carbohydrates	• Legumes/pulses, nuts and seeds • Wheat, rye • Onions, garlic, artichoke	• Laxation: weak laxative effect • Transit time: does not hasten transit time • Balance of bacteria: selective growth of certain microbiota, e.g., Bifidobacterium • SCFA: very rapidly fermented in terminal ileum and proximal colon to produce SCFA • Gas production: high	• In patients with IBS the rapid fermentation may contribute to gas, flatus and gastrointestinal symptoms • A number of studies have been undertaken in IBS—with mixed results [98]
Soluble highly fermentable "fiber" (e.g., RS, pectin, guar gum, and inulin)	Long-chain carbohydrates	• Legumes/pulses • Rye bread, barley • Firm bananas • Buckwheat groats (kashi), millet, oats • Cooked and cooled-pasta, potato and rice	• Laxation: mild laxative effect • Transit time: does not hasten gut transit. Can slow absorption from the small intestine • Balance of bacteria: increases overall bacterial species but not selective for bifidobacteria • SCFA: rapidly fermented in proximal colon to produce SCFA. RS is good an excellent substrate for the production of the SCFA butyrate • Gas production: moderate	• In patients with IBS the rapid fermentation may contribute to gas, flatus, and gastrointestinal symptoms • No well-designed studies have been undertaken in IBS

Continued

Table 8.4 Naturally occurring fibers according to solubility and fermentation properties—cont'd

Fiber type	Chain length	Sources	Potential benefits for IBS[a]	Potential risks for IBS[a]
Intermediate soluble fermentable "fiber" (psyllium/ispaghula) and oats 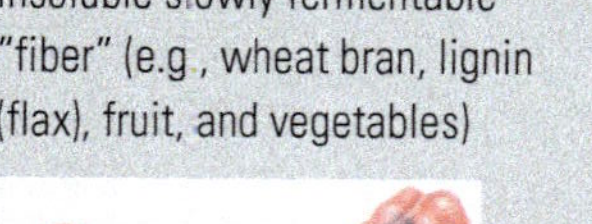	Long-chain carbohydrates	Seed of the plant *Plantago ovata*, and oats	• Laxation: good laxative effect • Transit time: does hasten transit time • Balance of bacteria: increases overall bacterial species but little evidence for selective growth • SCFA: moderately fermented along length of colon to produce SCFA • Gas production: moderate	• In patients with IBS studies have shown some positive effect on laxation • Side-effects of gas/flatus has produced mixed results for some patients with IBS [99]
Insoluble slowly fermentable "fiber" (e.g., wheat bran, lignin (flax), fruit, and vegetables) 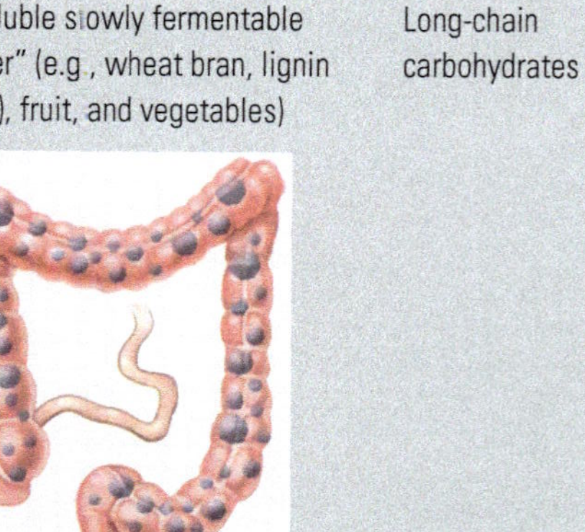	Long-chain carbohydrates	• Some vegetables and fruit • Wheat bran • Wholegrain cereal • Rye • Brown rice, wholemeal pasta, quinoa • Flax seed	• Laxation: good laxative effect • Transit time: does hasten transit time • Balance of bacteria: increases overall bacterial species but little evidence for selective growth • SCFA: slowly fermented to produce SCFA along the length of the colon • Gas production: moderate-high	• In patients with IBS wheat bran has not been shown to be effective. A major side-effect has been excessive gas/wind and bloating [100]. This may be due to the presence of high quantities of fructans also associated with the wheat bran [101] • Symptoms associated with wheat bran may not be acceptable to many patients

Insoluble, nonfermentable "fiber" (e.g., cellulose, sterculia, and methylcellulose) 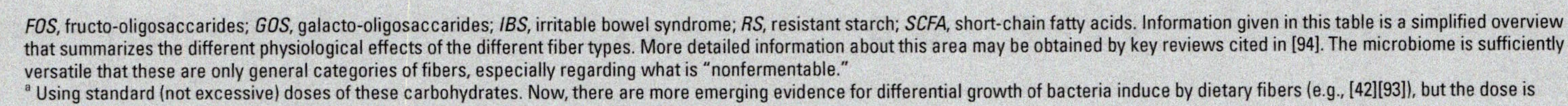	Long-chain carbohydrates	• High fiber grains and cereals • Nuts, seeds • Skins of fruit and vegetables	• Laxation: good laxative effect • Transit time: does hasten transit time • Balance of bacteria: no evidence for selective growth • SCFA: poorly fermented • Gas production: low	• Less gas/wind forming properties • This fiber type may have better characteristics for treating constipation in IBS patients. However, few well designed studies have been conducted

FOS, fructo-oligosaccarides; *GOS*, galacto-oligosaccarides; *IBS*, irritable bowel syndrome; *RS*, resistant starch; *SCFA*, short-chain fatty acids. Information given in this table is a simplified overview that summarizes the different physiological effects of the different fiber types. More detailed information about this area may be obtained by key reviews cited in [94]. The microbiome is sufficiently versatile that these are only general categories of fibers, especially regarding what is "nonfermentable."

[a] Using standard (not excessive) doses of these carbohydrates. Now, there are more emerging evidence for differential growth of bacteria induce by dietary fibers (e.g., [42][93]), but the dose is sometimes excessive compared to population-wide standard recommendations from nutritional experts.

Credit: Table 1 of Eswaran S, Muir J, Chey WD. Fiber and functional gastrointestinal disorders. Am J Gastroenterol 2013;108:718–27. https://doi.org/10.1038/ajg.2013.63.

microbiome composition and functional capacity. The association between milk consumption and fecal microbiome may have something to do with estrogen exposure from the cow milk, in addition to IBS (irritable bowel syndrome) risk in some individuals [90]. Microbiome associations with more local habits, such as the types of tea, soy milk, plants used for bathing, can also be validated with more volunteer participation.

As we take measures to better protect the ecosystems on our planet Earth, gases and particles in the air, metals and organic compounds in water and soils will also be changing [96,97]. Exposure to climate change-related events such as wildfires could have long-term effects on the microbiome and the immune system. The microbiome records now may then be historic.

8.2 From genetic risk to the prevention of diseases

We discussed metagenomics-based diagnosis and treatment in Chapter 7. For infants, airway microbiome at 1 month could predict asthma at 6 years old [13] (Table 8.3), and there might be ways to amend for preterm birth and Cesarean section early on (e.g., Probiotics [102], Facilitated microbial colonization [103]; Cortisols and catecholamine exposure, Chapter 6, Fig. 6.4). Large cohorts have been followed to understand the microbiome during the onset of Type 1 diabetes [22,104–106].

Trends for many diseases could be seen decades before clinical symptoms, and the microbiome adds a key dimension for visualizing such early trends. For example, fecal bacterial and plasma metabolomic markers for colorectal cancer, hyperuricemia (high level of urate in the blood, which does not necessarily lead to painful gout) and meat consumption are visible in a cohort with a mean age of 30 years old, with some individuals at a higher risk than others [53]. Poly-genetic risk scores (PRS) for periodontal diseases and dental caries already showed an area under the (receiver operating) curve (AUC) of more than 0.8, suggesting that some people would always have to take better care of their oral microbiome [84]. PRS for diseases such as breast cancer, cardiovascular diseases, Alzheimer's diseases are promising for preventive medicine [107], and the fecal or oral microbiome could add an important dimension. Progression of Parkinson's disease and Alzheimer's disease might be slowed with proper management of risk factors [108,109]. The intervention can simply be dietary changes, exercise, or quit smoking, which would be a wise thing to do before it is too late. Plasma biomarkers for metabolic syndrome and Alzheimer's disease, e.g. branched-chain amino acids (BCAAs) and acylcarnitines, could be metabolized by leg muscles and the kidneys [110–112]. Frequent moderate-intensity physical activity, e.g., brisk walking, is associated with lower risk and mortality for major diseases such as cancer and cardiovascular diseases [113,114]. The gut microbiome and inflammatory markers change after sprinting [115]. Dairy products are known as a protective factor for gout [90,116]. So if we compute from the fecal and/or oral microbiome that the urate level is not low [45,83,84], and the risk for Parkinson's disease is not high (fecal microbiome studies tend to be complicated by medication and disease duration, but useful biomarkers would still emerge) [117,118], maybe people with a genetic risk for Parkinson's disease can consume dairy products just fine [108].

Digitalizing the dietary information as nitrogen source (e.g., shifting from nitrite to amines), amino acids, metals, vitamins, etc. instead of crude questionnaires, would be a long-term effort for each culture. Sodium benzoate, a preservative that used to be widely present in pickled vegetables and in soda, is recently in clinical trials to see if it might help treat negative symptoms in schizophrenia patients [119] (the depression-like social exclusion, as opposed to the agitated positive symptoms). The incidence of vitamin A deficiency is decreasing globally (Chapter 7, Fig. 7.1). Firmicutes of the Clostridiale order suppress intestinal vitamin A production in mice, possibly finishing up a weaning reaction that induces T regulatory cells (Treg); and fecal Clostridiale species correlated with plasma vitamin A level in adult humans [23,53,120,121]. In addition to the gut microbiome, plasma vitamin A level also negatively associated with *L. iners* in the cervicovaginal microbiome (more *L. crispatus* in women who had never given birth), another potential explanation for the ethnical differences in the vaginal microbiome [122], in addition to the vitamin D and hormone explanation for bacterial vaginosis [54,123], or some genetic factors.

Body temperature and the number of breath taken per minute appeared to associate with the microbiome in the peritoneal cavity in women of reproductive age [67]. From long-term records of the menstrual cycles, scientists were able to match disturbances to the menstrual cycles with luminance (or gravitational) differences in the lunar cycle [124,125], which affects sleeping together with the hormones [124,126]; Fertile women had menses at the full moon and ovulate in the darkest days [125]. In the cervical microbiome, menstrual irregularity correlated with *L. vaginalis* [54], which correlated with *L. crispatus* and so possibly decreases with older age and decreased hormones.

Air pollution, which is a strong risk factor for cardiovascular diseases [96], lung cancer, etc. [127], can at least be recorded on the city scale. For developing countries, the burning of biomass is still a major source of pollution, instead of from factories or cars [97,128,129], which means different chemicals and particles that the human microbiome is exposed to.

Lung capacity showed association with the fecal microbiome [90]. Besides physical exercises, do we know how much talking and singing a person does? Salivary IgA and cortisol have been used as biomarkers to show favorable effects of enjoying music and choral singing [130,131], and oxytocin would also be a key marker [132].

Wearable devices are constantly being developed, and body fluids other than blood could be more readily analyzed in the future. For example, sweat can be assayed for glucose, electrolytes, lactate, uric acid, metals, etc. [133,134]. Oxygen saturation and heart rate readings are commonly provided by smartwatches [135]. Such disease-relevant and easily accessible measurements could fit in with the microbiome record.

Single-cell sequencing for detailed immune cell populations and T-cell receptor (TCR), B-cell receptor (BCR) sequences are as yet too expensive to replace cell counts from routine blood tests, and ELISA (Enzyme-linked Immunosorbent Assay) for cytokines. Human leukocyte antigen (HLA) types (MHC-I, MHC-II) that represent a strong genetic factor for many diseases can be available from whole genome sequences, without additional experimental procedures. Less understood major histocompatibility complex (MHC)-like proteins such as CD1 genes and MR1 (MHC class I-related protein 1) are also important for interaction with molecules that are relevant for the human microbiome [2,136–141]. It has been shown in mice that the development of mucosal-associated invariant T cells (MAIT), the predominant innate-like lymphocytes important for mucosal and skin homeostasis, took place during a time window early in life (2–3 weeks old for mice) [3]. Commensal bacteria can produce the vitamin B2 (riboflavin) derivative 5-(2-oxopropylideneamino)-6-D-ribitylaminouracil (5-OP-RU) [2]. When such bacteria or metabolites were applied to the skin or gavaged orally, 5-OP-RU is presented by MR1 to the T cell receptor (TCR) of MAIT cells which induce MAIT development in the thymus [2,3]. The number of these cells varies considerably among individuals.

Microbiome information may help relieve the ethical controversy over risks for complex diseases reported by commercial tests of the human genome [142]. Family history, a strong epidemiological factor for many complex diseases (colorectal cancer, breast cancer, hypertension, autoimmune disorders, etc.), could mean both genetic and microbiome similarity. By adding such modifiable information which together explains a larger portion of the risk, disease risks can be received with a more scientific understanding. The key question would always be, shall we do something about it?

Clinical trials are already being performed on high-risk individuals for diseases such as rheumatoid arthritis (Fig. 8.2). Microbiome information (Chapter 4, Fig. 4.4; Chapter 7) would make an important complement to the immune parameters. Probiotics are also being tested for a number of diseases [143–145], and it would probably be better to start in preclinical individuals.

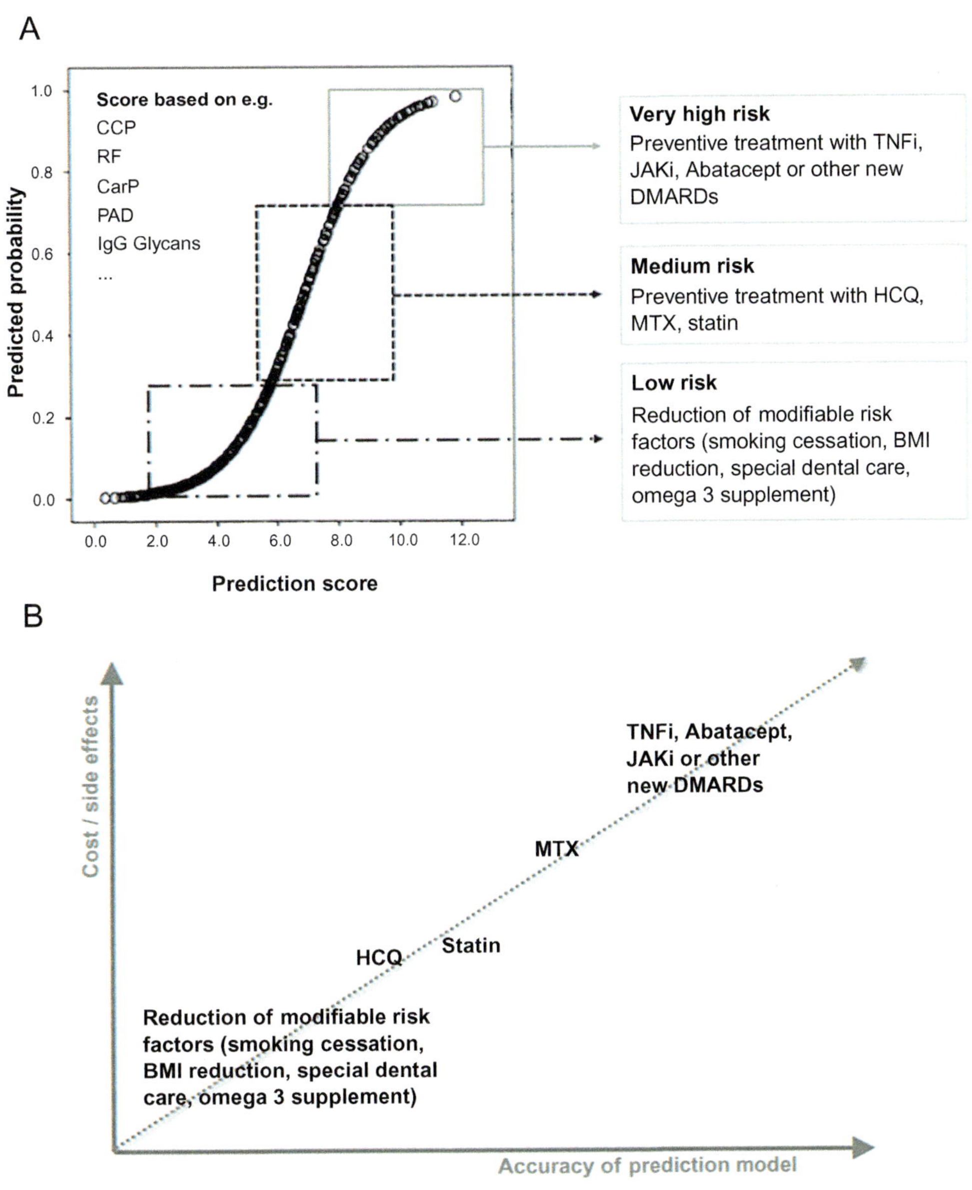

Fig. 8.2 Toward prevention of rheumatoid arthritis (RA). (A) Risk-based prevention model. (B) Linear relationship between cost/safety profile and required accuracy of the prediction model. More expensive and less safe treatment strategies will require higher accuracy of the prediction model to avoid unnecessary treatment of individuals that might not develop RA in the near future. *BMI*, body mass index; *DMARD*, disease modifying antirheumatic drug; *HCQ*, hydroxychloroquine; *MTX*, methotrexate; *TNFi*, tumor necrosis factor inhibitor. Credit: Fig. 2 of Mahler M, Martinez-Prat L, Sparks JA, Deane KD. Precision medicine in the care of rheumatoid arthritis: focus on prediction and prevention of future clinically-apparent disease. Autoimmun Rev 2020;19:102506. https://doi.org/10.1016/j.autrev.2020.102506.

8.3 Summary

From birth to old age, the microbiome in various body sites contains important information that could predict disease risks in the future (Fig. 8.1). Once-in-a-lifetime experiences and long-term exposures can all leave their mark in the human microbiome. The microbiome connects genetic and environmental factors, would greatly facilitate precision medicine and potentially allow many complex diseases to be prevented before the emergence of clinical symptoms. It will take engineers and scientists, and people from all walks of life to fill in a variety of information that is not currently recorded in microbiome studies. With demonstrated values for health, and as metagenomics become cheaper and bioinformatics becomes faster and more accessible, more people would be motivated to sample their microbiome. Facilities could be provided for educational purposes, and as part of a physical examination that are better aligned with occasional events, e.g., being locked down at home.

Worked sample 8.1

In your country or region, what do you think is the most needed microbiome test?

Where do you think the test should be offered?

How much do you think people are willing to pay for the test?

Worked sample 8.2

Prepare a shortlist of questions that you would like people to answer before, or after receiving results from the microbiome test you start to provide.

Are some people more willing to participate than others? What do they hope the microbiome test could do for them?

Do you see gaps in expectations? How would you like to improve, or refocus?

References

[1] Zegarra-Ruiz DF, Kim DV, Norwood K, Kim M, Wu W-JH, Saldana-Morales FB, et al. Thymic development of gut-microbiota-specific T cells. Nature 2021;1–5. https://doi.org/10.1038/s41586 021-03531-1.

[2] Legoux F, Bellet D, Daviaud C, El Morr Y, Darbois A, Niort K, et al. Microbial metabolites control the thymic development of mucosal-associated invariant T cells. Science 2019;366. https://doi.org/10.1126/science.aaw2719, eaaw2719.

[3] Constantinides MG, Link VM, Tamoutounour S, Wong AC, Perez-Chaparro PJ, Han S-J, et al. MAIT cells are imprinted by the microbiota in early life and promote tissue repair. Science 2019;366. https://doi.org/10.1126/science.aax6624, eaax6624.

[4] Zhang J, Branch DW, Ramirez MM, Laughon SK, Reddy U, Hoffman M, et al. Oxytocin regimen for labor augmentation, labor progression, and perinatal outcomes. Obstet Gynecol 2011;118:249–56. https://doi.org/10.1097/AOG.0b013e3182220192.
[5] Cho I, Yamanishi S, Cox L, Methé BA, Zavadil J, Li K, et al. Antibiotics in early life alter the murine colonic microbiome and adiposity. Nature 2012;488:621–6. https://doi.org/10.1038/nature11400.
[6] Cox LM, Yamanishi S, Sohn J, Alekseyenko AV, Leung JM, Cho I, et al. Altering the intestinal microbiota during a critical developmental window has lasting metabolic consequences. Cell 2014;158:705–21. https://doi.org/10.1016/j.cell.2014.05.052.
[7] Lynn MA, Tumes DJ, Choo JM, Sribnaia A, Blake SJ, Leong LEX, et al. Early-life antibiotic-driven dysbiosis leads to dysregulated vaccine immune responses in mice. Cell Host Microbe 2018;23:653–660.e5. https://doi.org/10.1016/j.chom.2018.04.009.
[8] Peñagarikano O, Lázaro MT, Lu X-H, Gordon A, Dong H, Lam HA, et al. Exogenous and evoked oxytocin restores social behavior in the Cntnap2 mouse model of autism. Sci Transl Med 2015;7:271ra8. https://doi.org/10.1126/scitranslmed.3010257.
[9] Lawson EA. The effects of oxytocin on eating behaviour and metabolism in humans. Nat Rev Endocrinol 2017;13:700–9. https://doi.org/10.1038/nrendo.2017.115.
[10] Ben-Ari Y. Is birth a critical period in the pathogenesis of autism spectrum disorders? Nat Rev Neurosci 2015;16:498–505. https://doi.org/10.1038/nrn3956.
[11] Russell SL, Gold MJ, Hartmann M, Willing BP, Thorson L, Wlodarska M, et al. Early life antibiotic-driven changes in microbiota enhance susceptibility to allergic asthma. EMBO Rep 2012;13:440–7. https://doi.org/10.1038/embor.2012.32.
[12] Pattaroni C, Watzenboeck ML, Schneidegger S, Kieser S, Wong NC, Bernasconi E, et al. Early-life formation of the microbial and immunological environment of the human airways. Cell Host Microbe 2018;24:857–865.e4. https://doi.org/10.1016/j.chom.2018.10.019.
[13] Thorsen J, Rasmussen MA, Waage J, Mortensen M, Brejnrod A, Bønnelykke K, et al. Infant airway microbiota and topical immune perturbations in the origins of childhood asthma. Nat Commun 2019;10:5001. https://doi.org/10.1038/s41467-019-12989-7.
[14] Lynn DJ, Benson SC, Lynn MA, Pulendran B. Modulation of immune responses to vaccination by the microbiota: implications and potential mechanisms. Nat Rev Immunol 2021. https://doi.org/10.1038/s41577-021-00554-7.
[15] Mansbach JM, Luna PN, Shaw CA, Hasegawa K, Petrosino JF, Piedra PA, et al. Increased Moraxella and Streptococcus species abundance after severe bronchiolitis is associated with recurrent wheezing. J Allergy Clin Immunol 2020;145:518–527.e8. https://doi.org/10.1016/j.jaci.2019.10.034.
[16] Bunyavanich S, Shen N, Grishin A, Wood R, Burks W, Dawson P, et al. Early-life gut microbiome composition and milk allergy resolution. J Allergy Clin Immunol 2016;138:1122–30. https://doi.org/10.1016/j.jaci.2016.03.041.
[17] Stephen-Victor E, Crestani E, Chatila TA. Dietary and microbial determinants in food allergy. Immunity 2020;53:277–89. https://doi.org/10.1016/j.immuni.2020.07.025.
[18] Rachid R, Stephen-Victor E, Chatila TA. The microbial origins of food allergy. J Allergy Clin Immunol 2021;147:808–13. https://doi.org/10.1016/j.jaci.2020.12.624.
[19] Bäckhed F, Roswall J, Peng Y, Feng Q, Jia H, Kovatcheva-Datchary P, et al. Dynamics and stabilization of the human gut microbiome during the first year of life. Cell Host Microbe 2015;17:690–703. https://doi.org/10.1016/j.chom.2015.04.004.

[20] Verma R, Lee C, Jeun E-J, Yi J, Kim KS, Ghosh A, et al. Cell surface polysaccharides of *Bifidobacterium bifidum* induce the generation of Foxp3 + regulatory T cells. Sci Immunol 2018;3:eaat6975. https://doi.org/10.1126/sciimmunol.aat6975.

[21] Henrick BM, Rodriguez L, Lakshmikantz T, Pou C, Henckel E, Olin A, et al. Bifidobacteria-mediated immune system imprinting early in life. BioRxiv 2021. https://doi.org/10.1101/2020.10.24.353250.

[22] Vatanen T, Franzosa EA, Schwager R, Tripathi S, Arthur TD, Vehik K, et al. The human gut microbiome in early-onset type 1 diabetes from the TEDDY study. Nature 2018;562:589–94. https://doi.org/10.1038/s41586-018-0620-2.

[23] Al Nabhani Z, Dulauroy S, Marques R, Cousu C, Al Bounny S, Déjardin F, et al. A weaning reaction to microbiota is required for resistance to immunopathologies in the adult. Immunity 2019;50:1276–1288.e5. https://doi.org/10.1016/j.immuni.2019.02.014.

[24] Knoop KA, Gustafsson JK, McDonald KG, Kulkarni DH, Coughlin PE, McCrate S, et al. Microbial antigen encounter during a preweaning interval is critical for tolerance to gut bacteria. Sci Immunol 2017;2. https://doi.org/10.1126/sciimmunol.aao1314, eaao1314.

[25] Lehtimäki J, Karkman A, Laatikainen T, Paalanen L, von Hertzen L, Haahtela T, et al. Patterns in the skin microbiota differ in children and teenagers between rural and urban environments. Sci Rep 2017;7:45651. https://doi.org/10.1038/srep45651.

[26] Ayeni FA, Biagi E, Rampelli S, Fiori J, Soverini M, Audu HJ, et al. Infant and adult gut microbiome and metabolome in rural Bassa and urban settlers from Nigeria. Cell Rep 2018;23:3056–67. https://doi.org/10.1016/j.celrep.2018.05.018.

[27] Depner M, Taft DH, Kirjavainen PV, Kalanetra KM, Karvonen AM, Peschel S, et al. Maturation of the gut microbiome during the first year of life contributes to the protective farm effect on childhood asthma. Nat Med 2020. https://doi.org/10.1038/s41591-020-1095-x.

[28] Stokholm J, Thorsen J, Blaser MJ, Rasmussen MA, Hjelmsø M, Shah S, et al. Delivery mode and gut microbial changes correlate with an increased risk of childhood asthma. Sci Transl Med 2020;12. https://doi.org/10.1126/scitranslmed.aax9929, eaax9929.

[29] Selma-Royo M, Calatayud Arroyo M, García-Mantrana I, Parra-Llorca A, Escuriet R, Martínez-Costa C, et al. Perinatal environment shapes microbiota colonization and infant growth: impact on host response and intestinal function. Microbiome 2020;8:167. https://doi.org/10.1186/s40168-020-00940-8.

[30] Kirjavainen PV, Karvonen AM, Adams RI, Täubel M, Roponen M, Tuoresmäki P, et al. Farm-like indoor microbiota in non-farm homes protects children from asthma development. Nat Med 2019;25:1089–95. https://doi.org/10.1038/s41591-019-0469-4.

[31] Barisic V, Abdelhadi A, Frank A, Riojas MA, Hazbón MH. Reclassification of the bifidobacterium and gardnerella genera. In: ATCC. ASM microbe 2019, San Francisco, California, United States; 2019.

[32] Ferretti P, Pasolli E, Tett A, Asnicar F, Gorfer V, Fedi S, et al. Mother-to-infant microbial transmission from different body sites shapes the developing infant gut microbiome. Cell Host Microbe 2018;24:133–145.e5. https://doi.org/10.1016/j.chom.2018.06.005.

[33] Tun HM, Konya T, Takaro TK, Brook JR, Chari R, Field CJ, et al. Exposure to household furry pets influences the gut microbiota of infant at 3-4 months following various birth scenarios. Microbiome 2017;5:40. https://doi.org/10.1186/s40168-017-0254-x.

[34] Song SJ, Lauber C, Costello EK, Lozupone CA, Humphrey G, Berg-Lyons D, et al. Cohabiting family members share microbiota with one another and with their dogs. Elife 2013;2. https://doi.org/10.7554/eLife.00458, e00458.

[35] Ardura J, Gutierrez R, Andres J, Agapito T. Emergence and evolution of the circadian rhythm of melatonin in children. Horm Res 2003;59:66–72. doi:68571.
[36] Prakash M, Johnny J. Whats special in a child's larynx? J Pharm Bioallied Sci 2015;7:S55–8. https://doi.org/10.4103/0975-7406.155797.
[37] Geddes DT, Chadwick LM, Kent JC, Garbin CP, Hartmann PE. Ultrasound imaging of infant swallowing during breast-feeding. Dysphagia 2010;25:183–91. https://doi.org/10.1007/s00455-009-9241-0.
[38] Pou C, Nkulikiyimfura D, Henckel E, Olin A, Lakshmikanth T, Mikes J, et al. The repertoire of maternal anti-viral antibodies in human newborns. Nat Med 2019. https://doi.org/10.1038/s41591-019-0392-8.
[39] Msallam R, Balla J, Rathore APS, Kared H, Malleret B, Saron WAA, et al. Fetal mast cells mediate postnatal allergic responses dependent on maternal IgE. Science 2020;370:941–50. https://doi.org/10.1126/science.aba0864.
[40] Le Rouzic V, Corona J, Zhou H. Postnatal development of hepatic innate immune response. Inflammation 2011;34:576–84. https://doi.org/10.1007/s10753-010-9265-5.
[41] Nakagaki BN, Mafra K, de Carvalho É, Lopes ME, Carvalho-Gontijo R, de Castro-Oliveira HM, et al. Immune and metabolic shifts during neonatal development reprogram liver identity and function. J Hepatol 2018;69:1294–307. https://doi.org/10.1016/j.jhep.2018.08.018.
[42] Patnode ML, Beller ZW, Han ND, Cheng J, Peters SL, Terrapon N, et al. Interspecies competition impacts targeted manipulation of human gut bacteria by fiber-derived glycans. Cell 2019;179:59–73.e13. https://doi.org/10.1016/j.cell.2019.08.011.
[43] De Filippo C, Cavalieri D, Di Paola M, Ramazzotti M, Poullet JB, Massart S, et al. Impact of diet in shaping gut microbiota revealed by a comparative study in children from Europe and rural Africa. Proc Natl Acad Sci U S A 2010;107:14691–6. https://doi.org/10.1073/pnas.1005963107.
[44] Yatsunenko T, Rey FE, Manary MJ, Trehan I, Dominguez-Bello MG, Contreras M, et al. Human gut microbiome viewed across age and geography. Nature 2012;486:222–7. https://doi.org/10.1038/nature11053.
[45] Liu X, Tang S, Zhong H, Tong X, Jie Z, Ding Q, et al. A genome-wide association study for gut metagenome in Chinese adults illuminates complex diseases. Cell Discov 2021;7:9. https://doi.org/10.1038/s41421-020-00239-w.
[46] Liu X, Tong X, Zou Y, Lin X, Zhao H, Tian L, et al. Inter-determination of blood metabolite levels and gut microbiome supported by Mendelian randomization. BioRxiv 2020. https://doi.org/10.1101/2020.06.30.181438. 2020.06.30.
[47] Hilty M, Burke C, Pedro H, Cardenas P, Bush A, Bossley C, et al. Disordered microbial communities in asthmatic airways. PLoS One 2010;5. https://doi.org/10.1371/journal.pone.0008578, e8578.
[48] Schnorr SL, Candela M, Rampelli S, Centanni M, Consolandi C, Basaglia G, et al. Gut microbiome of the Hadza hunter-gatherers. Nat Commun 2014;5:3654. https://doi.org/10.1038/ncomms4654.
[49] Lane AA, McGuire MK, McGuire MA, Williams JE, Lackey KA, Hagen EH, et al. Household composition and the infant fecal microbiome: the INSPIRE study. Am J Phys Anthropol 2019;169:526–39. https://doi.org/10.1002/ajpa.23843.
[50] Denger K, Weiss M, Felux A-K, Schneider A, Mayer C, Spiteller D, et al. Sulphoglycolysis in *Escherichia coli* K-12 closes a gap in the biogeochemical Sulphur cycle. Nature 2014;507:114–7. https://doi.org/10.1038/nature12947.
[51] Roy AB, Hewlins MJE, Ellis AJ, Harwood JL, White GF. Glycolytic breakdown of sulfoquinovose in bacteria: a missing Link in the sulfur cycle. Appl Environ Microbiol 2003;69:6434–41. https://doi.org/10.1128/AEM.69.11.6434-6441.2003.
[52] Gashu D, Nalivata PC, Amede T, Ander EL, Bailey EH, Botoman L, et al. The nutritional quality of cereals varies geospatially in Ethiopia and Malawi. Nature 2021;594:71–6. https://doi.org/10.1038/s41586-021-03559-3.

[53] Jie Z, Liang S, Ding Q, Li F, Tang S, Wang D, et al. A transomic cohort as a reference point for promoting a healthy gut microbiome. Med Microecol 2021. https://doi.org/10.1016/j.medmic.2021.100039.
[54] Jie Z, Chen C, Hao L, Li F, Song L, Zhang X, et al. Life history recorded in the vagino-cervical microbiome along with multi-omics. Genomics Proteomics Bioinformatics 2021. https://doi.org/10.1016/j.gpb.2021.01.005.
[55] Mueller JT, Gasteyer S. The widespread and unjust drinking water and clean water crisis in the United States. Nat Commun 2021;12:3544. https://doi.org/10.1038/s41467-021-23898-z.
[56] dos Santos Santiago GL, Tency I, Verstraelen H, Verhelst R, Trog M, Temmerman M, et al. Longitudinal qPCR study of the dynamics of *L. crispatus*, *L. iners*, *A. vaginae*, (sialidase positive) *G. vaginalis*, and *P. bivia* in the vagina. PLoS One 2012;7. https://doi.org/10.1371/journal.pone.0045281, e45281.
[57] Gajer P, Brotman RM, Bai G, Sakamoto J, Schutte UME, Zhong X, et al. Temporal dynamics of the human vaginal microbiota. Sci Transl Med 2012;4:132ra52. https://doi.org/10.1126/scitranslmed.3003605.
[58] Fettweis JM, Serrano MG, Brooks JP, Edwards DJ, Girerd PH, Parikh HI, et al. The vaginal microbiome and preterm birth. Nat Med 2019;25:1012–21. https://doi.org/10.1038/s41591-019-0450-2.
[59] Koedooder R, Singer M, Schoenmakers S, Savelkoul PHM, Morré SA, de Jonge JD, et al. The vaginal microbiome as a predictor for outcome of in vitro fertilization with or without intracytoplasmic sperm injection: a prospective study. Hum Reprod 2019;34:1042–54. https://doi.org/10.1093/humrep/dez065.
[60] Schoenmakers S, Laven J. The vaginal microbiome as a tool to predict IVF success. Curr Opin Obstet Gynecol 2020;32:169–78. https://doi.org/10.1097/GCO.0000000000000626.
[61] Pelzer ES, Allan JA, Waterhouse MA, Ross T, Beagley KW, Knox CL. Microorganisms within human follicular fluid: effects on IVF. PLoS One 2013;8. https://doi.org/10.1371/journal.pone.0059062, e59062.
[62] Fardini Y, Chung P, Dumm R, Joshi N, Han YW. Transmission of diverse oral bacteria to murine placenta: evidence for the oral microbiome as a potential source of intrauterine infection. Infect Immun 2010;78:1789–96. https://doi.org/10.1128/IAI.01395-09.
[63] Urbaniak C, Gloor GB, Brackstone M, Scott L, Tangney M, Reid G. The microbiota of breast tissue and its association with breast cancer. Appl Environ Microbiol 2016;82:5039–48. https://doi.org/10.1128/AEM.01235-16.
[64] Chambers SA, Townsend SD. Like mother, like microbe: human milk oligosaccharide mediated microbiome symbiosis. Biochem Soc Trans 2020;48:1139–51. https://doi.org/10.1042/BST20191144.
[65] Pannaraj PS, Li F, Cerini C, Bender JM, Yang S, Rollie A, et al. Association between breast milk bacterial communities and establishment and development of the infant gut microbiome. JAMA Pediatr 2017;171:647–54. https://doi.org/10.1001/jamapediatrics.2017.0378.
[66] Nayfach S, Rodriguez-Mueller B, Garud N, Pollard KS. An integrated metagenomics pipeline for strain profiling reveals novel patterns of bacterial transmission and biogeography. Genome Res 2016;26:1612–25. https://doi.org/10.1101/gr.201863.115.
[67] Chen C, Song X, Wei W, Zhong H, Dai J, Lan Z, et al. The microbiota continuum along the female reproductive tract and its relation to uterine-related diseases. Nat Commun 2017;8:875. https://doi.org/10.1038/s41467-017-00901-0.
[68] Chen C, Hao L, Wei W, Li F, Song L, Zhang X, et al. The female urinary microbiota in relation to the reproductive tract microbiota. Gigabyte 2020;2020:1–9. https://doi.org/10.46471/gigabyte.9.

[69] Wang J, Zheng J, Shi W, Du N, Xu X, Zhang Y, et al. Dysbiosis of maternal and neonatal microbiota associated with gestational diabetes mellitus. Gut 2018. https://doi.org/10.1136/gutjnl-2018-315988. gutjnl-2018-315988.
[70] Vuillermin PJ, O'Hely M, Collier F, Allen KJ, Tang MLK, Harrison LC, et al. Maternal carriage of Prevotella during pregnancy associates with protection against food allergy in the offspring. Nat Commun 2020;11:1452. https://doi.org/10.1038/s41467-020-14552-1.
[71] Cortés-Macías E, Selma-Royo M, Martínez-Costa C, Collado MC. Breastfeeding practices influence the breast milk microbiota depending on pre-gestational maternal BMI and weight gain over pregnancy. Nutrients 2021;13. https://doi.org/10.3390/nu13051518.
[72] Crusell MKW, Hansen TH, Nielsen T, Allin KH, Rühlemann MC, Damm P, et al. Gestational diabetes is associated with change in the gut microbiota composition in third trimester of pregnancy and postpartum. Microbiome 2018;6:89. https://doi.org/10.1186/s40168-018-0472-x.
[73] Zhang Y, Xiao C-M, Zhang Y, Chen Q, Zhang X-Q, Li X-F, et al. Factors associated with gestational diabetes mellitus: a Meta-analysis. J Diabetes Res 2021;2021:6692695. https://doi.org/10.1155/2021/6692695.
[74] Hewage SS, Aw S, Chi C, Yoong J. Factors associated with intended postpartum OGTT uptake and willingness to receive preventive behavior support to reduce type 2 diabetes risk among women with gestational diabetes in Singapore: an exploratory study. Nutr Metab Insights 2021;14. https://doi.org/10.1177/11786388211016827. 11786388211016828.
[75] Wang Q, Sun Q, Li X, Wang Z, Zheng H, Ju Y, et al. Linking gut microbiome to bone mineral density: a shotgun metagenomic dataset from 361 elderly women. Gigabyte 2021;2021:1–7. https://doi.org/10.46471/gigabyte.12.
[76] Ohlsson C, Sjögren K. Effects of the gut microbiota on bone mass. Trends Endocrinol Metab 2015;26:69–74. https://doi.org/10.1016/j.tem.2014.11.004.
[77] Yan J, Herzog JW, Tsang K, Brennan CA, Bower MA, Garrett WS, et al. Gut microbiota induce IGF-1 and promote bone formation and growth. Proc Natl Acad Sci U S A 2016;113:E7554–63. https://doi.org/10.1073/pnas.1607235113.
[78] Zhao H, Chen J, Li X, Sun Q, Qin P, Wang Q. Compositional and functional features of the female premenopausal and postmenopausal gut microbiota. FEBS Lett 2019;593:2655–64. https://doi.org/10.1002/1873-3468.13527.
[79] Bouslimani A, Porto C, Rath CM, Wang M, Guo Y, Gonzalez A, et al. Molecular cartography of the human skin surface in 3D. Proc Natl Acad Sci U S A 2015;112:E2120–9. https://doi.org/10.1073/pnas.1424409112.
[80] Kindschuh WF, Baldini F, Liu MC, Gerson KD, Liao J, Lee HH, et al. Preterm birth is associated with xenobiotics and predicted by the vaginal metabolome. BioRxiv 2021. https://doi.org/10.1101/2021.06.14.448190. 2021.06.14.448190.
[81] Gupta RS, Warren CM, Smith BM, Jiang J, Blumenstock JA, Davis MM, et al. Prevalence and severity of food allergies among US adults. JAMA Netw Open 2019;2. https://doi.org/10.1001/jamanetworkopen.2018.5630, e185630.
[82] Zhu F, Ju Y, Wang W, Wang Q, Guo R, Ma Q, et al. Metagenome-wide association of gut microbiome features for schizophrenia. Nat Commun 2020;11:1612. https://doi.org/10.1038/s41467-020-15457-9.
[83] Jie Z, Liang S, Ding Q, Li F, Tang S, Sun X, et al. Disease trends in a young Chinese cohort according to fecal metagenome and plasma metabolites. Med Microecol 2021. https://doi.org/10.1016/j.medmic.2021.100037.
[84] Liu X, Tong X, Zhu J, Tian L, Jie Z, Zou Y, et al. Metagenome-genome-wide association studies reveal human genetic impact on the oral microbiome. bioRxiv 2021. https://doi.org/10.1101/2021.05.06.443017.

[85] Hou D, Zhou X, Zhong X, Settles ML, Herring J, Wang L, et al. Microbiota of the seminal fluid from healthy and infertile men. Fertil Steril 2013;100:1261–9. https://doi.org/10.1016/j.fertnstert.2013.07.1991.

[86] Lundy SD, Sangwan N, Parekh NV, Selvam MKP, Gupta S, McCaffrey P, et al. Functional and taxonomic dysbiosis of the gut, urine, and semen microbiomes in male infertility. Eur Urol 2021;79:826–36. https://doi.org/10.1016/j.eururo.2021.01.014.

[87] Chen H, Luo T, Chen T, Wang G. Seminal bacterial composition in patients with obstructive and non-obstructive azoospermia. Exp Ther Med 2018. https://doi.org/10.3892/etm.2018.5778.

[88] Cavarretta I, Ferrarese R, Cazzaniga W, Saita D, Lucianò R, Ceresola ER, et al. The microbiome of the prostate tumor microenvironment. Eur Urol 2017;72:625–31. https://doi.org/10.1016/j.eururo.2017.03.029.

[89] Javurek AB, Spollen WG, Ali AMM, Johnson SA, Lubahn DB, Bivens NJ, et al. Discovery of a novel seminal fluid microbiome and influence of estrogen receptor alpha genetic status. Sci Rep 2016;6:23027. https://doi.org/10.1038/srep23027.

[90] Jie Z, Liang S, Ding Q, Li F, Tang S, Wang D, et al. Dairy consumption and physical fitness tests associated with fecal microbiome in a Chinese cohort. Med Microecol 2021.

[91] Svensson T, Saito E, Svensson AK, Melander O, Orho-Melander M, Mimura M, et al. Association of sleep duration with all- and major-cause mortality among adults in Japan, China, Singapore, and Korea. JAMA Netw Open 2021;4(9):e2122837. https://doi.org/10.1001/jamanetworkopen.2021.22837.

[92] Valenzuela PL, Carrera-Bastos P, Gálvez BG, Ruiz-Hurtado G, Ordovas JM, Ruilope LM, et al. Lifestyle interventions for the prevention and treatment of hypertension. Nat Rev Cardiol 2021;18(4):251–75. https://doi.org/10.1038/s41569-020-00437-9.

[93] Kovatcheva-Datchary P, Nilsson A, Akrami R, Lee YS, De Vadder F, Arora T, et al. Dietary fiber-induced improvement in glucose metabolism is associated with increased abundance of prevotella. Cell Metab 2015;22(6):971–82. https://doi.org/10.1016/j.cmet.2015.10.001.

[94] Eswaran S, Muir J, Chey WD. Fiber and functional gastrointestinal disorders. Am J Gastroenterol 2013;108:718–27. https://doi.org/10.1038/ajg.2013.63.

[95] Jie Z, Yu X, Liu Y, Sun L, Chen P, Ding Q, et al. The baseline gut microbiota directs dieting-induced weight loss trajectories. Gastroenterology 2021. https://doi.org/10.1053/j.gastro.2021.01.029.

[96] Al-Kindi SG, Brook RD, Biswal S, Rajagopalan S. Environmental determinants of cardiovascular disease: lessons learned from air pollution. Nat Rev Cardiol 2020;17:656–72. https://doi.org/10.1038/s41569-020-0371-2.

[97] Landrigan PJ, Fuller R, Acosta NJR, Adeyi O, Arnold R, Basu N, et al. The lancet commission on pollution and health. Lancet 2018;391:462–512. https://doi.org/10.1016/S0140-6736(17)32345-0.

[98] Bijkerk CJ, Muris JWM, Knottnerus JA, Hoes AW, de Wit NJ. Systematic review: the role of different types of fibre in the treatment of irritable bowel syndrome. Aliment Pharmacol Ther 2004;19:245–51. https://doi.org/10.1111/j.0269-2813.2004.01862.x.

[99] Biesiekierski JR, Rosella O, Rose R, Liels K, Barrett JS, Shepherd SJ, et al. Quantification of fructans, galacto-oligosacharides and other short-chain carbohydrates in processed grains and cereals. J Hum Nutr Diet 2011;24:154–76. https://doi.org/10.1111/j.1365-277X.2010.01139.x.

[100] Hunt R, Fedorak R, Frohlich J, McLennan C, Pavilanis A. Therapeutic role of dietary fibre. Can Fam Physician 1993;39:897–900 [903–10].

[101] Elia M, Cummings JH. Physiological aspects of energy metabolism and gastrointestinal effects of carbohydrates. Eur J Clin Nutr 2007;61(Suppl 1):S40–74. https://doi.org/10.1038/sj.ejcn.1602938.
[102] Morgan RL, Preidis GA, Kashyap PC, Weizman AV, Sadeghirad B, Chang Y, et al. Probiotics reduce mortality and morbidity in preterm, low birth weight infants: a systematic review and network meta-analysis of randomized trials. Gastroenterology 2020. https://doi.org/10.1053/j.gastro.2020.05.096.
[103] Korpela K, Helve O, Kolho K, Saisto T, Skogberg K, Dikareva E, et al. Maternal fecal microbiota transplantation in cesarean-born infants rapidly restores normal gut microbial development: a proof-of-concept study. Cell 2020;1–11. https://doi.org/10.1016/j.cell.2020.08.047.
[104] Paun A, Yau C, Meshkibaf S, Daigneault MC, Marandi L, Mortin-Toth S, et al. Association of HLA-dependent islet autoimmunity with systemic antibody responses to intestinal commensal bacteria in children. Sci Immunol 2019;4. https://doi.org/10.1126/sciimmunol.aau8125, eaau8125.
[105] Vatanen T, Kostic AD, D'Hennezel E, Siljander H, Franzosa EA, Yassour M, et al. Variation in microbiome LPS immunogenicity contributes to autoimmunity in humans. Cell 2016;165:842–53. https://doi.org/10.1016/j.cell.2016.04.007.
[106] Akil AA-S, Yassin E, Al-Maraghi A, Aliyev E, Al-Malki K, Fakhro KA. Diagnosis and treatment of type 1 diabetes at the dawn of the personalized medicine era. J Transl Med 2021;19:137. https://doi.org/10.1186/s12967-021-02778-6.
[107] Lambert SA, Abraham G, Inouye M. Towards clinical utility of polygenic risk scores. Hum Mol Genet 2019;28:R133–42. https://doi.org/10.1093/hmg/ddz187.
[108] Ascherio A, Schwarzschild MA. The epidemiology of Parkinson's disease: risk factors and prevention. Lancet Neurol 2016;15:1257–72. https://doi.org/10.1016/S1474-4422(16)30230-7.
[109] Nedergaard M, Goldman SA. Glymphatic failure as a final common pathway to dementia. Science 2020;370:50–6. https://doi.org/10.1126/science.abb8739.
[110] Toledo JB, Arnold M, Kastenmüller G, Chang R, Baillie RA, Han X, et al. Metabolic network failures in Alzheimer's disease: a biochemical road map. Alzheimers Dement 2017;13(9):965–84. https://doi.org/10.1016/j.jalz.2017.01.020.
[111] Overmyer KA, Evans CR, Qi NR, Minogue CE, Carson JJ, Chermside-Scabbo CJ, et al. Maximal oxidative capacity during exercise is associated with skeletal muscle fuel selection and dynamic changes in mitochondrial protein acetylation. Cell Metab 2015;21(3):468–78. https://doi.org/10.1016/j.cmet.2015.02.007.
[112] Jang C, Hui S, Zeng X, Cowan AJ, Wang L, Chen L, et al. Metabolite exchange between mammalian organs quantified in pigs. Cell Metab 2019;30(3):594–606.e3. https://doi.org/10.1016/j.cmet.2019.06.002.
[113] Ruiz-Casado A, Martín-Ruiz A, Pérez LM, Provencio M, Fiuza-Luces C, Lucia A. Exercise and the hallmarks of cancer. Trends Cancer 2017;3:423–41. https://doi.org/10.1016/j.trecan.2017.04.007.
[114] Fiuza-Luces C, Santos-Lozano A, Joyner M, Carrera-Bastos P, Picazo O, Zugaza JL, et al. Exercise benefits in cardiovascular disease: beyond attenuation of traditional risk factors. Nat Rev Cardiol 2018;15:731–43. https://doi.org/10.1038/s41569-018-0065-1.
[115] Motiani KK, Collado MC, Eskelinen J-J, Virtanen KA, LÖyttyniemi E, Salminen S, et al. Exercise training modulates gut microbiota profile and improves endotoxemia. Med Sci Sports Exerc 2020;52:94–104. https://doi.org/10.1249/MSS.0000000000002112.
[116] Kuo C-F, Grainge MJ, Zhang W, Doherty M. Global epidemiology of gout: prevalence, incidence and risk factors. Nat Rev Rheumatol 2015;11:649–62. https://doi.org/10.1038/nrrheum.2015.91.

[117] Hopfner F, Künstner A, Müller SH, Künzel S, Zeuner KE, Margraf NG, et al. Gut microbiota in Parkinson disease in a northern German cohort. Brain Res 2017;1667:41–5. https://doi.org/10.1016/j.brainres.2017.04.019.

[118] Bullich C, Keshavarzian A, Garssen J, Kraneveld A, Perez-Pardo P. Gut vibes in Parkinson's disease: the microbiota-gut-brain axis. Mov Disord Clin Pract 2019;6:639–51. https://doi.org/10.1002/mdc3.12840.

[119] Minichino A, Brondino N, Solmi M, Del Giovane C, Fusar-Poli P, Burnet P, et al. The gut-microbiome as a target for the treatment of schizophrenia: a systematic review and meta-analysis of randomised controlled trials of add-on strategies. Schizophr Res 2021;234:1–13. https://doi.org/10.1016/j.schres.2020.02.012.

[120] Grizotte-Lake M, Zhong G, Duncan K, Kirkwood J, Iyer N, Smolenski I, et al. Commensals suppress intestinal epithelial cell retinoic acid synthesis to regulate interleukin-22 activity and prevent microbial dysbiosis. Immunity 2018;49:1103–1115.e6. https://doi.org/10.1016/j.immuni.2018.11.018.

[121] Atarashi K, Tanoue T, Oshima K, Suda W, Nagano Y, Nishikawa H, et al. Treg induction by a rationally selected mixture of Clostridia strains from the human microbiota. Nature 2013;500:232–6. https://doi.org/10.1038/nature12331.

[122] Ravel J, Gajer P, Abdo Z, Schneider GM, Koenig SSK, McCulle SL, et al. Vaginal microbiome of reproductive-age women. Proc Natl Acad Sci U S A 2011;108(Suppl. 1):4680–7. https://doi.org/10.1073/pnas.1002611107.

[123] Jefferson KK, Parikh HI, Garcia EM, Edwards DJ, Serrano MG, Hewison M, et al. Relationship between vitamin D status and the vaginal microbiome during pregnancy. J Perinatol 2019;39:824–36. https://doi.org/10.1038/s41372-019-0343-8.

[124] Casiraghi L, Spiousas I, Dunster GP, McGlothlen K, Fernández-Duque E, Valeggia C, et al. Moonstruck sleep: synchronization of human sleep with the moon cycle under field conditions. Sci Adv 2021;7. https://doi.org/10.1126/sciadv.abe0465, eabe0465.

[125] Helfrich-Förster C, Monecke S, Spiousas I, Hovestadt T, Mitesser O, Wehr TA. Women temporarily synchronize their menstrual cycles with the luminance and gravimetric cycles of the moon. Sci Adv 2021;7. https://doi.org/10.1126/sciadv.abe1358, eabe1358.

[126] Taxier LR, Gross KS, Frick KM. Oestradiol as a neuromodulator of learning and memory. Nat Rev Neurosci 2020;21:535–50. https://doi.org/10.1038/s41583-020-0362-7.

[127] Liu Y, Ding H, Ting CS, Lu R, Zhong H, Zhao N, et al. Exposure to air pollution and scarlet fever resurgence in China: a six-year surveillance study. Nat Commun 2020;11:1–13. https://doi.org/10.1038/s41467-020-17987-8.

[128] Daellenbach KR, Uzu G, Jiang J, Cassagnes L, Leni Z, Vlachou A, et al. Sources of particulate-matter air pollution and its oxidative potential in Europe. Nature 2020;587. https://doi.org/10.1038/s41586-020-2902-8.

[129] Alotaibi R, Bechle M, Marshall JD, Ramani T, Zietsman J, Nieuwenhuijsen MJ, et al. Traffic related air pollution and the burden of childhood asthma in the contiguous United States in 2000 and 2010. Environ Int 2019;127:858–67. https://doi.org/10.1016/j.envint.2019.03.041.

[130] Mccraty R, Atkinson M, Rein G, Watkins AD. Music enhances the effect of positive emotional states on salivary IgA. Stress Med 1996;12:167–75. https://doi.org/10.1002/(SICI)1099-1700(199607)12:3<167::AID-SMI697>3.0.CO;2-2.

[131] Kreutz G, Bongard S, Rohrmann S, Hodapp V, Grebe D. Effects of choir singing or listening on secretory immunoglobulin A, cortisol, and emotional state. J Behav Med 2004;27:623–35. https://doi.org/10.1007/s10865-004-0006-9.

[132] Greenberg DM, Decety J, Gordon I. The social neuroscience of music: understanding the social brain through human song. Am Psychol 2021. https://doi.org/10.1037/amp0000819.

[133] Nyein HYY, Bariya M, Kivimäki L, Uusitalo S, Liaw TS, Jansson E, et al. Regional and correlative sweat analysis using high-throughput microfluidic sensing patches toward decoding sweat. Sci Adv 2019;5. https://doi.org/10.1126/sciadv.aaw9906, eaaw9906.
[134] Yang Y, Song Y, Bo X, Min J, Pak OS, Zhu L, et al. A laser-engraved wearable sensor for sensitive detection of uric acid and tyrosine in sweat. Nat Biotechnol 2019. https://doi.org/10.1038/s41587-019-0321-x.
[135] Bayoumy K, Gaber M, Elshafeey A, Mhaimeed O, Dineen EH, Marvel FA, et al. Smart wearable devices in cardiovascular care: where we are and how to move forward. Nat Rev Cardiol 2021;18(8):581–99. https://doi.org/10.1038/s41569-021-00522-7.
[136] Van Rhijn I, Godfrey DI, Rossjohn J, Moody DB. Lipid and small-molecule display by CD1 and MR1. Nat Rev Immunol 2015;15:643–54. https://doi.org/10.1038/nri3889.
[137] Donia MS, Fischbach MA. Small molecules from the human microbiota. Science 2015;349:1254766. https://doi.org/10.1126/science.1254766.
[138] Ma C, Han M, Heinrich B, Fu Q, Zhang Q, Sandhu M, et al. Gut microbiome—mediated bile acid metabolism regulates liver cancer via NKT cells. Science 2018;876. https://doi.org/10.1126/science.aan5931.
[139] Nicolai S, Wegrecki M, Cheng T-Y, Bourgeois EA, Cotton RN, Mayfield JA, et al. Human T cell response to CD1a and contact dermatitis allergens in botanical extracts and commercial skin care products. Sci Immunol 2020;5. https://doi.org/10.1126/sciimmunol.aax5430, eaax5430.
[140] An D, Oh SF, Olszak T, Neves JF, Avci FY, Erturk-Hasdemir D, et al. Sphingolipids from a symbiotic microbe regulate homeostasis of host intestinal natural killer T cells. Cell 2014;156:123–33. https://doi.org/10.1016/j.cell.2013.11.042.
[141] Linehan JL, Harrison OJ, Han S-J, Byrd AL, Vujkovic-Cvijin I, Villarino AV, et al. Non-classical immunity controls microbiota impact on skin immunity and tissue repair. Cell 2018;172:784–796.e18. https://doi.org/10.1016/j.cell.2017.12.033.
[142] Becker J, Burik CAP, Goldman G, Wang N, Jayashankar H, Bennett M, et al. Resource profile and user guide of the polygenic index repository. Nat Hum Behav 2021. https://doi.org/10.1038/s41562-021-01119-3.
[143] Pan H, Guo R, Ju Y, Wang Q, Zhu J, Xie Y, et al. A single bacterium resurrects the microbiome-immune balance to protect bones from destruction in a rat model of rheumatoid arthritis. Microbiome 2019;7:107. https://doi.org/10.1186/s40168-019-0719-1.
[144] Pan H, Li R, Li T, Wang J, Liu L. Whether probiotic supplementation benefits rheumatoid arthritis patients: a systematic review and meta-analysis. Engineering 2017;3:115–21. https://doi.org/10.1016/J.ENG.2017.01.006.
[145] O'Toole PW, Marchesi JR, Hill C, Na YC, Kim HS. Next-generation probiotics: the spectrum from probiotics to live biotherapeutics. Nat Microbiol 2017;2:17057. https://doi.org/10.1038/nmicrobiol.2017.57.

Index

Note: Page numbers followed by *f* indicate figures, *t* indicate tables, and *b* indicate boxes.

A

ABO blood group, 29
Absolute abundance, 69–71
Actinomyces odontolyticus, 110
Adhesion, 33–35
Adipose tissue, 63
Aggregatibacter actinomycetemcomitans, 89
Akkermansia muciniphila, 24*f*, 143–144
Alzheimer's disease, 163
Amplicon sequencing, 59–60
Aryl hydrocarbon receptor (AhR) agonists, 166–167*t*
Atherosclerotic cardiovascular disease (ACVD), 94
Atherosclerotic plaque-associated bacteria, 91–93*t*
Autoantigens, 146–147

B

Bacillus anthracis, 134*f*
Bacteria, 119*b*
Bacterial genomes, 111–112*f*
Bacterial species, 120–121*f*
Bacterial strains, 118
Bacteroides fragilis, 2*f*, 35–37
Bacteroides thetaiotaomicron, 24*f*
Bifidobacterium animalis subsp., 24*f*
Biomarkers, 159–160
Branched chain amino acids (BCAAs), 144–146
Bristol's stool score (BSS), 57

C

Cancer screening, 160
Cancer therapy, 168*f*
Candidate phyla radiation (CPR), 113*f*
Cardiometabolic diseases, 89–94
Cardiovascular diseases, 94
Causality, 149
Causal reasoning, 133–138
Cell-free DNA (cfDNA), 160*b*
Cell-free RNA (cfRNA), 160*b*
Christensenella minuta, 24*f*
Circadian rhythm, 43*b*
Circulation, 97
Colonoscopy, 159–160
Colorectal cancer, 175–176
Commensal microbes, 84–88
Commercial deoxyribonucleic acid (DNA) tests, 202
Corynebacterium spp., 10*f*
COVID-19 pandemic, 83–84
  epidemiology, 83
CPR. *See* Candidate phyla radiation (CPR)
Culturomics, 110
Cutibacterium acnes, 29–31

D

Dental bacteria, 3*b*
Deoxyribonucleic acid (DNA)
  extraction, 57–58
    metagenomic samples, 67–68
  sequencing amount, 57–58
Diabetes, 144–146
Dietary fibers, 195–200
Drug metabolism, 168*f*

E

Enterotypes, 38–46
Escherichia coli, 2*f*

F

Faecalibacterium prausnitzii, 24*f*, 135–136
Fecal microbiome, 29*b*
Fungal taxonomy, 114*f*
Fusobacterium spp., 175–176

G

Gut microbiome, 33–38, 169–170*t*
  circadian rhythm, 43*b*

H

Habitats, 29–31
Helicobacter pylori, 141–142
Hematogenous spread, 86
Higher-taxonomic resolution, 118–121
Histone deacetylase 3 (HDAC3), 43

I

Immunoglobulin A (IgA), 33–35
Indoleamine 2,3-dioxygenase 1(IDO1) inhibitors, 164–165*t*
Infant gut microbiota, 84
Interkingdom interactions, 94–96

K

Koch's postulates, 133

L

Lachnospiraceae, 33–35
Lactobacillus crispatus, 94
LCT alleles, 29
Liver diseases, 157–159
Low biomass samples, 59–65
Lymphatic drainage, 86

M

Machine learning, 118–119
Macroecology, 21
Mendelian randomization (MR), 138–141, 139*b*
Metabolomics, 97
Metagenome, 6*b*
Metagenome-assembled genomes (MAGs), 109
Metagenome-wide association studies (MWAS), 71–73
Metagenomics, 1–5
  deoxyribonucleic acid, 67–68
  disease screening, 157–161
  personalized treatment, 161–163
  vs. traditional microbiology, 5*f*

Metagenomic shotgun sequencing, 67
Methanobrevibacter, 40
Methanobrevibacter smithii, 24*f*, 40*b*
Methanogenic archaea, 40*b*
Microbial cells, 5–12
Microbial counts, 7
Microbial diversity, 46
Microbial genomes, 120–121*f*
Microbial growth, 65–67
Microbiology, 5*f*
Microbiome, 6*b*
  from ancient times, 13–16
  cancer therapies, 162–163*t*
  causal evidence, 140–141*t*
  diseases, 138–143, 175–177
  diversity, 21–26, 27*t*
  female reproductive tract, 11*f*
  immediate and historical events, 192–200
  interkingdom interactions, 94–96
  omics, 97–101
  record from birth, 189–192
  richness, 21–26
  species, 13
  stability, 21–26
Microbiota, 6*b*
  de novo assembly, 26–28
Microflora, 21
Microscopy, 1
Mucosal lining, 26*b*
Multiple effective molecules, 143–144
MWAS. *See* Metagenome-wide association studies (MWAS)
Mycobacterium abscessus, 174–175
Mycobiota, 87*f*
Mycoplasma genitalium, 124–126, 125–126*f*

N
Newton's simplicity rule, 135*b*

O
Occult sepsis, 86
Ockham's razor, 135*b*
Opportunistic pathogen, 133
Oral hygiene, 142
Oral microbiome, 31–33
Organ-specific metabolite production, 97–98*t*
Outer membrane vesicles, 147–149

P
Paleogenomics, 13–16
Pancreatic ductal adenocarcinoma (PDAC), 12*f*
Path diagrams, 135–136
Peristalsis, 35–38
Permutational analyses of variances (PERMANOVA), 72
Phages, 147–149
Placenta microbiome, 62–63
Pollution, 201
Poly-genetic risk scores (PRS), 200–201
Polymerase chain reaction (PCR), 84–88
Prevotella copri, 122*f*
Probiotics, 200
Prokaryotic traits, 15*f*

R
Randomized controlled trials (RCTs), 139
Rheumatoid arthritis (RA), 88–89
  prevention, 203*f*
Ribosomal RNA (rRNA), 62*f*
Robustness, 26–28
Ruminococcaceae, 33–35

S
Sequencing amount, 68–69
The Serengeti rules, 38–46
Shotgun metagenomic data, 59*t*
Single-cell sequencing, 202
Skin microbiome, 29–31
Sparsity, 118
Streptococcus parasanguinis, 139
Supra-gingival dental calculus, 15*f*

T
Treponema spp., 23*b*
Treponema succinifaciens, 24*f*
Trimethylamine (TMA), 40
Trimethylamine N-oxide (TMAO), 40*b*
Trophic levels, macroecology, 21

V
Viral particles, 12–13
Virus-like particles (VLPs), 12
Vitamins, 201

W
Whole-cell modeling, 124–127

9780323913690